Food Contamination: Qualitative and Quantitative Analysis

Food Contamination: Qualitative and Quantitative Analysis

Edited by Cindy Featherstone

SYRAWOOD
PUBLISHING HOUSE
New York

Published by Syrawood Publishing House,
750 Third Avenue, 9th Floor,
New York, NY 10017, USA
www.syrawoodpublishinghouse.com

Food Contamination: Qualitative and Quantitative Analysis
Edited by Cindy Featherstone

© 2019 Syrawood Publishing House

International Standard Book Number: 978-1-68286-685-6 (Hardback)

Cataloging-in-Publication Data

Food contamination : qualitative and quantitative analysis / edited by Cindy Featherstone.
    p. cm.
Includes bibliographical references and index.
ISBN 978-1-68286-685-6
1. Food contamination. 2. Food adulteration and inspection.
I. Featherstone, Cindy.
RA601 .F66 2019
363.192--dc23

TABLE OF CONTENTS

Permissions

List of Contributors

Index

PREFACE

Food contamination is the degradation of food quality due to the presence of harmful chemicals and microorganisms. This field encompasses the scientific study of food contaminants, their types and methods of entry in food. Some of the common food contaminants include pesticides, mycotoxins, arsenic, benzene, etc. The chapters in this book are compiled to provide detailed information about multiple aspects of food contamination analysis. From theories to research to practical applications, case studies related to all contemporary topics of relevance to this field have been included in this book. Food scientists, researchers, experts and students who want to broaden the expanse of their knowledge will find this book immensely beneficial.

The researches compiled throughout the book are authentic and of high quality, combining several disciplines and from very diverse regions from around the world. Drawing on the contributions of many researchers from diverse countries, the book's objective is to provide the readers with the latest achievements in the area of research. This book will surely be a source of knowledge to all interested and researching the field.

In the end, I would like to express my deep sense of gratitude to all the authors for meeting the set deadlines in completing and submitting their research chapters. I would also like to thank the publisher for the support offered to us throughout the course of the book. Finally, I extend my sincere thanks to my family for being a constant source of inspiration and encouragement.

Editor

Bacterial contaminations of raw cow's milk consumed at Jigjiga City of Somali Regional State, Eastern Ethiopia

Melese Abate Reta[1*], Tesfaye Wolde Bereda[2] and Ayalew Nigusie Alemu[3]

Abstract

Background: Milk is a compensatory part of daily diet especially for the expectant mothers as well as growing children. It is virtually a sterile fluid when secreted into alveoli of udder. However, beyond this stage of production, microbial contamination might generally occur from different sources.

Methods: A cross-sectional study was carried out from March 2013-January 2014 in Jigjiga city to assess bacterial contamination of raw milk meant for human consumption and to determine antimicrobial susceptibility patterns of the isolates. A total of 120 raw milk samples were aseptically collected from different sampling points that were hypothesized to be a source of potential contaminations. Data were analyzed using SPSS version 17 computer software. P-value of <0.05 was taken as statistical significance.

Results: Overall, the organisms identified and their prevalence rates were *Escherichia coli* 70(58 %), *Staphylococcus aureus* 29(24.2 %), *Shigella Sp.* 21 (17.5 %), *Proteus sp.* 9 (7.5 %) and *Salmonella sp.* 4 (3.3 %). The isolation rates of these identified bacteria from each sampling points are statistically significant in *E. coli* and *Proteus sp.* ($P < 0.05$). High antibiotic resistance for *E. coli* isolates were observed to Doxycycline (42.3 %) and Ampicillin (30 %). *Shigella sp.* was resistant to Ampicillin (38.1 %). *Salmonella sp.* isolates were highly resistant to Amoxicillin (50 %). Out of a total of 29 *S.aureus* isolates, high resistance rate was observed to penicillin G 27(93.1 %) followed by tetracycline 20(69 %), and very low level of resistance to vancomycin 2(6.9 %) and rifampicin 1(3.4 %). Multidrug resistance was also observed in 55.2 % of the total isolates.

Conclusions: Considering the high rate of raw milk contamination with the above isolated bacteria, sanitary practice during collecting, transporting and vending is recommended since the consumption of unpasteurized milk may inflict an important public health risk.

Keywords: Bacterial contamination, Critical sampling points, Raw milk, Antibiotic, Jigjiga

Background

Milk is used throughout the world as a human food at least one form or more. It is virtually a sterile fluid when secreted into alveoli of udder. However, beyond this stage of production, microbial contamination might generally occur from different sources (Mennane et al. 2007). Conditions for contamination of raw milk at different critical points are due to less hygienic practices in pre-milking udder preparation, sub-optimal hygiene of milk handlers, and poor sanitation practices associated with milking and storage equipments (Garedew et al. 2012). Milk is largely made up of water, within which a wide range of nutrients including vitamins, proteins, fats and carbohydrates are suspended. These rich nutritional contents, the production and processing procedures in commercial milk production render it susceptible to contamination by a host of pathogenic microbes that could cause diseases in humans. Therefore, milk is known to be an efficient vehicle for transmission of disease causing agents to humans (Garedew et al. 2012). The demand of consumers for safe and high quality milk has placed a significant responsibility on dairy producers, retailers and

---
* Correspondence: melese1985@gmail.com
[1]Faculty of Health Science, Department of Nursing, Woldia University, P.O.Box 400, Woldia, Ethiopia
Full list of author information is available at the end of the article

manufacturers to produce and market safe milk and milk products (Adesiyun et al. 1995; Mennane et al. 2007). Milk and milk products have important role in feeding the rural and urban population of Ethiopia owing to its high nutritional value. It is produced daily, sold for cash or readily processed. It is a cash crop in the milkshed areas that enables families to buy other foodstuffs and significantly contributing to the household food security (Abebe et al. 2012). Lack of refrigeration facilities at farm and household level in developing countries of tropical regions with high ambient temperature implies that raw milk will easily be spoiled during storage and transportation (Godefay and Molla 2000). Milk and milk products may carry toxic metabolites of different pathogenic organisms growing in it. Ingestion of such products contaminated with these metabolites cause food poisoning for consumers. On the other hand the ingestion of viable pathogenic bacteria along with the food product leads to food borne infection (Aneja et al. 2002). The disease causing bacteria in the milk are *Salmonella sp. Mycobacterium bovis, Corynebacterium sp., Clostridium perfringens, Yersinia enterocolitica, Coxiella burnetii, Brucella, Staphylococcus sp., Campylobacter jejuni, Mycobacterium avium, Listeria sp., Escherichia coli,* and coliforms (Fadaei 2014; Olatunji et al. 2009). The total coliforms, *E. coli* and other enteric bacteria are reliable indicators of fecal pollution generally in insanitary conditions of water, food, milk and other dairy products. Recovery of *E. coli* from food is an indicative of possible presence of enteropathogenic and/or toxigenic microorganism which could constitute a public health hazard (Soomro et al. 1996). These microorganisms are usually associated with food borne diseases and outbreaks, as recorded by official health organizations (Bouazza et al. 2012). The presence of these pathogenic bacteria in milk appeared as main public health concerns, especially for those people who still drink unpasteurized raw milk (Claeys et al. 2013). Despite this, the aim of this study was to determine the presence of contaminating microorganisms and their antibiotic resistance patterns in the raw milk produced by individual farmers, collectors and milk vendors in Jigjiga city, eastern Ethiopia.

Methods
Study area, design and study period
A cross-sectional study was conducted in Jigjiga city from March 2013-January 2014. Jigjiga is the capital city of Ethiopian Somali Regional State located at 628 km east of Addis Ababa at 9° 20' north latitude and 42° 47' east longitude. The altitude of the district ranges from 900–1600meters above sea level and receives an annual rainfall of 300–500 mm with the mean minimum and maximum annual temperatures of 20°c and 28°c respectively (CSA 2003). The community in this region is pastoral and agro-pastoralist and there is large milk

production from cows, camels and goats. The study populations were raw cow's milk from individual farmers' cows, milk collectors, and milk venders in Jigjiga city.

Collection of raw milk samples at critical sampling points and transportation
Milk samples were collected from points considered to be associated with contamination (critical sampling points). The sampling points were the teat during milking, milking buckets at farm level, transport containers, and selling point up on arrival at the market. Overall, 120 raw milk samples were analyzed: of these, 30 raw milk samples were from teat, 30 from milking buckets, 30 from storage containers, and 30 from selling point up on arrival at the markets. During sampling of raw milk directly from teats, the udder and teats were cleaned and dried before sampling; each teat end was scrubbed gently with cotton swabs moistened with 70 % ethyl alcohol. The first 3-4 streams of milk were discarded, and approximately 10 ml of milk was collected into sterile sampling bottles. The other raw milk samples were collected in the morning following standard safety procedures. Prior to sampling from milking buckets and transport containers, the milk was thoroughly mixed by shaking and 25 ml of milk was transferred into a sterile screw capped bottle. Transportation of samples to the Ethiopian Somali Regional Laboratory was immediately conducted for further processing using ice packs following the standard safety procedures (Robinson 2002).

Bacterial identification and isolation from milk samples
Detection of E.coli: All the samples positive for *E. coli* contamination were confirmed using Gram's staining, cultural and biochemical examinations. The samples were inoculated on MacConkey Agar (Difco laboratories, USA) and incubated aerobically at 37°c for 24 h. The plates were observed for the growth of *E. coli*. A single, isolated colony was picked and sub-cultured again on MacConkey agar for purification of the isolate. Simultaneously another single colony with similar characters was picked for the preparation of smear and stained with Gram's stain for the examination of staining and morphological characters of the isolate using bright field microscope. The cultural characteristics of the isolates were confirmed by inoculating the pure colonies on Blood Agar (Oxoide, Germany), Nutrient Agar (Oxoid CM0003, Basingstoke, England), Nutrient Broth and Violet Red Bile Agar (Oxoid CM107). Biochemical tests were performed to confirm the *E. coli* using catalase test, Simmon's Citrate Agar, sugar fermentation on Triple Sugar Iron Agar(Oxoid CM0277, Basingstoke, England), Gelatin liquefaction, Indole Production, Nitrate reduction, Urease production, Voges proskaur, Methyl red and Presumptive test. **Detection of Salmonella Sp:** The

isolation and identification involves three steps; 1 ml of milk was pre-enriched with 9 ml of buffered peptone water (Oxoid CM509, Basingstoke, England) and incubated for 24 h at 37°c. A portion (0.1 ml) of the pre-enriched cultured was transferred to 10 ml of selenite cysteine broth (Merck) and incubated at 37°c for 24 h respectively. Finally, from the selective enrichment media the sample was inoculated on to xylose lysine deoxycholate (XLD) agar (Oxoid CM0469, Basingstoke, England) and incubated at 37°c for 24 h. Characteristic *Salmonella* colonies, having a slightly transparent zone of reddish color and a black center were sub-cultured on nutrient agar and confirmed biochemically using triple sugar iron agar (TSI)(Oxoid CM0277, Basingstoke, England), Christensen's urea agar (Oxoid CM53, Basingstoke, England), lysine iron agar (LIA) (Oxoid CM381, Basingstoke, England), Voges Proskauer (VP), methyl red (MR)(Micromaster Thane, India), and Indole tests (Becton Dickinson, USA) (Hendriksen 2003). **Detection of** *S.aureus*: Gram staining was performed (Cruikshank et al. 1975) and Gram-positive cocci that occurred in clusters under the microscope were subjected to preliminary biochemical tests (the catalase and oxidase tests). The identities of the isolates were confirmed based on positive results for the DNase test, beta haemolytic patterns on blood agar enriched with 5 % (v/v) sheep blood and the coagulase slide test for *S. aureus* using the (PROLD Diagnostics, Canada). The slide agglutination test was performed according to the manufacturer's instructions. Briefly, cells from a pure colony were placed on the clean area of the slide using a sterile toothpick and a drop of the PROLD reagent was added. These were mixed using the toothpick and the isolates were identified based on the formation of agglutination. An isolates that formed agglutination were recorded as *S. aureus* and maintained at 4°c in 30 % glycerol for further characterization by antibiotic susceptibility testing. **Detection of** *Shigella Sp:* Specimens were plated directly on primary media: Salmonella-Shigella agar (Merck) and Selenite F broth (Mast Diagnostics DM 210, Mast Diagnostics, UK). For those negative samples on primary sold media, sub-culturing from enrichment broth to primary media was performed to improve recovery of the isolates. All of the inoculated media were incubated at 37°c for 18-24h. The non-black colonies observed on the center were suspected positive test for *Shigella sp.* and Klingler Iron Agar (KIA) was used for biochemical differentiation of *Shigella* from other coliform bacteria. Colonies of suspected *Shigella* was inoculated on Salmonella-shigella Agar plate (Merck), deoxycholate citrate agar (DCA) (Oxoid CM 35; Oxoid Ltd, UK) and incubated at 37°c for 24 h. Growth of suspected *Shigella sp.* change in color butt of media its color(red) to yellow and red slope remained as it is because *Shigella sp.* is lactose fermenter in anaerobic condition. **Detection of** *Proteus*

Sp: One (1 ml) of milk sample was enriched in 10 ml of Buffer peptone water aseptically and incubated at 37°c for 24 h. Inoculum from the enrichment broth was streaked on Hektoen Enteric Agar (HEA) and MacConkey Agar (Difco laboratories, USA) and incubated at 37°c for 24 h. The cultures were identified on the basis of their morphological, and biochemical characteristics.

Antimicrobial susceptibility testing

The antimicrobial susceptibility patterns of the above detected bacteria were carried out following the Kirby-Bauer disc diffusion method on Mueller Hinton agar (Oxoid CM0337 Basingstoke, England) as described by the Clinical and Laboratory Standards Institute (CLSI 2008). The criteria used to select the antimicrobial agents tested were based on the availability and frequency of prescription for the management of bacterial infections in animals as well as for human in Ethiopia and on the basis of their different structures and mechanisms of action. Antimicrobial susceptibility test was performed for all *S. aureus* isolates according to the criteria of the Clinical and Laboratory Standards Institute (CLSI 2008). For susceptibility test for *S. aureus*, one anti-microbial from each subclass of antimicrobials which were commonly used for treatment of bovine mastitis or considered as important antimicrobial agents for human were selected for antibiogram based on the criteria of Clinical and Laboratory Standards Institute (CLSI 2008). Thus, antimicrobials used for treatment of bovine mastitis included in this study were erythro-mycin (E/15 µg), cephalothin (KF/30 µg), penicillin-G(10unit), sulphoxazole-trimethoprim (SXT/25 µg), amoxicillin-clavulinic acid (AMC/30 µg), chloroamphenicol (C/30 mg), (Oxoid), tetracycline (TE/30 µg) and gentamicin (CN/10 µg) (Biomerioux). Antimicrobials not used for treatment of bovine mastitis but important for human were oxacillin (OX/1 µg), vancomycin (VA/30 µg), clindamycin (DA/10 µg) and rifampicin (RD/5 µg) (Oxoid). Finally, the diameters of the zone of inhibition around the disks were measured to the nearest millimeter using rulers, and the isolates were classified as susceptible, intermediate and resistant (CLSI 2008). *E. coli ATCC 25922* was used as a quality control organism for the antimicrobial susceptibility test (Hendriksen, 2002). Moreover, isolates showing resistance to three or more antimicrobial subclass were considered as multidrug resistant.

Statistical analysis

The collected data for bacterial contamination analysis were entered and analyzed using SPSS version 17 computer software. Accordingly, descriptive statistics such as percentages and frequency distribution was used to describe/present bacterial isolates and antimicrobial susceptibility which was expressed as percent of resistant, intermediate and susceptible. *P*-value <0.05 was taken as cut-off for statistical significance.

Results

Isolated bacterial species

Overall, five bacterial targets were identified in the milk sampled in the study area. The bacteria so identified and their isolation rate were *E.coli* 70(58 %), *Salmonella sp.* 4(3.3 %), *Shigella sp.* 21(17.5 %), *Staphylococcus aureus* 29 (24.2 %) and *Proteus Sp.* 9(7.5 %). These are indicative of significant contamination of milk and important human pathogens. The most prevalent organism overall was *E. coli*, while the least prevalent was *Salmonella sp.* In this study, the contamination degree of milk by the isolated bacteria is utterly worsened at each critical sampling point. High contamination level was observed at market point sampled milk. The difference in isolation rate across market chain (critical sampling points) is statistically significant in *E. coli.* ($P = 0.00$) and *Proteus sp.* ($P = 0.016$) (Table 1).

Results of the present study revealed that 49 (40.8 %) of milk sampled had at least two different bacterial organisms, 6 (12 %) from the udder, 10(20.4 %) milking bucket, 12 (24.5 %) from storage container and 21 (42.9 %) from the market point (Fig. 1).

Antimicrobial susceptibility of the bacterial isolates

The antimicrobial susceptibility tests of the bacterial isolates were grossly very variable. About 76.1 % *E. coli* isolates were resistant and it had the highest resistance rates to Doxycycline, Ampicillin and Gentamycin (42.3 %, 30 % and 30 %) respectively. A quarter of *E. coli* isolates (25.4 %) were multidrug resistant (≥3drugs). Similarly higher antimicrobial resistance (74.6 %) was recorded against *Salmonella sp.* isolates as well. The highest resistance rate to *Salmonella sp.* was observed in Amoxicillin (50 %). The highest resistance rate to *shigella* was observed in Ampicillin (38.1 %). All *shigella* isolates were highly susceptible to Co-trimoxazole(81 %). About 14 % of *Shigella* isolates were multidrug resistant-fairly better than *E.coli* isolates. All *Proteus sp.* isolates were 66.7 % sensitive to Ciprofloxacin and showed (55.6 %) resistance to Ampicillin (Table 2).

The observations made in the present study clearly proved that *S. aureus* showed resistance to all antimicrobials tested except for Rifampicin and Vancomycin. These indicate that the problem is highly distributed and disseminated. Moreover, the overall resistance of *S. aureus* isolates, to Vancomycin, Rifampicin, Clindamycin and Gentamycin showed less than 25 % of resistance. The highest resistance rate was observed in Penicillin (93.1 %), followed by Tetracycline (69 %). On the other hand, about 55.2 % (16/29) of *S.aureus* isolates were found to be multidrug resistant (Table 3).

MAR phenotypes of *S. aureus*

Multiple antibiotic resistance (MAR) phenotypes were determined for *S. aureus* (Table 4). The predominant MAR phenotypes for *S. aureus* isolated from this study area were PG-TE -Ox and PG-TE-AC-E-SXZ-Ox in 24.1 % and 17.2 % of the isolates, respectively. Furthermore, MAR phenotypes PG-TE- AC-Ox, PG-TE-AC-E-SXZ-Ox-CN, PG-TE-AC-E-SXZ-Ox- CN-CH and PG-TE-AC-E-SXZ-Ox- CN- VA were obtained in 3.4 % of the isolates. Also PG-TE-AC 6.9 %, PG-TE-Ox- KF 13.8 % and PG-TE-AC-E-Ox 6.9 % were the MAR phenotypes for *S. aureus* isolated from this study area (Table 4).

It is thus evident that MAR *S. aureus* was isolated from all critical sampling points. However, among the isolates from this study area 55.2 % of the isolates develop MAR. Among all MAR phenotypes of *S. aureus*, 40.3 % of them were resistance to six different antibiotics and 7.2 % were resistance to seven antibiotics. Fifty four percent (54 %) of them were resistance to 3 or 4 antibiotics.

Discussion

The outcome of our study revealed that 84.1 % of milk samples were contaminated with at least one bacterium that comprised of *E. coli*, *Salmonella sp.*, *Shigella sp.*, *S. aureus*, and *Proteus Sp.*, with isolation rates of 70(58 %), 4(3.3 %), 21(17.5 %), 29(24.2 %), and 9(7.5 %), respectively.

The levels of contamination with each isolated bacteria were higher across critical sampling points (from teats, Milking bucket, transportation container, and at market points). Similar findings reported by Daka et al.(2012) revealed that the level of contamination with *S.aureus*

Table 1 Occurrence of isolated bacteria across critical sampling points ($n = 120$) of milk collected from Jigjiga city

Bacteria isolated	Isolate No(%)	No. of positive sample (%)				P- value
		Udder ($n = 30$)	Collection Bucket($n = 30$)	Storage Material($n = 30$)	Market point ($n = 30$)	
E. coli	70(58 %)	9(30 %)	16(53.3 %)	22(73.3 %)	23(76.7 %)	0.00
Salmonella sp.	4(3.3 %)	1(3.3 %)	0(0.0 %)	1(3.3 %)	2(6.7 %)	0.140
Shigella sp.	21(17.5 %)	4(13 .3 %)	3(10 %)	4 (13.3 %)	10(33.3 %)	0.069
S. aureus.	29(24.2 %)	6(20 %)	7(23.3 %)	6(20 %)	10(33.3 %)	0.727
Proteus sp.	9(7.5 %)	0(0.0 %)	2 (6.7 %)	3(10 %)	4(13.3 %)	0.016

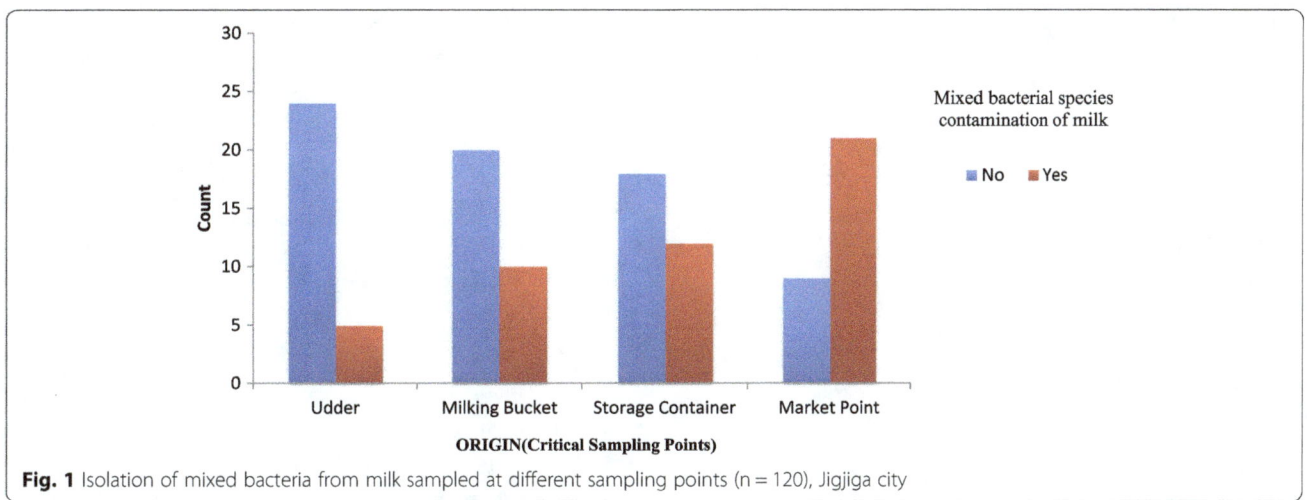

Fig. 1 Isolation of mixed bacteria from milk sampled at different sampling points (n = 120), Jigjiga city

were higher in milk obtained from teat(17.9 %), Milking bucket at farm level(25.7 %), storage containers at milk collection center(26.9 %) and from transportation container(21.8 %). Conditions for contamination of raw milk at different critical points are due to less hygienic practices in pre-milking udder preparation, sub-optimal hygiene of milk handlers, and poor sanitation practices associated with milking and storage equipments, higher environmental contamination during transportation or contamination during waiting along the roadside (Garedew et al. 2012). Based on observations made during the collection of samples, we therefore report that improper hygiene and poor farm management practices contributed to the presence of these isolated bacteria in the milk. In this study area milk was obtained from animals by washing their hands and/or the utensils and containers used. In certain cases, untreated groundwater was used to wash the containers that were used for milking. This may have contributed to the high level of enteric bacteria and S.aureus isolated. Improving the hygienic conditions of the milking environment and/or utensils may reduce the prevalence of entropatogenic as well as S.aureus in milk and prevent its transmission to humans. Olatunji et al (2009) in Nigeria had reported that higher isolation frequencies of E. coli

(24.4 %), S. aureus (38.2 %), and Salmonella sp. (2 %), from apparently normal milk samples from different critical sampling points. It is known that even when drawn under aseptic condition, milk always contains microorganisms which are derived from the milk ducts in the udder, in addition contaminants coming from milking utensils and human handlers (Solomon et al. 2013). Higher isolation frequencies, especially for E. coli across market chain was observed in the current study as compared to similar studies performed to assess bacteriological quality of raw milk in Ethiopia (Tassew and Seifu 2011; Tiruneh 1996). This might be due to poor and unhygienic bedding condition in the majority of farms and absence of teat dipping and disinfection practices in the current study. These practices have been known as critical components of mastitis prevention and control program in dairy herds (Galton et al. 1986). Other findings by different researchers confirm that E. coli grow well in milk and hence endanger its keeping milk quality (Frazeir and Westhoff 1988). E. coli and coliforms are often used as indicator microorganisms, and the presence of E. coli in milk samples implies a risk that other enteropathogenic bacteria may be present in the sample (Najib 2003; Olatunji et al. 2009; Arafa and Soliman 2013).

Table 2 Antibiotic sensitivity pattern of bacterial isolates in milk samples collected from Jigjiga city

Antimicrobial	E. coli(n = 70)			Salmonella sp.(n = 4)			Shigella sp.(n = 21)			Proteus sp.(n = 9)		
	R (%)	I (%)	S (%)	R(%)	I (%)	S (%)	R (%)	I (%)	S (%)	R (%)	I (%)	S (%)
Amoxicillin	15(21.4)	31(44.3)	24(34.3)	2(50)	1((25)	1(25)	2(9.5)	11(52.4)	8(38.1)	1(11.1)	6(66.7)	2(22.2)
Ampicillin	21(30)	28(40)	21(30)	1(25)	1(25)	2(50)	8(38.1)	8(38.1)	5(23.8)	5(55.6)	3(33.3)	1(11.1)
Ciprofloxacin	7(10)	17(24.3)	46(65.7)	1(25)	1(25)	2(50)	2(9.5)	6(28.6)	13(61.9)	0(0.0)	3(33.3)	6(66.7)
Co-trimoxazole	12(17.1)	22(31.4)	36(51.4)	1(25)	2(50)	1(25)	0(0.0)	4(19.0)	17(81)	1(11.1)	3(33.3)	5(55.6)
Chloramphenicol	15(21.4)	26(37.1)	29(41.4)	1(25)	2(50)	1(25)	3(14.3)	7(33.3)	11(52.4)	1(11.1)	5(55.6)	3(33.3)
Gentamycin	21(30)	35(50)	14(20)	1(25)	2(50)	1(25)	4(19.0)	10(47.6)	7(33.3)	2(22.2)	4(44.4)	3(33.3)
Doxycycline	30(42.9)	27(38.6)	13(18.5)	1(25)	2(50)	1(25)	6(28.6)	10(47.6)	5(23.8)	2(22.2)	3(33.3)	4(44.4)

R Resistance, I Intermediate, S Sensitive, n number

Table 3 Antimicrobial susceptibility pattern of *S. aureus* isolates (n = 29) from milk samples collected from Jigjiga city

Antimicrobial	Susceptible number (%)	Intermediate number (%)	Resistant number (%)
Pencillin (P)	1(3.4)	1(3.4)	27(93.1)
Chloroamphenicol (CH)	11(37.9)	10(34.5)	8(27.6)
Cephalothin (KF)	15(51.7)	5(17.2)	9(31.0)
Gentamycin (CN)	10(34.5)	12 (41.4)	7(24.1)
Erythromycin (E)	5(17.2)	17(58.6)	7(24.1)
Clindamycin (DA)	9(31.0)	16(55.2)	4(13.8)
Tetracycline (TE)	5(17.2)	4 (13.8)	20(69)
Amoxicillin + clavulanic	19(65.5)	0(0.0)	10(34.5)
Rifampicin (RD)	27(93.1)	1 (3.4)	1(3.4)
Oxacillin (OX)	18(62.1)	2(6.9)	9(31.0)
Vancomycin (VA)	24(82.8)	3(10.3)	2(6.9)
Sulphamethoxazole-timethoprim (SXZ)	19(65.5)	3 (10.3)	7(24.1)

Our results indicated that 24.2 % of the samples were positive for *S.aureus*. This is a favorable finding because, for human health some strains of *S. aureus* are capable of producing heat stable enterotoxins (Asperger 1994). A comparable finding to our result was reported by Abebe et al. (2013) that *S.aureus* prevalence was 15.5 % in raw milk samples. In contrast to this, different literatures revealed a very significant isolation rate of *S. aureus* from raw milk samples (Olatunji *et al*. 2009; Pourhassan and Taravat-Najafabadi 2011; Mohanty *et al*. 2013; Sanaa et al. 2005). Although the prevalence of *S. aureus* has been reported to vary with the size and geographic region of the area sampled, a high proportion of these bacteria in milk relates to poor hygiene practices. Based on observations made during the collection of samples, we therefore report that improper hygiene and poor farm management practices contributed to the

Table 4 The predominant MAR phenotypes for *S. aureus* isolated from milk samples (n = 29) collected from Jigjiga city

MDR patterns	Phenotype	Number observed	Percentage
Three	PG-TE-AC	2	6.9
	PG-TE -Ox	7	24.1
Four	PG-TE-Ox- KF	4	13.8
	PG-TE- AC-Ox	1	3.4
Five	PG-TE-AC-E-Ox	2	6.9
Six	PG-TE-AC-E-SXZ-Ox	5	17.2
Seven	PG-TE-AC-E-SXZ-Ox- CN	1	3.4
Eight	PG-TE-AC-E-SXZ-Ox- CN- CH	1	3.4
	PG-TE-AC-E-SXZ-Ox- CN- VA	1	3.4

The percentage representations of the phenotypes were obtained by dividing the number of a particular phenotype by the total number of multiple antibiotic resistant isolates identified in a given area. *VA* Vancomycin, *PG* Penicillin G, *SXZ* Sulphamethoxazole-timethoprim, *E* Erythromycin, *Ox* Oxacillin, *AC* Amoxicillin-Clavulanic Acid, *TE*Tetracycline, *KF* Cephalothin; *CN* Gentamycin, *CH* Chloroamphenicol

presence of *S. aureus* in the milk. In this study low *salmonella sp.* isolation rate with 3.3 % was found. Junaidu et al (2011), Forough et al. (2012) and Sanaa et al.(2005) had reported comparable findings with 2.17 %, 4 % and 1.43 % prevalence respectively. Addis et al. (2011) reported a prevalence of 10.7 % from raw milk which is higher than the present report. In the other study by Addis et al. (2011) from 195 dairy cows tested 28.6 % were positive from milk samples. Akoachere et al.(2009) in Cameroon reported a high prevalence (27 %) of *Salmonella* among cattle. This may be due to the difference in the living condition, like housing conditions, feeding habits, types of feed given for the cattle, of the two cattle populations. The detection of *Salmonella* in 3.3 % of the samples tested indicates that the degree of prevalence of the pathogen in raw milk in jigjiga is relatively higher than originally believed. Although contamination of dairy products currently accounts for a small percentage of foodborne illness, it is clear that raw milk consumption and the consumption of products made with raw milk present some risk. Although proper pasteurization minimizes these risks to the public, there is a small but growing group of people that consume unpasteurized milk or milk products, either for practical or cultural reasons, or because of perceived health benefits (Karns et al. 2005). Although the levels of *Salmonella* in the milk samples tested here seemed to be very low and the infectious dose for this organism is low, the potential for this organism to grow in improperly stored raw milk and in products made from raw milk presents a public health risk, particularly to susceptible members of the population.

The isolation rates of *proteus Sp.* in this study (7.5 %) is comparable with the report by Junaidu et al (2011) with 8.69 % prevalence. Most of the organisms identified in this study were enteric bacteria indicating probable faecal contamination of the milk as a result of poor hygiene. The practice of pooling milk from different sources by traders, and the absence of pasteurization generally observed among them could increase the risk posed by such organisms.

In the present study, Doxycycline had the highest resistance rates in *E. coli*. In contrast to this, fairly higher resistant rate was recorded in Ampicillin (100 %) and Amoxicillin (42.11 %) (Thaker et al. 2012). On the other hand the highest resistance rate for *Salmonella sp.* in this report was observed to Amoxicillin (50 %). Different researchers reported antimicrobial resistant *Salmonella* isolates of milk in their previous studies from Ethiopia (Molla et al. 2003; Mekonnen et al. 2005) and from other countries (White et al. 2001). Forough et al. (2012) reported that *salmonella sp.* isolates were resistance to Ampicillin (42.58 %), Tetracycline (42.58 %) and Nalidixic acid (78.57 %). Addis et al. (2011) reported a high resistance rate *salmonella* isolates to ampicillin (100 %). The remarkable degree of resistance to many drugs represents public health hazard due to the fact that food borne outbreaks would be difficult to treat and this pool of MDR *Salmonella* in food supply represents a reservoir for the transferable resistant genes (Diaze De Aguayo et al. 1992). The reasons for the recovery of antimicrobial resistance *Salmonella* isolates were most likely due to the indiscriminate use of antimicrobials (WHO 1988), self-medication and administration of sub therapeutic dose of antimicrobials to livestock for prophylactic purpose (Acha and Szyfers 2001). Antimicrobial use in animal production systems has long been suspected to be a cause of the emergence and dissemination of antimicrobial resistant *Salmonella* (Forough et al. 2012).

In this study the highest resistance rate for *shigella sp.* was observed to Ampicillin (38.1 %) followed by Doxycycline (28.6 %). In contrast to our finding Sanaa et al. (2005) reported that *shigella* isolated from raw milk were sensitive to Gentamycin (64.3 %) followed by Chloramphenicol (92.1 %), and the highest antimicrobial resistant pattern was observed in Ampicillin and Amoxicillin (92.9 %, 92.9 %) followed by penicillin (42.9 %). In agreement with our result Ayalu et al. (2011) reported that *shigella* isolates were 100 % resistant to Ampicillin and Amoxicillin but sensitive to Chloramphenicol, Gentamicin, and Norfloxacin(41.2 %, 88.2 %, and 94.1 %) respectively. Shiferaw et al. (2012) reported that 74 % *shigella* isolates were resistant to Ampicillin, and 58 % to Streptomycin. On the other hand, All the *Shigella* isolates were resistant to Ampicillin, 94 % to Tetracycline, and 82 % to Ciprofloxacin in a report by Debdutta et al. (2012).

The observations made in the present study clearly proved that *S. aureus* showed resistance to all antimicrobials tested except for rifampicin and Vancomycin (Table 3). These indicate that the problem is highly distributed and disseminated. Moreover, the overall resistance of

S. aureus isolates to vancomycin, rifampicin, clindamycin and gentamycin showed less than 25 % of resistance and this is similar with the report of Ma et al. (2006) from the dairy farm in Taiwan. The reason why these antimicrobials were less resistant might be they are not frequently used in the study area in veterinary services, and perhaps in human medicine. Similar suggestion was given by Jaims et al.(2002) that the development of antimicrobial resistance is nearly always as a result of repeated therapeutic and/or indiscriminate use of them. However, the present study has demonstrated the existence of alarming level of resistance of *S. aureus* to commonly used antimicrobials (pencillin G and tetracycline) in the study area. This is due to the fact that tetracycline and penicillin are frequently and improperly used antimicrobials in animal and human treatment. The results were in accordance with reports from earlier studies in other countries (Jakee et al. 2008; Edward et al. 2002; Gentilini et al. 2002) suggesting a possible development of resistance from prolonged and indiscriminate usage of some antimicrobials. This is in contrast with the report of Ma et al. (2006) on his report with respect to pencillin and tetracycline in Taiwan. This is not surprising because penicillin G and tetracycline are the most commonly used antimicrobials for the treatment of infection or mastitis in veterinary practice in Ethiopia. Moreover, penicillin resistance is plasmatic and, it spread out very quickly to several other strains. Pereira et al. (2009) showed that 70 to 73 % of *S. aureus* strains isolated from various foods were resistant to β-lactam such as pencillin and ampicillin. Staphylococci are frequently isolated from bovine mastitis which is one of the most common causes for the use of antimicrobial in lactating dairy cows. Similarly, the present investigation indicated that the resistance pattern of penicillin was found to be 93.1 % (Table 3) which is similar to the finding made by Tariku et al.(2011) (87.2 %) in Ethiopia, Landin (2006) (80 %) in Sweden, Gooraninejad et al. (2007) (57 %) in Iran and Myllys et al.(1998) (50 %) in Finland. This is in contrast to findings observed by Adesiyun (1994) who reported 23 % of resistance to pencillin G in West India.

Moreover, the present study showed the resistance of *S. aureus* to tetracycline (69 %), amoxicillin-clavulinic acid (34.5 %), oxacillin (31 %), cephalothin (31 %), chloramphenicol (27.6 %), sulphamethoxazole-trimethoprim (24.1 %), erythromycin (24.1 %), gentamycin (24.1 %), clindamycin (13.8 %) observed in milk samples taken from dairy cows in jigjiga city. This is in accordance with the findings of Tariku et al.(2011) who reported resistance of *S. aureus* to amoxicillin-clavulinic acid (46 %), chloroamphenicol (16 %), vancomycin (3 %), but it disagree with the observation made by Tariku et al. (2011) in the case of tetracycline(0 %), Co-trimoxazole (0 %) and clindamycin (4 %) in dairy farms in Jimma town. The probable explanation could be that *S. aureus* strains have the capacity to change their resistance behavior to the exposed antimicrobials.

With a particular emphasis to tetracycline, the present observation agrees with preliminary finding conducted by Bayhun (2008) (55.3 %). However, apparent difference was observed in the report of Tariku et al.(2011) (0 %). This is due to the fact that tetracycline is the most commonly used antimicrobial in the treatment of infections in the livestock sector in Ethiopia. Moreover, tetracycline is widely used as growth factors in veterinary medicine for livestock rearing as well in the treatment of bacterial infection occurring in human medicine (Ardic et al. 2005). Furthermore, the resistance profile of S. aureus to amoxicillin-clavulinic acid and oxacillin in milk samples was found to be high. This is due to the fact that resistance of S. aureus to pencillin G, amoxicillin and oxacillin may be attributed to the production of β-lactamase, an enzyme that inactivates pencillin and closely related antimicrobial. It is believed that about 50 % of mastitis causing S. aureus produces β-lactamase (Green and Bradely 2004). Likewise, S. aureus showed resistance to vancomycin and clindamycin. This might indicate transfer of resistant strain among environment, livestock and human since this antimicrobials are not used in veterinary practice.

The MAR phenotypes (Table 4) obtained in the study correlated with the percentage of antibiotic resistance. Although the development of resistance to a particular antibiotic depends on the level of exposure to the antimicrobials, (Rychlik et al. 2006) there are many other factors that are involved. We are therefore suggesting that molecular methods be used to characterize these isolates for the presence of antibiotic-resistance determinants, which may provide data to support our conclusions. S. aureus is normally resident in humans; therefore, the S. aureus present in the cow's milk may have resulted from transmission from humans, which raises questions regarding the hygiene practices followed.

Conclusions

This study revealed that raw cow's milk in the study area could be an important source of infection with a wide range of organisms, particularly enteropathogens. An important source of microbial contamination of the milk is faecal pollution probably from cow dung. There is the need for instituting effective control measures to protect public health. This includes mandatory milk pasteurization by traders and improved hygienic handling of the commodity during milking, ensuring milking is not done on cow dung. The occurrence of multidrug resistance S. aureus should be under consideration during selection of antimicrobials for treatment of mastitis especially if the possibility exists in the transfer of resistance in or between microbial species. Moreover, S. aureus is a common

human commensal, and multidrug resistant S. aureus may present without clinical illness. However, when they cause infection they are extremely serious. Furthermore, dairy cows become infected with multidrug resistant S. aureus, therefore diagnosis of S. aureus does not have implication for treatment only but also it indicates zoonotic transmission since it becomes reservoir for human infection. In practice of indiscriminate use of drugs should be controlled. Further studies that could incorporate isolation of milk contaminating bacteria to the species level should be done to evaluate the imminent danger posed by microbes from milks.

Acknowledgements
Jigjiga University is greatly acknowledged for funding and providing all rounded technical assistance for the smooth accomplishment of this research. The authors are also highly indebted to Ethiopian Somali Regional Health Bureau, Regional Public Health and Research Laboratory for laboratory facilities.

Authors' contributions
MA carried out the conception of the research concept and designed the methodology, data analysis and interpretation and preparation of the manuscript for publication. TW carried out the laboratory work, sample collection and revision of the manuscript. AN critically revised the proposal, designed the methodology, and reviewed the manuscript. All authors read and approved the final manuscript.

Competing interests
The authors declare that there is no financial or non-financial competing interest from anybody or institute. We also want to assure that we did not receive any technical assistant in developing the research concept or preparation of the manuscript.

Author details
[1]Faculty of Health Science, Department of Nursing, Woldia University, P.O.Box 400, Woldia, Ethiopia. [2]College of Natural Sciences, Department of Biology, Wolkite University, P.O.Box 07, Wolkite, Ethiopia. [3]College of Veterinary Medicine, Department of Veterinary Microbiology and Public Health, Jigjiga University, P.O.Box 1020, Jigjiga, Ethiopia.

References
Abebe B, Zelalem Y, Ajebu N. Hygienic and microbial quality of raw whole cow's milk produced in Ezha district of the Gurage zone, Southern Ethiopia. Wudpecker J Agric Res. 2012;1(11):459–65.
Abebe M, Daniel A, Yimtubezinash W, Genene T. Identification and antimicrobial susceptibility of S. aureus isolated from milk samples of dairy cows and nasal swabs of farm workers in selected dairy farms around Addis Ababa, Ethiopia. Afr J Microbiol Res. 2013;7(27):3501–10.
Acha PN, Szyfres B. Zoonoses and Communicable Diseases Common to Man and Animals: Bacteriosis and Mycosis. 3rded.Vol I. Washington DC: Pan American Health Organization; 2001. p. 233–46.
Addis Z, Kebed N, Worku Z, Gezahegn H, Yirsa A, Kassa T. Prevalence and antimicrobial resistance of Salmonella isolated from lactating cows and in contact humans in dairy farms of Addis Ababa. BMC Infect Dis. 2011;11:222–8.
Adesiyun A. Characteristics of S. aureus strains isolated from bovine mastitic milk: Bacteriophage and antimicrobial agent susceptibility and enterotoxigenecity. J Vet Med. 1994;42:129–39.
Adesiyun AA, Webb L, Rahman S. Microbiological quality of raw cow's milk at collection centers in Trinidad. J Food Prot. 1995;58:139–46.
Akoachere TKJ, Tanih FN, Ndip ML, Ndip RN. Phenotypic Characterization of Salmonella Typhimurium Isolates from Food-animals and Abattoir Drains in Buea, Cameroon. J Health Popul Nutr. 2009;27(5):612–8.
Aneja RP, Muthur BN, Chandan RC, Banerejee AK. Technology of Indian milk products. New Delhi: Dairy Indian Yearbook; 2002. p. 183–96.

Arafa M, Soliman M. Bacteriological Quality and Safety of Raw Cow's Milk and Fresh Cream. Slov Vet Res. 2013;50(1):21–30.

Ardic N, Ozyurt M, Sareyyupoglu B. Investigation of erythromycin and tetracycline resistance genes in Methicillin-resistant Staphylococci. Int J Antimicrob Agents. 2005;26:213–8.

Asperger H. *Stapylococcus aureus*. In: The Significance of Pathogenic Microorganisms in Raw Milk, International Dairy Federation. Brussels: IDF; 1994. p. 24–42.

Ayalu AR, Berhanu S, Jemal Y, Gizachew A, Sisay F, Jean MV. Antibiotic susceptibility patterns of Salmonella and Shigella isolates in Harar, Eastern Ethiopia. J Infect Dis Immun. 2011;3(8):134–9.

Bayhun S. Beta-lactamase activities and antibiotic resistance comparison of *Staphylococcus aureus* isolated from clinic and food material (M.Sc Thesis). Gazi University, Institute of Science Technology;2008. p.126

Bouazza H, Hassikou R, Ohmani F, Hmmamouchi J, Ennadir J, Qasmaoui A, et al. Hygienic quality of raw milk at Sardi breed of sheep in Morocco. Afr J Microbiol Res. 2012;6(11):2768–72.

Claeys WL, Cardoen S, Daube G, Block JD, Dewettinck K, Katelijne Dierick K, et al. Herman Raw or heated cow milk consumption. Rev Risks benefits Food Control. 2013;31:251–62.

Clinical and Laboratory Standards Institute (CLSI). Performance Standards for Antimicrobial Disk and Dilution Susceptibility Tests for Bacteria Isolated From Animals. Clin Lab Stand Inst. 2008;28:M31–A3.

Cruikshank R, Duguid JP, Marmoin BP, Swain RH. Medical microbiology. 12th ed. New York: Longman Group Limited; 1975. p. 34.

CSA (Central Statistical Authority). Central Statistical Authority, Federal Democratic Republic of Ethiopia, Central Statistical Investigatory, Statistical Report. Addis Ababa: Central Statistical Authority of Ethiopia; 2003

Daka D, Gebresilassie S, Yihdego D. Antibiotic-resistance Staphylococcus aureus isolated from cow's milk in the Hawassa area, South Ethiopia. Ann Clin Microbiol Antimicrob. 2012;11:26.

Debdutta B, Sugunan AP, Haimanti B, Thamizhmani R, Sayi DS, Thanasekaran K, et al. Antimicrobial resistance in *Shigella* -rapid increase & widening of spectrum in Andaman Islands, India. Indian J Med Res. 2012;135(3):365–70.

Diaze De Aguayo ME, Duarte AB, Montes De Oca Canastillo F. Incidence of multiple antibiotic resistant organisms isolated from retail milk products in Hermosillo, Mexico. J Food Prot. 1992;55(5):370–3.

Edward N, Anna K, Michal K, Henryka L, Krystyna K. Antimic-robial susceptibility of staphylococci isolated from mastitic cows. Bull Vet Inst. 2002;46:289–94.

Fadaei A. Bacteriological Quality of Raw Cow Milk in Shahrekord, Iran. Veterinary World. 2014;7(4):240–3.

Forough T, Elahe T, Manochehr M, Ebrahim R, Rafie S. Occurrence and Antibiotic Resistance of Salmonella spp Isolated from Raw Cow's Milk from Shahahrekord, Iran. Int J Microbiol Res. 2012;3(3):242–5.

Frazeir WC, Westhoff DT. Food Microbiology (4thed). Singapore: Mcgraw-Hill Book Company; 1988. p. 419–28.

Galton DM, Petersson LG, Merril WG. Effects of Pre- Milking Udder Preparation on Bacterial Counts of in Milk and on Teat. J Dairy Sci. 1986;69:260–6.

Garedew L, Berhanu A, Mengesha D, Tsegay G. Identification of gram-negative bacteria from critical control points of raw and pasteurized cow milk consumed at Gondar town and its suburbs, Ethiopia. BMC Public Health. 2012;12:950.

Gentilini E, Danamiel A, Betancor M, Rebuelto MR, Fermepin RM, Detorrest RA. Antimicrobial susceptibility of coagulase-negative staphylococci isolated from bovine Mastitis in Argentina. J Dairy Sci. 2002;85:1913–7.

Godefay B, Molla B. Bacteriological quality of raw milk from four dairy farms and milk collection center in and around Addis Ababa, Berl. Münch Tierarztl Wschr. 2000;113:1–3.

Gooraninejad S, Ghorbanpoor M, Salati AP. Antibiotic Susceptibility of Staphylococci isolated from bovine sub-clinical mastitis. Pak J Biol Sci. 2007;10:2781–3.

Green M, Bradely A. Clinical Forum- S. aureus mastitis in cattle UK. VET. 2004;9:4.

Hendriksen RS. A global Salmonella surveillance and laboratory support project of the World Health Organization: Laboratory Protocols (Susceptibility testing of Salmonella using disk diffusion). 2002. p. 3.

Hendriksen RS. A global Salmonella surveillance and laboratory support project of the World Health Organization: Laboratory Protocols (Isolation of Salmonella). 2003. p. 4.

Jaims E, Montros LE, Renata DC. Epidemiology of drug resistance; the case of *Staphylococcus aureus* and *Coagulase negative Staphylococci* infections. Salud Publica Mex. 2002;44:108–12.

Jakee J, Ata S, Nagwa M, Bakry SA, Zouelfakar EE, Gad El-Said WA. Characteristics of S. aureus strains isolated from human and animal sources. Am-Euras J Agric Environ Sci. 2008;4:221–9.

Junaidu AU, Salihu MD, Tambuwal FM, Magaji AA, Jaafaru S. Prevalence of Mastitis in Lactating Cows in some selected Commercial Dairy Farms in Sokoto Metropolis. Adv Appl Sci Res. 2011;2(2):290–4.

Karns JS, Van Kassel JS, McKluskey BJ, Perdue M. Prevalence of Salmonella enteric in bulk tank milk from US dairies as determined by Polymerase Chain Reaction. J Dairy Sci. 2005;88:3475–9.

Landin H. Treatment of mastitis in Swedish dairy production (in Swedish with English summary). Svensk Veterinärtidning. 2006;58:19–25.

Ma Y, Chang SK, Chou CC. Characterization of bacterial susceptibility isolates in sixteen dairy farms in Taiwan. J Dairy Sci. 2006;1:55–6.

Mekonnen H, Workineh S, Bayleyegne M, Moges A, Tadele K. Antimicrobial susceptibility profile of mastitis isolates from cows in three major Ethiopian dairies. Med Vet. 2005;176(7):391–4.

Mennane Z, Ouhssine M, Khedid K, Elyachioui M. Hygienic Quality of Raw Cow's Milk Feeding from Domestic. Int J Agri Biol. 2007;9(1):1560–8530.

Mohanty NN, Das P, Pany SS, Sarangi LN, Ranabijuli S, Panda HK. Isolation and antibiogram of *Staphylococcus*, Streptococcus and E. coli isolates from clinical and subclinical cases of bovine mastitis. Vet World. 2013;6(10):739–43.

Molla B, Alemayehu D, Salah W. Source and distribution of *Salmonella serovars* isolated from food animals, slaughter house personnel and retail meat products in Ethiopia: 1997-2002. Ethiopia J Health Dev. 2003;17:63–70.

Myllys V, Asplund K, Brofeld E, Hirevela-Koski V, Honkanen-Buzalski T. Bovine Mastitis in Filand in 1988 and 1995. Changes in Prevalence and Antimicrobial resistance. Acta Vet Scand. 1998;39:119–26.

Najib G. Risk assessment of dairy products, Consumer committee seminar,Msca Tour Municipality MM Service, Columbia USA. 2003.

Olatunji EA, Ahmed I, Ijah UJ. Evaluation of microbial qualities of skimmed milk (nono) in Nasarawa State, Nigeria. Proceeding of the 14th Annual Conf. of Ani.Sc. Asso. of Nig. (ASAN) LAUTECH Ogbomoso, Sept. 14th-17th, 2009. 2009.

Pereira V, Lopes C, Castro A, Silva J, Gibbs P, Teixeira P. Characterization for Enterotoxins production, virulence factors and antibiotic susceptibility of S. aureus isolates from various food in Portugal. Food Microbiol. 2009;26:278–82.

Pourhassan M, Taravat-Najafabadi ART. The spatial distribution of bacterial pathogens in raw milk consumption on Malayer City, Iran. Shiraz E Med J. 2011;12:2–10.

Robinson RK. Dairy Microbiology Handbook: The Microbiology of Milk and Milk Products. 3rd ed. USA: John Wiley & Sons, Inc; 2002. p. 51–305.

Rychlik I, Gregorova D, Hradecka H. Distribution and function of plasmids in Salmonella enterica. Vet Microbiol. 2006;112(1):1–10.

Sanaa OY, Nazik EA, Ibtisam EM, Zubeir EL. Incidence of Some Potential Pathogens in Raw Milk in Khartoum North (Sudan) and Their Susceptibility to Antimicrobial Agents. J Anim Vet Adv. 2005;4(3):356–9.

Shiferaw B, Solghan S, Palmer A, Joyce K, Barzilay EJ, Krueger A, Cieslak P. Antimicrobial susceptibility patterns of Shigella isolates in Foodborne Diseases Active Surveillance Network (FoodNet) sites, 2000-2010. Clin Infect Dis. 2012;54 Suppl 5:S458–63.

Solomon M, Mulisa M, Yibeltal M, Desalegn G, Simenew. Bacteriological quality of bovine raw milk at selected dairy farms in Debre Zeit town, Ethiopia. Compr J Food Sci Technol Res. 2013;1(1):1–8.

Soomro AH, Arain MA, Khaskheli M, Bhuto B. Isolation of E. coli from raw milk and milk products in relation to public health sold under market conditions at Tandonjam, Pakistan. J Nutr. 1996;1(3):151–2.

Tariku S, Jemal H, Molalegne B. Prevalence and susceptibility assay of *Staphylococcus aureus* isolated from bovine mastitis in dairy farms in Jimma town South West Ethiopia. J Anim Vet Adv. 2011;10:745–9.

Tassew A, Seifu E. Microbial Quality of Raw Cow's Milk Collected from Farmers and Dairy Cooperatives in Bahir Dar Zuria and Mecha District, Ethiopia. Agric Biol J N Am. 2011;2(1):29–33.

Thaker HC, Brahmbhatt MN, Nayak JB. Study on occurrence and antibiogram pattern of E. coli from raw milk samples in Anand, Gujarat, India. Vet World. 2012;5(9):556–9.

Tiruneh Z. A Study on Bovine Subclinical Mastitis at Stella Dairy Farm, DVM Thesis. Addis Ababa: Faculty of Veterinary Medicine, Addis Ababa University; 1996. p. 30–45.

White DG, Zhao S, Sudler R, Ayers S, Friedman S, Chen S, et al. The isolation of antibiotic-resistant Salmonella from retail ground meats. New Engl J Med. 2001;345(16):1147–54.

World Health Organization. Salmonellosis Control: The Role of Animal and Product Hygiene, Technical Report Series, 774. Geneva: WHO; 1988.

2

Assessing the potential for *Salmonella* growth in rehydrated dry dog food

Ruth A. Oni[1], Elisabetta Lambertini[1,2,3]* and Robert L. Buchanan[1,2]*

Abstract

Background: A substantial percentage of dog owners add water to dry dog food to increase its palatability. The recent association of *Salmonella* contamination of dry pet foods with salmonellosis cases in both dogs and their owners has generated a need to determine the ability of *Salmonella* to grow in eight commercial brands of rehydrated dry dog food.

Results: Eight brands of commercial dry dog food were rehydrated to 20, 35 and 50% added moisture, inoculated with two *S. enterica* strains (~10^5 CFU/g) and incubated for 72 h at 18 °C, 22 °C, or 28 °C. Dog food brand, moisture content, and temperature affected pathogen growth/survival patterns. Rehydration to 20% moisture did not support growth of *S. enterica*, and in general there was a 0.5–2.0 Log decline. At 35% moisture and 28 °C, 4 of 8 brands supported up to 3.4 Log(CFU/g) of growth, while *Salmonella* levels declined in three brands, and remained unchanged in one. Rehydration to 50% moisture at 28 °C supported increases of up to 4.6 Log(CFU/g) in 5 of 8 brands. Growth kinetics determinations with two of the brands that supported growth had calculated lag times, generation times, and maximum population densities of 4.4 and 2.2 h, 1.4 and 10.8 h, and 7.3 and 6.9 Log(CFU/g) when rehydrated to 35% moisture and held at 30 °C.

Conclusions: Results of this study establish that the rehydration of dry dog food with sufficient amounts of water may support the growth of *S. enterica*. Based on the most rapid observed lag times, growth of *Salmonella*, if present, in rehydrated dog food could be avoided by discarding or refrigerating uneaten portions within 2–3 h of rehydration. These data allow accidental or intentional rehydration of dry dog food to be factored into predictive microbiology models and exposure assessments.

Keywords: Pet food, Pet owners, Rehydration, Salmonellosis, Home environment, Growth kinetics

Background

Many of the forty-three million U.S. households who own dogs rely on dry pet food as their pet's primary source of nutrition (AVMA American Veterinary Medical Association 2012). Investigations of two related, multi-state outbreaks of human salmonellosis in 2006 and 2008 identified dry dog food as the source of infection (CDC U.S. Centers for Disease Control and Prevention 2008). With seventy-nine case-patients identified in 21 states of the U.S., this outbreak underscored the importance of proper handling and storage of pet foods in the home to prevent human infection (Behravesh et al. 2010). A third multi-state outbreak

of salmonellosis was linked to a different manufacturer (CDC U.S. Centers for Disease Control and Prevention. Multistate outbreak of human *Salmonella* Infantis infections linked to dry dog food final update 2012; FDA U.S. Food and Drug Administration. Investigation of multistate outbreak of human infections linked to dry pet food 2012). These outbreaks, ongoing recalls of pet foods and treats, and surveys of pet foods and animal feeds have increased consumer concerns about the safety of these products (Finley et al. 2006; Adley et al. 2011; Buchanan et al. 2011; Li et al. 2012; Lambertini et al. 2016a). This, in turn, has led to a need to better understand the microbiological characteristics of dry pet food and treats.

Salmonella is noted for its ability to survive for extended periods in dry food products (Tamminga et al. 1977, Juven et al. 1984, Hiramatsu et al. 2005), including dry pet food (Lambertini et al. 2016b). During the

---
* Correspondence: rbuchana@umd.edu
[1]Department of Nutrition and Food Science, University of Maryland, College Park 20742, MD, USA
Full list of author information is available at the end of the article

development of "what-if" scenarios for an exposure assessment for *S. enterica* in dry dog food (Lambertini et al. 2016c), the purposeful or incidental wetting of the product was evaluated as a potential factor affecting the levels of *Salmonella* in the product in the home environment, and hence potential human and pet exposure. Based on a evaluation of questions asked by pet owners posted on the Internet, the instructions for moistening on dog food packages, and the veterinary literature (Laflamme et al. 2008), it appears that a substantial, though unquantified portion of dog owners moisten dry dog food and/or mix it with wet food prior to feeding. Furthermore, rehydrated pet food can remain at room temperature for substantial periods before being consumed or discarded, thereby potentially increasing levels of *S. enterica* in the home environment. However, no data were available to quantify the degree and extent of *S. enterica* growth under such circumstances.

Accordingly, the objective of the current study was to preliminarily characterize the growth of *S. enterica* in a range of dry dog foods after rehydration, and provide a means of including this factor in risk assessments.

Methods

A complete factorial design (3 × 3 × 8 × 3) with three variables – moisture level, storage temperature and brand of dog food - was used to examine *Salmonella* survival/growth in eight commercial dog food brands. Temperatures of 18, 22, and 28 °C were selected to mimic the range of temperatures that might be encountered during different seasons. Added moisture levels were set at 20, 35, and 50% added water.

Analysis of background microflora of commercial dry dog foods

A 5-lb bag of each of the eight brands of dog food was purchased at a local supermarket after careful visual examination to ensure the packages were undamaged. Upon opening of the bags in the laboratory, samples of the dog food were analyzed for *Salmonella* and aerobic plate counts using the methods described below in conjunction with the standard cultural and confirmatory techniques from the FDA Bacteriological Analytical Manual were used to screen for *Salmonella* (Andrews et al. 2016). The brands were also tested for water activity and pH (see below).

Bacterial strains used and preparation of inocula

In all tests, dry dog food was inoculated with a cocktail of two *S. enterica* strains: *Salmonella enterica* serovar Typhimurium CVM98 (animal isolate) and *Salmonella enterica* serovar Enteritidis KPL13076 (clinical isolate originally obtained from CDC). These strains were selected based on their ability to survive for extended periods in dry environments (Oni et al. 2015). An exception was one growth kinetics study at 30 °C which was conducted using *S.* Typhimurium CVM 98 and *S.* Typhimurium LT-2, due to the loss of *S.* Enteritidis KPL13076 stock culture during a power failure in our −80 °C freezer. *S.* Typhimurium LT-2 is a laboratory strain used extensively for studying *Salmonella* detection and behavior in various foods including dry foods undergoing extended dry storage. The strains were acquired from the culture collection of the Department of Nutrition and Food Science, University of Maryland. After activating individual stock cultures of both *Salmonella* strains by streaking onto Brain Heart Infusion Agar (BHIA) plates (Becton Dickinson, Sparks, MD) and incubating at 37 °C for 18 to 24 h, a single colony of each strain was selected from each plate, streaked onto separate plates of Xylose Lysine Desoxycholate Agar (XLDA) (Becton Dickinson), and incubated at 37 °C for 24 h. Single black colonies were selected from the XLDA plates and used to inoculate five 10-ml tubes of BHI broth (Becton Dickinson), which were then incubated at 37 °C for 24 h. The five 10-ml tubes were combined in a sterile 50-ml centrifuge tube (BD Falcon, Franklin Lake, NJ), and centrifuged at 3,000 × *g* for 10 min at 7 °C. Cell pellets were washed three times with 5 ml of sterile 0.1% peptone water and re-centrifuged, and the final cell pellet was re-suspended in 3 ml of sterile 0.1% peptone water. Equal volumes of each strain were combined, re-centrifuged, and re-suspended in 1 ml of sterile 0.1% peptone water to produce the two-strain cocktail with a final concentration of approximately 10^9 CFU/ml.

Preparation of samples and measurement of *Salmonella* growth/survival at specified moisture levels

The pH and water activity (a_w) of each dry dog food brand was measured upon opening each bag. pH was determined at 25 °C by weighing 1 g portions, pulverizing with a wooden mallet, hydrating with distilled water (1:2.5 g/v), and using a pH meter (Orion pH electrode 9165 BN, Orion Research, Boston, MA, USA) to take measurements. A water activity meter (Novasina IC-500, AW-LAB, Switzerland) was used to measure water activity using the manufacturer's specifications.

Three 140-g portions of each brand of dog food, one for each target moisture level, were weighed into sterile plastic bags. The appropriate amounts of sterilized distilled/deionized water were added to rehydrate the dog food samples to 20, 35, 50% added water as calculated based on the initial weight of the dog food. No attempt was made to have all the brands have the same percent moisture or water activity after rehydration. Working under a biosafety hood, 10 μl (0.01 ml) of the concentrated *Salmonella* cocktail was transferred to the water to be added to the dog food. After vortexing for 20 s, the diluted inoculum was gradually added to the corresponding dog food sample. With each addition, the bag was gently massaged and shaken to ensure a homogeneous distribution. Portions (~10 g) of inoculated dog food were transferred to triplicate labeled plastic containers and stored for 72 h at 18, 22 and 28 °C. Three 10-g samples of each moisture level were analyzed immediately to determine the initial *Salmonella* population densities. After incubation, each 10-g sample was transferred to a Whirlpak bag (Nasco, Fort Atkinson, WI), mixed with 90 ml of sterile 0.1% PW, and stomached for 30 s. The supernatant was used to make serial dilutions, after which 50 μl aliquots of appropriate dilutions which were spiral-plated in duplicate on BHIA (total aerobic bacteria) and XLDA (*S. enterica*). This dual media plating system was used to allow estimation of the degree of injury of salmonellae recovered from the samples (Oni et al. 2015). Plates were incubated at 35 °C for 48 h, and

enumerated at 24 and 48 h using an automated colony counter (Neutec Group Inc., Farmingdale, NY).

Growth kinetics of *Salmonella* in dry dog food

Two additional studies to more closely examine the growth kinetics of the *S. enterica* were carried out with dog food brands #2 and #4 at the 35% rehydration level. A rehydration level of 35% was selected as the level most likely used by consumers, based on an informal survey of pet owners in our laboratory. In the first study, the strong temperature dependency of *S.* Typhimurium CVM98/*S.* Enteritidis KPL13076 growth in brand #4 (Fig. 1b) was followed over 72 h using four temperatures: 15, 20, 25, and 30 °C. Growth was measured over 72 h by periodically taking duplicate samples and quantifying *Salmonella* population densities as described above. The second related study examined the growth kinetics of *S.* Typhimurium CVM98/*S.* Typhimurium LT-2 in brand #2 rehydrated to 35% and incubated at 30

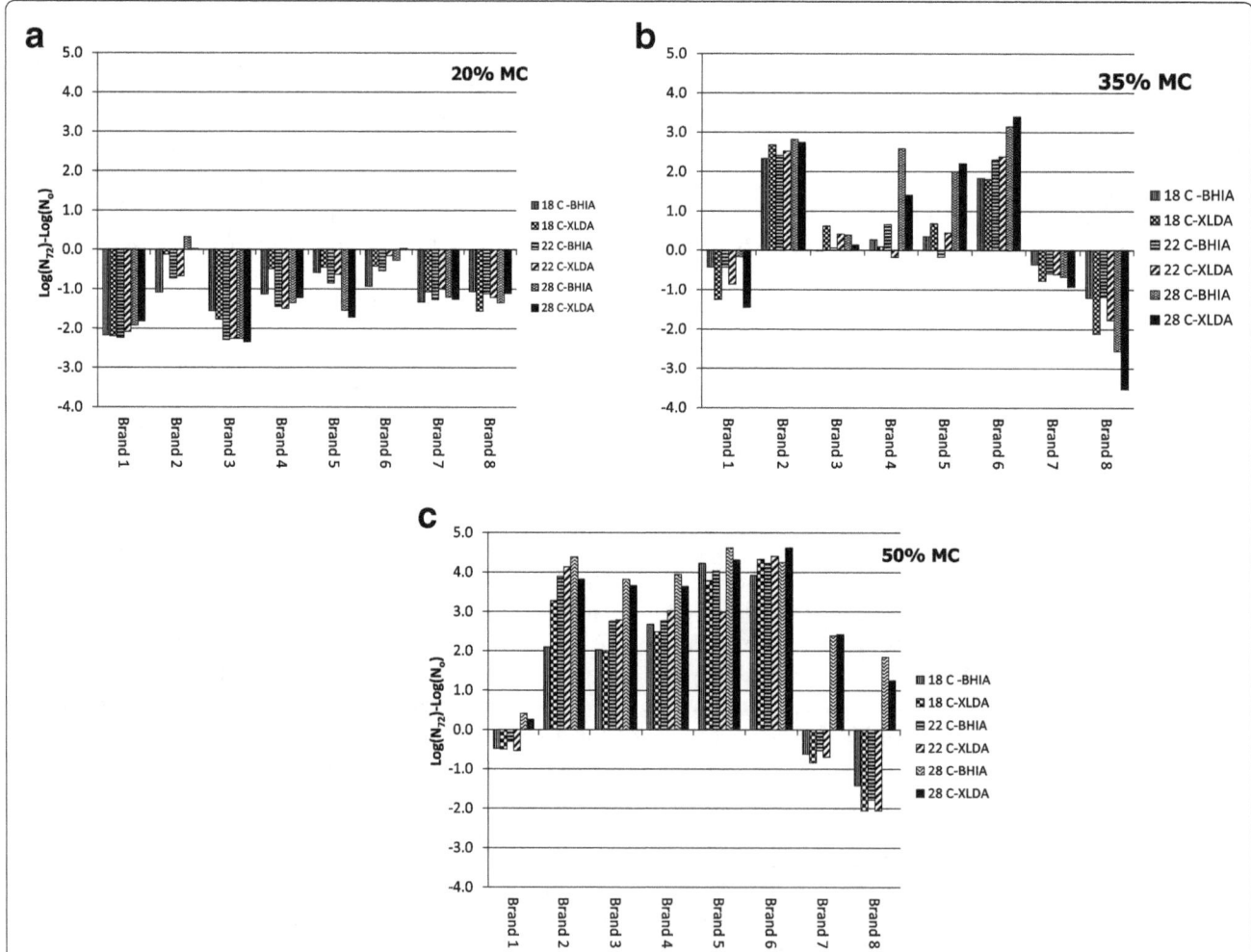

Fig. 1 Comparison of normalized growth and/or survival [Log (N_{72}) − Log (N_0)] of a cocktail of *S.* Typhimurium CVM98 and *S.* Enteritidis KPL13076 in eight brands of dry dog food at added moisture levels of (**a**) 20%, (**b**) 35%, and (**c**) 50%, and incubated at 18, 22 and 28 °C for 72 h

Table 1 Water activity and pH values of dry dog food brands

	Dog Food Brand							
	#1	#2	#3	#4	#5	#6	#7	#8
pH	5.30 ± 0.02^a	5.52 ± 0.01	5.70 ± 0.02	5.83 ± 0.02	6.13 ± 0.02	6.12 ± 0.04	5.39 ± 0.04	5.02 ± 0.02
a_w (no added water)	0.495 ± 0.012	0.486 ± 0.010	0.401 ± 0.008	0.492 ± 0.009	0.459 ± 0.003	0.434 ± 0.009	0.661 ± 0.009	0.653 ± 0.010

[a]Values represent the mean and standard deviation of three samples

°C as described above. This brand was selected because it was one of the brands that supported substantial growth at this rehydration level. *Salmonella* counts obtained from XLDA plates were log-transformed to Log(CFU/g), and the time series fitted to the three-phase linear model (Buchanan et al. 1997) using the IPMP-2014 software (USDA/ARS, Wyndmoor, PA).

Results and Discussion

The measured pH and a_w of the eight brands of dry dog food used in this study are provided in Table 1. No *Salmonella* were detected in the uninoculated dog food (lower limit of detection ~50 CFU/g). The aerobic plate counts (BHIA plates) of the uninoculated dog food were generally below the limit of detection (~50 CFU/g), with those showing growth having counts on average of ~3.8 Log(CFU/g), and consisting largely of molds and yeast (data not shown). The measured a_w of the eight brands after addition of 20, 35, and 50% water is provided in Table 2. Visual observations of the effect of rehydration on the appearance and texture of dog food indicated that with 20% added moisture there was hardly any noticeable effect. Conversely, a moisture level of 50% resulted in the dog food having a soggy appearance and texture. Rehydration to 35% moisture yielded food pellets that were moist yet firm.

All three variables—dog food brand, moisture content, and storage temperature—influenced pathogen growth and survival patterns (Fig. 1). Clear differences were observed among the eight dog food brands. At the lowest added moisture level of 20%, a 0.5–2.0 Log decline in *S. enterica* levels was observed after the 72 h incubation (Fig. 1a). The extent of the reduction was similar at cool (18 °C), room (22 °C) or warm (28 °C) temperatures. At the 35% rehydration level, four of the eight dog food

brands supported *Salmonella* increases up to 3.4 Log(CFU/g), while *Salmonella* levels declined in three brands and remained largely unchanged in brand #3 (Fig. 1b). In three of the four brands that supported growth, the extent of growth was enhanced by incubation at 28 °C. In the three brands that declined at 35% moisture, there was evidence of approximately 1 Log of injury based on the differences in BHIA and XLDA counts. Injury seemed to be enhanced at the warmer incubation temperatures. When rehydrated to 50% moisture, increases up to 4.6 Log(CFU/g) were observed in five of eight brands, with the extent of growth being enhanced by the warmer incubation temperatures (Fig. 1c). Brand #7 and brand #8, which showed substantial reductions in *Salmonella* levels at 35% moisture, supported substantial growth at 50% rehydration and 28 °C, but continued to show 1–2 Log reductions at the lower temperatures. Brand #1 remained largely unchanged over the 72-h incubation period, though to a much lesser degree it showed a response pattern similar to brands #7 and #8.

The decreases observed across all brands at the 20% rehydration level suggest a general effect. One possibility is that this level of moisture was sufficient to support increased metabolic activity but was insufficient to support growth. This hypothesis is supported by pH and a_w values after rehydration (Table 2). After 20% rehydration, the observed a_w values ranged from 0.92 to 0.97. In such a situation, the increased stress of active metabolism under the pH/a_w conditions could lead to physiological damage. However, when the moisture content was

Table 2 Water activity of dog food brands after the addition of 20, 35, and 50% water

Water Added (%)[a]	Dog Food Brand							
	#1	#2	#3	#4	#5	#6	#7	#8
20	0.94^b	0.96	0.94	0.97	0.95	0.97	0.93	0.92
35	0.98	0.97	0.98	0.99	0.99	0.99	0.97	0.95
50	>0.99	>0.99	>0.99	>0.99	>0.99	>0.99	>0.99	>0.99

[a]Proportion of dry weight
[b]Mean of two samples

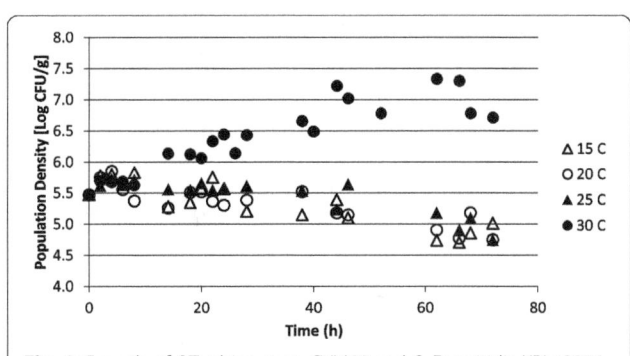

Fig. 2 Growth of *S.*Typhimurium CVM98 and *S. Enteritidis* KPL13076 in brand #4, rehydrated to 35% moisture and incubated at 15, 20, 25, and 30 °C

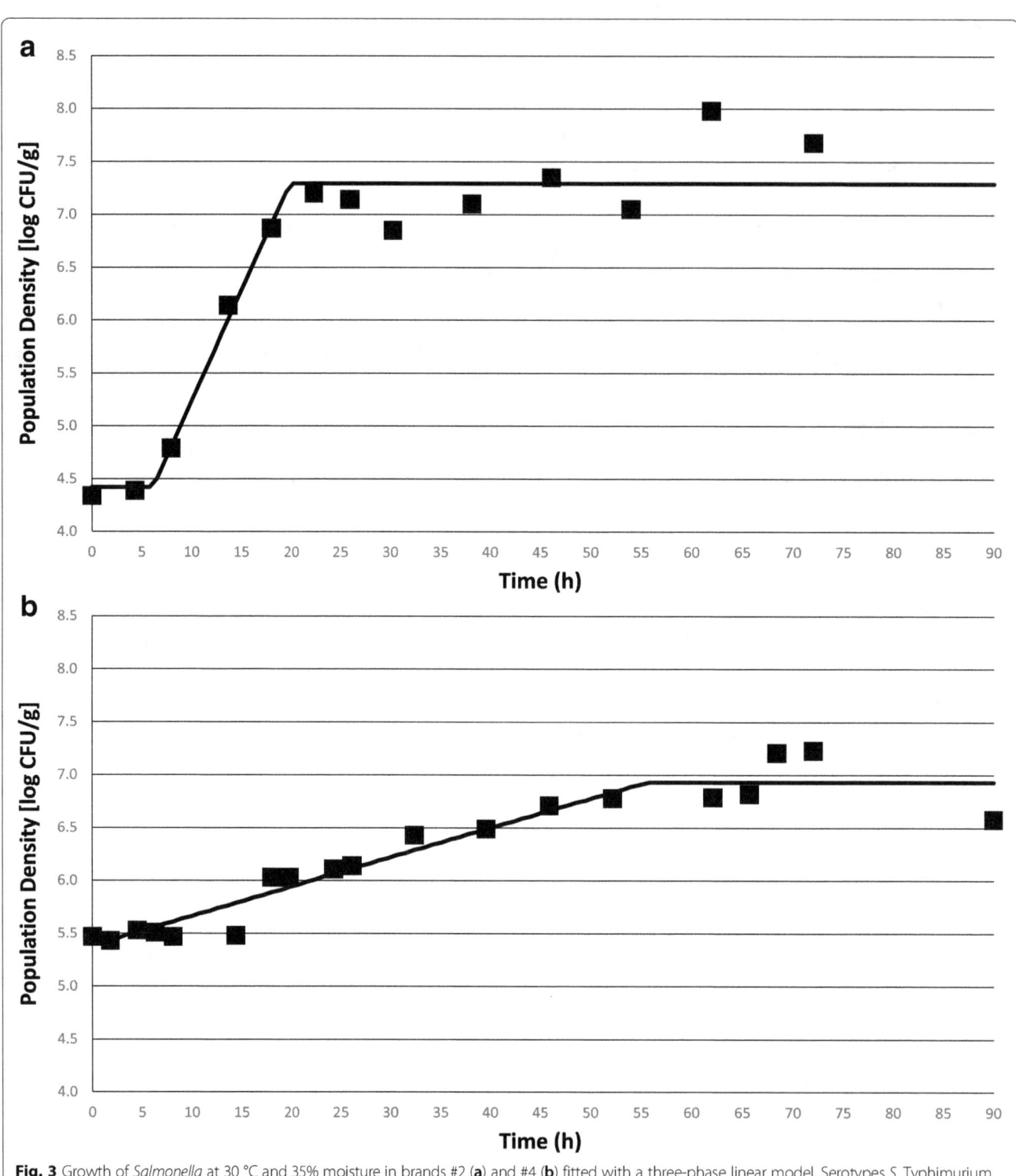

Fig. 3 Growth of *Salmonella* at 30 °C and 35% moisture in brands #2 (**a**) and #4 (**b**) fitted with a three-phase linear model. Serotypes *S.* Typhimurium CVM98 and *S.* Enteritidis KPL13076 were used with brand #4 and *S.* Typhimurium CVM98 and *S.* Typhimurium LT-2 were used with brand #2

further increased to 35% and 50%, the a_w was closer to the optimal for *S. enterica*. At these higher moisture levels, three of the brands continued to display reductions in *Salmonella* levels (Fig. 1b, c). This suggests that these brands may have an antimicrobial ingredient that is released or activated with increased water availability. However, comparison of the ingredients labels for each of the brands did not provide any insights into the specific component(s) of the formulations that would account for the putative antimicrobial response. In general, bacterial counts on non-selective BHIA and selective XLDA plates were similar (Fig. 1a-c), suggesting that no other microorganisms were growing when the *Salmonella* were not. This also suggests that in most instances when *S. enterica* grew, there was little injury. However, as mentioned above, this was not the case in brands #7 and #8 where *S. enterica* levels declined (Fig. 1b).

Two preliminary studies were undertaken to characterize the growth kinetics of *Salmonella* in 35% rehydrated dog food. In the first, the strong temperature effect on *S.* Typhimurium CVM98/*S.* Enteritidis KPL13076 in brand #4 was evaluated at 15, 20, 25, and 30 °C (Fig. 2). Growth was only observed at 30 °C which is consistent with the earlier results based on a 72 h incubation at 28 °C.

In the second preliminary study, the growth kinetics of *S.* Typhimurium CVM98/*S.* Typhimurium LT-2 were determined at 30 °C after rehydration of brand #2 (Fig. 3a). These conditions supported growth to levels similar to those observed with brand #2 with 72 h incubation at 28 °C (Fig. 1b). The 30 °C growth curve data for brands #2 and #4 were fitted to the three-phase linear growth model (Fig. 3a, b) to generate commonly used growth kinetics metrics (Table 3). While the lag phase duration was similar for the two brands, there was a substantial difference in the generation times. This resulted in brand #4 taking a substantially longer time to reach its maximum population density. The reason(s) for the difference in growth rates is unclear, particularly considering that brand #4 had a slightly higher a_w after rehydration (Table 2). Additional research is currently underway to explore the reasons underlying the differences noted among the brands in regard to the survival or growth of *Salmonella*.

The current study clearly demonstrates that rehydration of dry dog food may support the growth of *S. enterica* if present. The specific role that rehydration plays in outbreaks of salmonellosis among pet owners has not been considered in past outbreak investigations (CDC U.S. Centers for Disease Control and Prevention 2008, 2012; Behravesh et al. 2010), but should be in the future. Likewise, consideration of rehydration as a contributing

factor in exposure assessments (Lambertini et al. 2016c) and future risk assessments may be critical for accurately estimating risks and identifying practical risk mitigation strategies. However, this could be a significant risk management challenge considering the substantial differences among brands observed in the current study.

Conclusions

The results of this study establish that the rehydration of dry dog food with sufficient amounts of water can support the growth of *S. enterica*. Thus, if *S. enterica* is present, allowing rehydrated dog food to sit too long uneaten, particularly at warmer temperatures, is likely to increase the risk of salmonellosis for both the dog and its owners. Based on the observed lag time, the present work also indicates that these risks could be effectively managed by ensuring that uneaten food is discarded or refrigerated within approximately 2–3 h of rehydration. The data in this study also help inform a quantitative exposure assessment for *S. enterica* in dry dog food (Lambertini et al. 2016c).

Acknowledgements
This research is part of a risk assessment project funded by Mars Petcare® in partnership with the Center for Food Safety and Security Systems (CFS³), University of Maryland, College Park. The authors would like to thank Joseph Park, Janet Akinduro and Yinzhi Qu for their assistance in the laboratory.

Authors' contributions
RAO was involved in the design of the study and experiment protocol, carried out laboratory experiments, data collection and drafted the manuscript. RLB was involved in the study design, data analysis, and manuscript review. EL assisted in data analysis and manuscript review. All authors read and approved the final manuscript.

Competing interests
The authors declare that they have no competing interest.

Author details
[1]Department of Nutrition and Food Science, University of Maryland, College Park 20742, MD, USA. [2]Center for Food Safety and Security Systems, University of Maryland, College Park 20742, MD, USA. [3]Current address: Environmental and Health Sciences, RTI International, Rockville, MD 20852, USA.

Table 3 Growth kinetics parameters obtained by fitting the three-phase linear model to *Salmonella* growth kinetics data for brands #2 and #4, rehydrated to 35% moisture and incubated at 30 °C (see Fig. 3)

Kinetics Parameters	Brand #2	Brand #4
Lag Phase Duration	4.4 h	2.2 h
Exponential Growth Rate	0.209 Log(CFU/g)/h	0.028 Log(CFU/g)/h
Generation Time	1.4 h	10.8 h
Maximum Population Density	7.3 Log(CFU/g)	6.9 Log(CFU/g)
Time to N_{max} (TN_{max})	20 h	56 h

References
Adley C, Dillon C, Morris CP, Delappe N, Cormican M. Prevalence of *Salmonella* in pig ear pet treats. Food Res Int. 2011;44:193–7.

Andrews WH, Jacobson A, Hammack T. Chapter 5 – *Salmonella*. FDA Bacteriological Analytical Manual. 2016; http://www.fda.gov/Food/FoodScienceResearch/LaboratoryMethods/ucm070149.htm. Accessed 10 Nov 2016.

AVMA (American Veterinary Medical Association). (2012) U.S. Pet Ownership Statistics. *U.S. Pet Ownership and Demographics Sourcebook*. www.avma.org/KB/Resources/Statistics/Pages/Market-research-statistics-US-pet-ownership.aspx. (Accessed August 19, 2015)

Behravesh CB, Ferraro A, Deasy III M, Dato V, Moll M, Sandt C, Rea NK, Rickert R, Marriott C, Warren K, Urdaneta V, Salehi E, Villamil E, Ayers T, Hoekstra RM, Austin JL, Ostroff S, Williams IT, Salmonella Schwarzengrund Outbreak Investigation Team. Human *Salmonella* infections linked to contaminated dry dog and cat food, 2006–2008. Pediatrics. 2010;126(3):477–83.

Buchanan RL, Whiting RC, Damert WC. When is simple good enought: a comparison of the Gompertz, Baranyi, and thre-phase linear models for fitting bacterial groth curves. Food Microbiol. 1997;14:313–26.

Buchanan RL, Baker RC, Charlton AJ, Riviere JE, Standaert RF. Pet food safety – a shared concern. Brit J Nut. 2011;106:S78–84.

CDC (U.S. Centers for Disease Control and Prevention). Multistate outbreak of human *Salmonella* infections caused by contaminated dry dog food, United States, 2006–2007. Morb Mortal Wkly Rep. 2008;57(19):521–4.

CDC (U.S. Centers for Disease Control and Prevention). Multistate outbreak of human *Salmonella* Infantis infections linked to dry dog food (final update). 2012; www.cdc.gov/salmonella/dog-food-05012/index.html. (Accessed August 19, 2015)

Finley R, Reid-Smith R, Weese JS. Human health implications of *Salmonella*-contaminated natural pet treats and raw pet food. Clin Infec Dis. 2006;42: 686–91.

FDA (U.S. Food and Drug Administration). Investigation of multistate outbreak of human infections linked to dry pet food. 2012; www.fda.gov/Food/RecallsOutbreaksEmergencies/Outbreaks/ucm302904.htm. (Accessed August 19, 2015)

Hiramatsu R, Matsumoto M, Sakae K, Miyazaki Y. Ability of Shiga toxin-producing *Escherichia coli* and *Salmonella* spp. to survive in a desiccation model system and in dry foods. Appl Environ Microbiol. 2005;71:6657–63.

Juven BJ, Cox NA, Bailey JS, Thomson JE, Charles OW, Shutze JV. Survival of *Salmonella* in dry food and feed. J Food Prot. 1984;6:445–8.

Laflamme DP, Abood SK, Fascetti AJ, Fleeman LM, Micchel KE, Bauer C, Kemp BL, Doren JR, Willoughby KN. Pet feeding practices of dog and cat owners in the United States and Australia. J Am Vet Med Assoc. 2008;232(5):687–94.

Lambertini E, Buchanan RL, Narrod C, Pradhan AK. Transmission of bacterial zoonotic pathogens between pets and humans: The role of pet food. Crit Rev Food Sci Nutr. 2016a;56:364–418.

Lambertini E, Mishra A, Guo M, Cao H, Buchanan RL, Pradhan AK. Modeling the long term kinetics of *Salmonella* survival on dry dog food. Food Microbiol. 2016b;58:1–6.

Lambertini E, Buchanan RL, Narrod C, Ford RM, Baker RC, Pradhan AK. Quantitative assessment of human and pet exposure to *Salmonella* associated with dry pet foods. Int J Food Microbiol. 2016c;216:79–90.

Li X, Bethune LA, Jia Y, Lovell RA, Proescholdt TA, Benz SA, Schell TC, Kaplan G, McChesney DG. Surveillance of *Salmonella* prevalence in animal feeds and characterization of the *Salmonella* isolates by serotyping and antimicrobial susceptibility. Foodborne Pathogens Dis. 2012;9:692–8.

Oni R, Sharma M, Buchanan RL. Survival of *Salmonella enterica* in dry manure and persistence on spinach leaves. J Food Prot. 2015;78:1791–9.

Tamminga SK, Beumer RR, Kampelmacher EH. Survival of *Salmonella eastbourne* and *Salmonella typhimurium* in milk chocolate prepared with artificially contaminated milk powder. J Hyg Cambridge. 1977;79:333–7.

3

Determination of pesticide residues in honey: a preliminary study from two of Africa's largest honey producers

Janet Irungu[*], Suresh Raina and Baldwyn Torto

Abstract

Background: The presence of pollutants in honey can influence honey bee colony performance and devalue its use for human consumption. Using liquid chromatography tandem mass spectrometry (LC-MS/MS), various clean-up methods were evaluated for efficient determination of multiclass pesticide contaminants in honey. The selected clean-up method was optimized and validated and then applied to perform a preliminary study of commercial honey samples from Africa.

Results: The most efficient method was primary-secondary amine (PSA) sorbent which was significantly different from the others ($P < 0.05$; average recovery ~94 %) and was applied to analyze 96 pesticide residues in 28 retail honey samples from Kenya and Ethiopia. From our preliminary data, a total of 17 pesticide residues were detected at ~10-fold below maximum residue limit (MRL) established for food products except for malathion which was detected at almost 2-fold above its acceptable MRL.

Conclusions: A highly efficient approach for determining pesticide residues in honey with good recoveries was developed. All residue contaminants were detected at levels well below their acceptable MRLs except malathion suggesting that the retail honey analyzed is safe for human consumption. Although PSA clean-up method was selected as the most efficient for cleaning honey samples, omitting the clean-up step was the most economical approach with potential applicability in the food industry.

Keywords: Pesticide residues, Honey bees, Liquid chromatography-tandem mass spectrometry (LC-MS/MS), Honey, Method development

Background

The recent sudden decline of honey bee colonies is of global concern not only because of pollination services they provide in food production process, but also due to honey production among other benefits. While there are multiple variables, including poor nutrition, pests, diseases, and loss of natural bee habitat, negatively affecting bee health, it is becoming increasingly clear that the widespread use of pesticides on agricultural crops is a major factor (Vanengelsdorp and Meixner 2010; Gill et al. 2012; Brodschneider and Crailsheim 2010). As such, to preserve honey bee health which is inextricably integrated with human health and to preserve the quality of bee by-products especially honey, requires regular monitoring using rigorous analytical methods to confirm product quality (Muli et al. 2014; Kujawski and Namiesnik 2008).

Honey is composed of over 300 compounds, mostly carbohydrates (>75 %) and water (~18 %), with minor components comprising of proteins, amino acids, vitamins, antioxidants, minerals, essential oils, sterols, pigments, phospholipids, and organic acids (Bogdanov et al. 2008; Kujawski and Namiesnik 2008). Whereas these diverse ranges of compounds make it a nutrient rich food commodity, they also make it a highly complex analytical matrix especially when analysing the presence of trace compounds such as toxins, pesticide residues and other environmental pollutants (Kujawski and Namiesnik 2008). The presence of pesticide residues and other contaminants in honey can have adverse health effects on bees and humans, decrease the quality of honey

* Correspondence: jirungu@icipe.org
African Reference Laboratory for Bee Health, International Centre of Insect Physiology and Ecology (icipe), P.O. Box 30772-00100, Nairobi, Kenya

and devalue its beneficial properties (Bogdanov et al. 2008). Typically, pesticide residues in honey occurs when bees in search for food, visit crops that have been treated with various agro-chemicals and/or when bee-keepers use chemicals to control bee pests or diseases (Bogdanov 2006). So far, several researchers have reported various residues of pesticides in honey at varying concentrations (De Pinho, et al. 2010; Irani 2009; Barganska et al. 2013; Blasco et al. 2011; Garcia-Chao et al. 2010; Herrera et al. 2005; Rissato et al. 2007; Weist et al. 2011; Fontana et al. 2010; Kujawski and Namiesnik 2011; Wang et al. 2010; Campillo et al. 2006; Choudhary and Sharma 2008; Martel et al. 2007; Erdoğrul 2007; Blasco et al. 2003) confirming the need to constantly monitor the presence of pesticide residues in honey to assess any potential health risk and to ensure that its quality, whether as food or as a therapeutic, is not compromised. However, to date, only few studies have been carried out to monitor pesticide residues in honey produced from Africa (Eissa et al. 2014). A recent study conducted in Kenya in 2010 detected four pesticides from beeswax and bee bread at very low concentrations (Muli et al. 2014). However, the cumulative levels and presence of pesticides in hive products over time can pose health problems for both honeybees and humans. Therefore there is the need to develop highly sensitive and selective analytical techniques that have the ability to analyze multiple pesticides simultaneously in hive products.

Since honey is a complex analytical matrix, it is often necessary to clean-up the sample prior to instrumental analysis (Kujawski and Namiesnik 2008). This facilitates removal of matrix co-extractives that could result in enhancement or suppression of the signal of the targeted analytes during analysis (Ferrer et al. 2011; Kittlaus et al. 2011; Kruve et al. 2008). Conversely, this clean-up step is usually the most expensive, time consuming and laborious sample preparation step with the highest probability of introducing errors on recovery and method repeatability. Conventional extraction/clean-up methods such as liquid-liquid (LLE) or solid-phase extractions (SPE), require large volumes of organic solvents and usually target pesticides from a single chemical class (Fontana et al. 2010; Fernández and Simal 1991; Wang et al. 2010; Martel et al. 2007). Recently, extensive research has been geared towards finding more economical and environmental friendly methods that can yield good recoveries for a diverse range of pesticides. For instance, a recent study compared four different methods for extracting 12 organophosphates and carbamates from honey and concluded that the choice of the method depends on the targeted analytes (Blasco, et al. 2011). In another example (Kujawski et al. 2014), two methods; solid supported liquid-liquid extraction(SLE) and a modified Quick, Easy, Cheap, Effective and Safe (QuEChERS) method for multiresidue analysis were compared using

extraction efficiencies for determination of 30 LC-amenable pesticides in honey at their MRLs. These authors concluded that in terms of recovery (ranged from 34 to 96 %) the methods had no significant difference but in terms of costs and time, the modified QuEChERS was better (Kujawski et al. 2014). In this study, an ultra-high performance liquid chromatography coupled to tandem mass spectrometry (LC-MS/MS) was employed to analyze multiclass chemical contaminants in African honey at parts per billion (ppb) levels. Four different clean-up methods including PSA plus graphitized carbon (GCB), PSA plus C18, PSA alone, and a no clean-up approach were investigated using 96 LC-amenable pesticides to determine their applicability in a multiclass residue analysis in honey by comparing their recoveries. The method was validated and applied to conduct a preliminary study of pesticide residues in commercial honey samples obtained from Kenya and Ethiopia which are among the major producers of honey in Africa. Previous data on honey production in Africa indicates that Ethiopia is the largest producer with an estimate of 41,233 tons of honey followed by Tanzania at 28,678 tons and Kenya at 25,000 tons in 2004- 2006 (FAOSTAT). To the best of our knowledge, this is the first in-depth multiclass pesticide residue analysis of commercial honey from Africa. These results provide some insights in the safety of honey from Africa and some baseline information for future studies on other components of the hive matrix in relation to honey bee colony losses.

Methods
Chemicals and reagents
All pesticide standards were of high purity (>94 %) and were obtained from Sigma-Aldrich (Chemie GmbH, Germany) and Dr Ehrenstorfer (Augsburg, Germany) and were stored according to manufacturer's recommendations until use. Pesticide stock solutions were prepared in acetonitrile at 1 μg/mL and stored in amber screw-capped glass vials at −20 °C.

LC-MS/MS instrumentation
An Agilent 1290 ultra high performance liquid chromatography (UHPLC) series coupled to a 6490 model triple quadrupole mass spectrometer (Agilent technologies) with an ifunnel JetStream electrospray source operating in the positive ionization mode was applied using dynamic multi-reaction monitoring (DMRM) software features. The electrospray ionization settings were gas temperature, 120 °C; gas flow, 15 L/min; nebuliser gas, 30 psi; sheath gas temperature, 375 °C; sheath gas flow, 12 L/min; capillary voltage, 3500 V; nozzle voltage, 300 V. The ifunnel parameters were high pressure RF 150 V and low pressure RF 60 V. Nitrogen was used both as a nebuliser and as the collision gas. Mass Hunter Data Acquisition; Qualitative and Quantitative analysis

software (Agilent Technologies, Palo Alto, CA, v.B.06 and v.B.07) were used for method development, data acquisition and data processing for all the analyses.

The chromatographic separation was performed on a Rapid Resolution reverse phase column-C18 1.8 μm, 2.1 × 150 mm column (Agilent Technologies). The mobile phases comprised of 100 % water in 5 mM ammonium formate containing 0.1 % formic acid for solvent A and acetonitrile in 5 mM ammonium formate containing 0.1 % formic acid for solvent B. A gradient elution at a flow rate of 0.4 mL/min was used.

Optimization of LC-MS/MS parameters

Pesticide standard solutions, individually or as mixes, were used for method development and instrument parameters optimization. To ensure that the maximum sensitivity for identification and quantification of the targeted pesticides is obtained, careful optimization of all MS parameters was performed by infusing the standard solutions directly into the MS followed by infusion through the column to establish their respective retention times (RT). The parameters optimised included collision energy (CE), gas temperature; gas flow, sheath gas temperature and flow, high and low pressure radio-frequency. Table 1 demonstrates the parameters developed and optimised for the 96 pesticide residues targeted in this study.

Data analysis

Targeted analytes were identified by monitoring two transition ions where possible, for each analyte as recommended by SANCO guidelines for LC-MS/MS analysis (SANCO/12571/2013). The most dominant transition ion was used for quantification whereas the second most intense ion as a qualifier for confirmation purposes. Calibration standard solutions were prepared at seven calibration levels covering a concentration range of 0.1 to 100 parts per billion (ppb), including the zero point. The resulting calibration curve was used to determine the instrument's limit of reporting (LOR) and limits of detection (LOD). These were set as calibration standard concentrations producing signal to noise ratio of 3 and 10 respectively. The LOR was set as the minimum concentration that could be quantified with acceptable accuracy and precision. The LC-MS/MS system's linearity was evaluated by assessing the signal responses of the calibration standards.

Sample preparation

Prior analysis of a honey sample, obtained from the local organic farmer from Kenya, was performed to ensure that it did not contain any of the studied compounds. This sample was selected as a blank during method development for spiking, preparing matrix matched calibration curves and recovery purposes. Samples were

prepared following the QuEChERS method (Anastassiades et al. 2003) with some modifications. Briefly, 5 g of this sample was weighed into a 50 ml falcon tube and 10 ml of water were added and the mixture homogenized. Aceto-nitrile (10 ml) plus a mixture of salts (4 g magnesium sulphate, 1 g sodium chloride, 1 g of trisodium citrate dehydrate and 0.5 g of disodium hydrogen citrate sesqui-hydrate) were added and the samples were vortexed for 1 min and centrifuged at 4200 rpm for 5 min. Aliquots of the supernatant were transferred to separate eppendorf tubes and subjected to either no clean-up or to various QuEChERS clean-up methods. A portion of 1 mL of the final solution was then transferred to an auto-sampler vial for LC-MS/MS analysis.

Extraction efficiency

A series of spiked samples were used to assess extraction efficiency of the method. These samples were prepared as follows: blank honey samples fortified at 10 times LOQ (10 ng/g) were dissolved in appropriate amounts of water and homogenized. Extractions of the spiked residues were performed following QuEChERS methods. Honey samples were spiked with a mixture of pesticide residues possessing different physic-chemical properties. After extraction, aliquots of the extract were subjected to three QuEChERS clean-up methods (PSA plus GCB or PSA plus C18 or PSA alone). Figure 1 represents a schematic diagram illustrating the workflow that was employed during method development. Extraction efficiencies of these clean-up methods were compared to extraction efficiencies of no clean-up methods to evaluate which of those methods will be best suited for our analysis. Instead, these samples were subjected to high centrifugation (12,000 rpm held at 4 °C) for 10 min and filtered through 0.22 μm PTFE filters on a Samplicity system (Merck Millipore, Germany). Each test was replicated three times.

Matrix effects

The effect of matrix co-extractives was performed by assessing ion suppression or enhancement effects of signals from chromatograms of matrix matched standard solutions compared to spiked extracts at the same concentration levels as per DG SANCO guidelines for LC-MS/MS analysis (SANCO/12571/2013). These were prepared using the extract of blank matrix (honey) covering a target analyte concentration range of 0.1 to 100 ng/g. Detection and quantification limits of the method were determined as described previously.

Validation of the analytical procedure

Analytes to be validated were spiked into the blank honey sample at LOR (1 ng/g) and at the lowest MRL level (0.01 mg/kg or 10 ng/g). Analysis was performed as

Table 1 Instrumental parameters of the MS/MS detector and retention times (RT) of the 96 pesticides standard mixture used for method development

Compound name	RT (min)	Parent ion (m/z)	[a]Trans1	CE1(V)	[a]Trans2	CE2(V)
Omethoate	2.72	214	125	20	109	25
Acetamiprid	2.84	223	126	20	90	35
Acephate	2.84	184	143	5	125	15
Propamocarb	3.19	189	144	5	102	15
Oxamyl	3.58	237	90	0	72	15
Methomyl	3.84	163	106	5	88	0
Thiamethoxam	3.95	292	211	5	181	20
Monocrotophos	3.95	224	193	0	127	10
Aldicarb	3.98	208	116	0	89	10
Imidacloprid	4.42	256	209	10	175	15
Thiabendazol	4.45	202	175	25	131	35
Cymiazole	4.70	219	171	25	144	35
Dimethoate	4.82	230	199	0	125	20
Thiacloprid	5.13	253	126	20	90	40
Propagite	5.25	368	231	5	175	10
Aldicarb fragment	5.43	116	89	4	70	4
Pirimicarb	5.90	239	182	10	72	20
Dichlorvos	6.13	221	109	12	79	24
Carbofuran	6.36	222	165	5	123	20
Nicosulfuron	6.40	411	213	12	182	16
Metsulfuron-methyl	6.51	382	199	20	167	15
Metribuzin	6.54	215	187	15	84	20
Malathion	6.64	331	126	5	99	10
Carbaryl	6.93	202	145	0	127	25
Fosthiazole	7.16	284	228	5	104	20
Thiodicarb	7.16	355	108	10	88	10
Amidosulfuron	7.22	370	261	10	218	20
DEET	7.75	192	119	16	91	32
Molinate	7.75	188	126	25	98	12
Tribenuron-methyl	7.87	396	155	5		
Metalaxyl	7.89	280	220	10	160	20
Flutriafol	8.01	302	70	15	123	30
Diuron	8.02	233	72	20	72	20
Isoxafluote	8.08	360	251	20	220	35
Methidathion	8.46	303	145	0	85	15
Flazasulfuron	8.73	408	182	15		
Fenobucarb	8.79	208	152	5	95	10
Azoxystrobin	9.01	404	372	10	344	25
Linuron	9.19	249	182	10	160	15
Fludioxonil	9.30	247	169	32	126	32
Promecarb	9.64	208	151	0		
Bosclid	9.67	343	271	28	307	12
Triadimefon	10.01	294	197	10	69	20

Table 1 Instrumental parameters of the MS/MS detector and retention times (RT) of the 96 pesticides standard mixture used for method development *(Continued)*

Bromuconazole	10.02	378	159	35	70	20
Bifenazate	10.09	301	170	15		
Cyproconazole	10.16	292	70	15	125	35
Fluquinconazole	10.27	376	349	16	307	24
Iprovalicarb	10.27	321	203	0	119	20
Triadimenol	10.36	296	70	5	99	10
Flufenacet	10.38	364	194	5	152	15
Bupirimate	10.42	317	166	20	108	25
Tetraconazole	10.45	372	159	30	70	20
Ethoprophos	10.48	243	131	15	97	30
Epoxyconazol	10.65	330	121	20	101	45
Cyazofamid	10.68	325	261	5	108	10
Cyprodinil	10.81	226	93	40	77	45
Fenbuconazole	10.85	337	125	35	70	15
Metolachlor	10.94	284	252	10	176	20
Fenamiphos	10.95	304	217	20	202	35
Flusilazole	10.97	316	247	15	165	25
Picoxystrobin	11.05	368	205	0	145	20
Tebufenozid	11.10	353	297	0	133	15
Diflubenzuron	11.17	311	158	10	141	35
Rotenone	11.24	395	213	20	192	20
Fipronil	11.25	435	330	12	250	28
Kresoxim-methyl	11.53	314	267	0	222	10
Tebuconazole	11.53	308	125	40	70	20
Procymidon	11.64	284	67	12	256	28
Benalaxyl	11.71	326	294	5	148	15
Diazinon	11.71	305	169	20	153	20
Coumaphos	11.76	363	307	16	227	28
Prochloraz	11.76	376	308	5	266	10
Chlorfenvinphos	11.77	359	170	40	155	8
Hexaconazole	11.93	314	159	30	70	15
Pyraclostrobin	12.04	388	194	5	163	20
Clofentezin	12.06	303	138	10	102	40
Pirimiphos-methyl	12.21	306	164	20	108	30
Spinosyn A	12.23	732	142	30	98	45
Metconazole	12.30	320	125	40		
Bitertenol	12.38	338	269	0	70	0
Chlorpyrifos-methyl	12.41	322	290	10	125	25
Trifloxystrobin	12.78	409	186	10	145	45
Spinosyn D	12.88	747	142	35	98	55
Ipconazole	12.97	334	125	45	70	25
Indoxacarb	12.99	528	203	45	150	20
Novaluron	13.32	493	158	20	141	45
Buprofezin	13.45	306	201	5	116	10

Table 1 Instrumental parameters of the MS/MS detector and retention times (RT) of the 96 pesticides standard mixture used for method development *(Continued)*

Profenofos	13.48	375	347	5	305	15
Ethion	13.93	385	199	4	143	20
Temephos	14.02	467	419	20	125	44
Chlorpyrifos	14.08	350	200	15	198	15
Pyriproxyfen	14.17	322	185	20	96	10
Lufenuron	14.19	511	158	20	141	45
Hexythiazox	14.46	353	228	10	168	25
Fenazaquin	15.35	307	161	10	57	25
Pyridaben	15.44	365	309	10	147	25
Bifenthrin	16.47	440	181	5	166	20
Etofenprox	16.57	394	177	10	107	45

[a]Transition ions used to quantify and qualify the targeted analytes

described previously. The recoveries and precision of the extraction method were determined as the average of five replicates. The method linearity was evaluated by assessing the signal responses of the targeted analytes from matrix-matched calibration solutions prepared by spiking blank extracts at seven concentration levels, from 0.1 to 100 ng/g, including the zero point or the blank. The method precision was expressed as percent relative standard deviation (%RSD) of the intra-day and inter-day analyses ($n = 5$). Blank matrices along with reagent blank were run during validation to ensure minimal risk of interferences, guarantee specificity of the method and to check for potential solvent contamination.

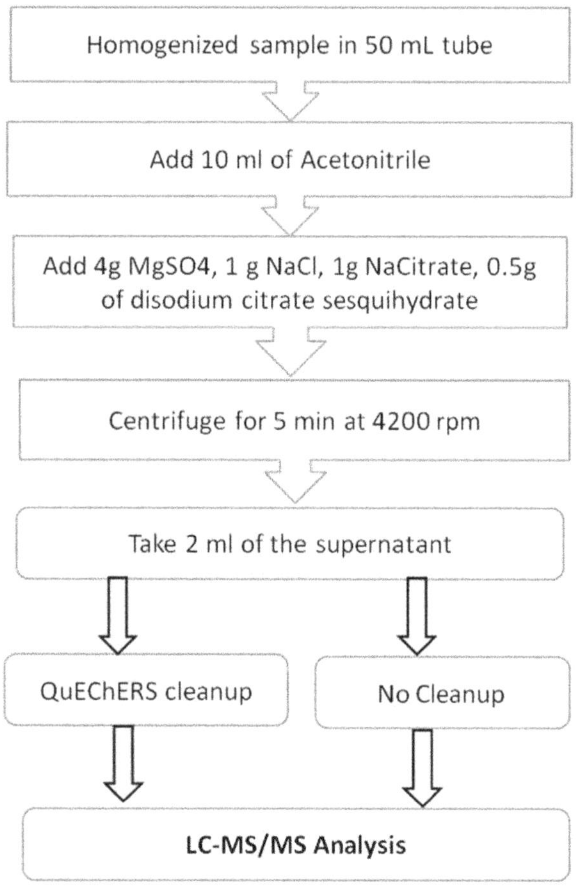

Application to real samples

The developed method was applied to conduct a preliminary study on chemical contaminants present in commercial honey in Africa. Ethiopia and Kenya were selected for this study as they are among the major producers of honey in Africa. From each country, 14 commercial honey samples were collected from local markets/farmers. These samples consisted of five honey samples from stingless (*Apis meliponina*) and nine honey bee (*Apis mellifera*) samples from various regions in each country. A total of 28 samples were analyzed at the African Reference Laboratory for Bee Health, International Centre of Insect Physiology and Ecology (*icipe*), Duduville Campus, Nairobi, Kenya at two different seasons (November 2014 and July 2015). All samples were stored in their original packaging under the recommended conditions prior to use and were prepared as previously described. The same calibration curve described above was run at the end of the sample series to check the stability of the detector after data acquisition of the unknown samples.

Statistical analysis

Data were analyzed using R version 3.1.1 (R Core Team 2014). For each pesticide or compound, the four cleanup methods were compared using one-way Analysis of Variance (ANOVA) and the means separated using the Student-Newman-Kuels (SNK) test. All tests were performed at 5 % significance level. Means with the same letter across are not significantly different.

```
Homogenized sample in 50 mL tube
              ↓
Add 10 ml of Acetonitrile
              ↓
Add 4g MgSO4, 1 g NaCl, 1g NaCitrate, 0.5g
of disodium citrate sesquihydrate
              ↓
Centrifuge for 5 min at 4200 rpm
              ↓
Take 2 ml of the supernatant
        ↓                 ↓
QuEChERS cleanup      No Cleanup
        ↓                 ↓
         LC-MS/MS Analysis
```

Fig. 1 Schematic diagram representing sample preparation workflow

Results and discussion

LC-MS/MS analysis

In this study, the methods investigated were selected based on the known matrix interferences expected from honey. Since sugars constitute the greatest proportion of honey (>75 %), three of the four methods investigated included PSA, as it removes sugars, along other interferences. Samples were spiked with a mixture of 96 pesticide standards at the default MRL value (0.01 mg/kg) since it provided great recoveries with the best reproducibility across multiple analytes during method development. Figure 2 shows representative chromatograms of honey extract processed using the four clean-up methods. Although the chromatographic profiles appeared similar for the four clean-up methods, the lowest recoveries were obtained from pesticides subjected to PSA combined with GCB clean-up with recoveries ranging from 5 to 117 % (Table 2). The use of GCB was important in removing pigment in honey; however, it also resulted in significant analyte losses during sample clean-up which could potentially lead to false negative results. Out of the 96 pesticides evaluated, 51 pesticides had the lowest recoveries from this method compared to the other methods (Table 2). Additionally, more than 45 % of the pesticides subjected to this method did not meet the minimum recommended criteria (>70 %) as indicated in the Guidance document on analytical quality control and validation procedures for pesticide residues analysis in food and feed (SANCO/12571/2013). On the other hand, for most pesticides, the best recoveries were obtained when PSA was used as a clean-up method. When compared to PSA plus C18 clean-up method, there were significant (P <0.05) differences in more than 10 % of the pesticides evaluated.

Results from this study also indicate that out of the 96 pesticides studied, only three pesticides, nicosulfuron (43 %), procymidon (58 %) and propamocarb (58 %), had recoveries that were below the acceptable limit when PSA was used alone. There was no significant (P <0.05) difference in recoveries for procymidon cleaned using C18 plus PSA (78 %) and PSA alone (58 %). Therefore, to improve recoveries for nicosulfuron and propamocarb, other alternatives must be considered. For instance, for nicosulfuron, based on the data provided in Table 2, the clean-up step can be omitted to yield 100 % recovery. This suggests that in the absence of clean-up resources, satisfactory information on levels of residue contamination in honey can still be achieved with minimal sample manipulations as found in other studies (Kujawski et al. 2014). Although omitting the clean-up step offers time savings in sample processing and is more economical, further precaution must be taken to avoid any potential clogging of the LC-MS system or eventual contamination of the MS ionization source. Based on the findings highlighted in Table 2, the use of PSA was selected as the best method for our analysis but was complemented with the no clean-up method to maximize on recoveries of all targeted pesticides.

Analytes eluted in 17 min followed by a short high-organic rinse to maintain the column and also in avoiding matrix carryover into the next sample. Elution of the remaining matrix material during subsequent analysis can cause unexpected matrix effects resulting in significant ionization inefficiencies. Matrix effects may either result to signal enhancement leading to recoveries >100 % or signal suppression resulting in poor recoveries. Aside from polar pesticides, other pesticides were well distributed across the elution window facilitating proper scan rate for scheduled

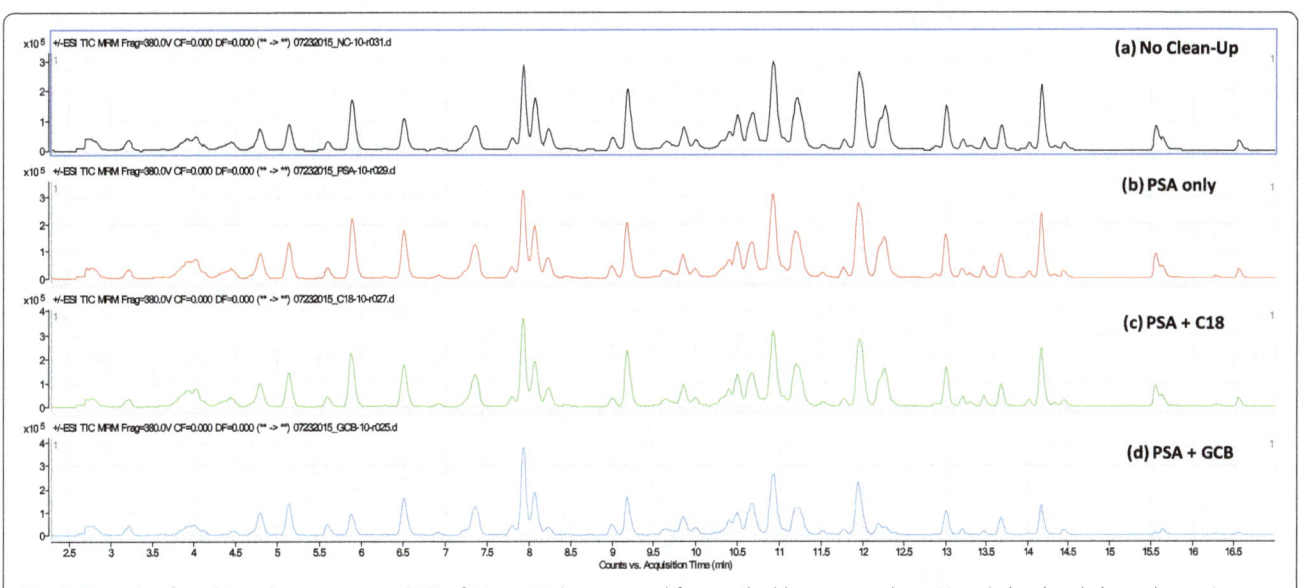

Fig. 2 Example of total ion chromatograms (TIC) of 96 pesticides extracted from spiked honey sample at 10 ng/g level and cleaned up using (**a**) No clean-up (**b**) PSA only (**c**) PSA+C18 (**d**) PSA+GCB

Table 2 Percentage recoveries (±SD) of 96 pesticides subjected to either QuEChERS clean-up methods or no clean-up

Compound name	% recovery at 10LOR (10 ng/g) ± SD			
	GCB+PSA	C18+PSA	PSA	No clean-up
Acephate	72.8 ± 0.8[b]	85.1 ± 0.6[a]	76.1 ± 0.8[ab]	52.5 ± 0.7[c]
Acetamiprid	98.1 ± 0.6[a]	99.8 ± 0.03[a]	99.6 ± 0.3[a]	74.8 ± 0.0[b]
Aldicarb fragment	104.5 ± 0.4[a]	100.5 ± 0.3[b]	97.9 ± 0.2[b]	70.6 ± 0.1[c]
Amidosulfuron	87.0 ± 0.8[b]	74.3 ± 0.4[c]	89.5 ± 0.1[b]	94.3 ± 0.2[a]
Azoxystrobin	77.0 ± 0.7[b]	108.8 ± 0.8[a]	106.5 ± 0.8[a]	101.8 ± 0.5[a]
Benalaxyl	88.9 ± 0.6[a]	97.3 ± 1.0[a]	97.5 ± 0.4[a]	97.7 ± 0.6[a]
Bifenazate	23.7 ± 0.8[b]	117.5 ± 0.3[a]	111.6 ± 1.2[a]	121.6 ± 0.4[a]
Bifenthrin	45.7 ± 1.1[b]	92.5 ± 0.9[a]	79.8 ± 0.8[a]	90.2 ± 0.02[a]
Bitertanol	88.9 ± 0.6[b]	105.6 ± 0.4[a]	99.4 ± 0.01[a]	100.6 ± 0.7[a,b]
Bosclid (Nicobifen)	39.6 ± 1.0[b]	113.1 ± 1.3[a]	106.3 ± 0.4[a]	115.3 ± 0.7[a]
Bromuconazole	85.2 ± 1.6[b]	96.9 ± 0.1[ab]	103.0 ± 0.4[a]	92.3 ± 0.4[ab]
Bupirimate	61.3 ± 0.6[b]	104.6 ± 0.1[a]	102.5 ± 0.6[a]	110.7 ± 1.4[a]
Buprofezin	84.6 ± 0.5[c]	104.0 ± 0.4[ab]	106.9 ± 0.4[a]	102.9 ± 0.8[b]
Carbaryl	98.2 ± 1.2[a]	110.9 ± 0.9[a]	102.5 ± 0.2[a]	72.6 ± 0.1[b]
Carbofuran	108.3 ± 0.8[b]	120.4 ± 0.9[a]	119.9 ± 0.1[ab]	64.1 ± 0.6[c]
Chlorfenvinphos	78.1 ± 0.7[c]	93.6 ± 0.1[b]	103.6 ± 0.2[a]	94.9 ± 0.7[b]
Chlorpyrifos	21.4 ± 0.6[b]	93.4 ± 0.6[a]	94.2 ± 0.3[a]	87.6 ± 1.0[a]
Chlorpyrifos-methyl	26.4 ± 0.8[c]	105.3 ± 0.5[a]	99.5 ± 0.1[a]	95.5 ± 0.4[b]
Clofentezin	6.2 ± 0.7[b]	97.3 ± 0.3[a]	98.5 ± 0.3[a]	91.1 ± 0.8[a]
Coumaphos	5.4 ± 0.4[b]	102.6 ± 1.3[a]	105.2 ± 0.5[a]	109.2 ± 0.5[a]
Cyazofamid	79.1 ± 0.2[c]	102.2 ± 0.1[a]	100.6 ± 0.2[a]	92.3 ± 0.2[b]
Cymiazol	56.0 ± 0.9[c]	92.2 ± 0.4[a]	89.5 ± 0.1[a]	75.2 ± 0.5[a]
Cyproconazole	100.3 ± 0.2[b]	87.8 ± 1.0[b]	106.8 ± 0.2[a]	92.3 ± 0.5[b]
Cyprodinil	10.5 ± 0.6[c]	90.8 ± 0.2[b]	104.5 ± 0.4[a]	105.0 ± 1.0[ab]
DEET	109.1 ± 0.4[a]	100.4 ± 1.2[a]	96.0 ± 0.0[a]	83.5 ± 0.5[b]
Diazinon	81.4 ± 0.01[b]	98.7 ± 0.3[a]	99.0 ± 0.3[a]	99.4 ± 1.1[a]
Dichlorvos	107.0 ± 0.8[a]	97.3 ± 0.6[b]	99.4 ± 0.3[b]	85.7 ± 0.2[c]
Diflubenzuron	18.9 ± 4.8[b]	101.9 ± 0.3[a]	106.3 ± 1[a]	104.8 ± 0.6[a]
Dimethoate	99.2 ± 1.0[a]	101.7 ± 0.2[a]	94.3 ± 0.3[a]	62.8 ± 0.1[b]
Diuron	33.0 ± 0.8[c]	100.7 ± 0.6[a]	108.1 ± 0.2[a]	92.1 ± 0.4[b]
Epoxyconazol	38.8 ± 2.6[b]	91.2 ± 0.2[a]	96.4 ± 1.1[a]	89.9 ± 0.6[a]
Ethion	78.7 ± 0.2[b]	98.3 ± 0.1[a]	103.0 ± 0.5[a]	95.1 ± 0.1[a]
Ethoprophos	87.7 ± 0.9[a]	94.1 ± 0.4[a]	98.3 ± 1.0[a]	90.8 ± 0.7[a]
Etofenprox	24.2 ± 0.5[b]	98.5 ± 0.4[a]	99.5 ± 0.1[a]	92.2 ± 0.0[a]
Fenamiphos	56.4 ± 1.0[b]	107.8 ± 0.2[a]	111.6 ± 0.3[a]	107.8 ± 0.5[a]
Fenazaquin	9.9 ± 1.7[d]	93.5 ± 0.5[b]	98.4 ± 0.1[a]	89.4 ± 0.1[a]
Fenbuconazole	40.5 ± 1.2[b]	107.8 ± 0.3[a]	109.1 ± 0.9[a]	107.0 ± 0.3[a]
Fenobucarb	94.7 ± 2.0[b]	80.6 ± 0.4[c]	101.6 ± 0.1[a]	90.4 ± 0.2[b]
Fipronil	107.9 ± 1.0[a]	111.4 ± 0.1[a]	108.3 ± 0.3[a]	111.1 ± 0.4[a]
Flazasulfuron	81.1 ± 1.5[b]	46.7 ± 0.5[d]	70.3 ± 0.3[c]	96.7 ± 0.4[a]
Fludioxonil	34.1 ± 2.9[b]	105.9 ± 0.5[a]	104.7 ± 0.1[a]	110.8 ± 0.5[a]
Flufenacet	102.8 ± 1.9[a]	117.5 ± 0.9[a]	103.8 ± 0.7[a]	100.9 ± 0.5[a]
Fluquinconazole	43.6 ± 2.0[b]	92.7 ± 1.1[a]	99.3 ± 0.9[a]	90.5 ± 0.9[a]

Table 2 Percentage recoveries (±SD) of 96 pesticides subjected to either QuEChERS clean-up methods or no clean-up *(Continued)*

Flusilazole	97.8 ± 0.8[b]	117.4 ± 0.7[a]	108.3 ± 0.8[ab]	98.6 ± 0.5[ab]
Flutriafol	94.3 ± 0.2[b]	97.9 ± 0.4[ab]	101.3 ± 0.1[a]	96.8 ± 0.5[ab]
Fosthiazate	101.8 ± 0.2[a]	107.1 ± 0.9[a]	103.6 ± 0.1[a]	70.1 ± 0.1[b]
Hexaconazole	90.0 ± 1.0[b]	99.3 ± 1.1[b]	110.5 ± 0.5[a]	97.9 ± 0.1[b]
Hexythiazox	77.2 ± 0.6[c]	99.8 ± 0.1[a]	99.6 ± 0.2[a]	94.1 ± 0.4[b]
Imidacloprid	80.3 ± 0.2[a]	88.3 ± 0.2[a]	87.6 ± 0.3[a]	66.4 ± 0.3[b]
Indoxacarb	56.4 ± 1.6[b]	103.0 ± 0.5[a]	102.1 ± 0.1[a]	96.1 ± 0.3[a]
Ipconazole	57.7 ± 0.9[b]	103.7 ± 0.2[a]	102.7 ± 0.1[a]	98.9 ± 0.5[a]
Iprovalicarb	58.6 ± 6.5[a]	95.6 ± 0.1[a]	99.1 ± 1.3[a]	74.5 ± 1.2[a]
Isoxaflutole	99.0 ± 0.3[a]	98.9 ± 0.5[a]	105.6 ± 0.4[a]	120.8 ± 2.0[a]
Kresoxim-methyl	74.7 ± 0.2[a]	96.0 ± 1.1[a]	93.1 ± 0.3[a]	92.3 ± 0.8[a]
Linuron	39.7 ± 3.4[b]	103.3 ± 0.03[a]	107.9 ± 0.5[a]	97.5 ± 0.1[a]
Lufenuron	5.9 ± 3.2[d]	105.1 ± 0.4[a]	98.6 ± 0.2[b]	95.4 ± 0.2[c]
Malathion	102.9 ± 0.2[a]	113.6 ± 0.2[a]	109.4 ± 0.3[a]	98.1 ± 0.1[a]
Metalaxyl	100.3 ± 0.1[a]	102.8 ± 0.4[a]	108.1 ± 0.2[a]	99.3 ± 0.5[a]
Metconazole	56.8 ± 1.7[b]	101.8 ± 0.8[a]	109.9 ± 0.1[a]	101.2 ± 0.2[a]
Methidathion	76.4 ± 0.7[b]	98.2 ± 0.4[a]	99.8 ± 0.5[a]	77.7 ± 0.2[b]
Methomyl	63.9 ± 7.9[a]	111.1 ± 0.6[a]	105.6 ± 0.3[a]	86.0 ± 0.4[a]
Metolachlor	88.6 ± 0.3[a]	100.2 ± 0.1[a]	97.7 ± 0.3[a]	98.4 ± 0.9[a]
Metribuzin	106.3 ± 0.1[a]	103.9 ± 0.5[a]	106.0 ± 0.5[a]	48.7 ± 0.3[b]
Metsulfuron-methyl	72.7 ± 1.7[b]	46.6 ± 0.7[c]	72.8 ± 0.5[b]	122.1 ± 0.4[a]
Monocrotophos	86.4 ± 0.1[a]	98.4 ± 0.2[a]	86.3 ± 0.3[a]	14.7 ± 7.1[b]
Nicosulfuron	43.9 ± 2.5[b]	19.0 ± 1.4[c]	42.6 ± 0.3[b]	100.6 ± 2.1[a]
Novaluron	16.5 ± 2.5[c]	104.4 ± 0.2[a]	107.4 ± 0.3[a]	92.1 ± 0.6[b]
Omethoat	88.0 ± 0.2[b]	90.6 ± 0.2[a]	86.4 ± 0.3[b]	83.5 ± 0.1[b]
Oxamyl	90.3 ± 0.1[a]	94.7 ± 0.0[a]	94.4 ± 0.1[a]	67.8 ± 0.2[b]
Picoxystrobin	76.7 ± 0.1[b]	94.3 ± 0.5[a]	92.6 ± 1.1[a]	79.1 ± 0.3[b]
Pirimicarb	37.1 ± 2.2[c]	103.6 ± 1.1[a]	102.3 ± 0.5[a]	81.2 ± 0.5[b]
Pirimiphos-methyl	44.6 ± 1.1[b]	99.2 ± 1.1[a]	99.1 ± 0.2[a]	91.2 ± 0.2[a]
Prochloraz	34.1 ± 3.7[b]	106.5 ± 0.4[a]	111.4 ± 0.4[a]	103.6 ± 0.1[a]
Procymidon	54.1 ± 1.0[a]	78.8 ± 0.7[a]	58.3 ± 1.7[a]	66.5 ± 0.7[a]
Profenofos	31.7 ± 2.6[b]	95.9 ± 0.7[a]	97.2 ± 0.2[a]	89.1 ± 0.4[a]
Promecarb	106.6 ± 0.4[a]	107.8 ± 0.1[a]	102.6 ± 0.2[a]	95.1 ± 0.1[a]
Propamocarb	75.9 ± 0.0[a]	49.3 ± 0.7[c]	57.9 ± 0.2[b]	73.4 ± 0.1[a]
Propargit	70.1 ± 1.4[ab]	98.7 ± 0.4[a]	99.4 ± 0.7[a]	64.2 ± 0.2[ab]
Pyraclostrobin	5.2 ± 0.9[c]	111.8 ± 0.0[a]	106.7 ± 0.2[a]	100.3 ± 0.3[b]
Pyridaben	53.1 ± 2.0[b]	100.0 ± 0.8[a]	102.6 ± 0.3[a]	89.9 ± 1.3[a]
Pyriproxyfen	36.9 ± 2.6[b]	98.1 ± 0.5[a]	97.9 ± 0.3[a]	92.3 ± 0.1[a]
Rotenone	46.7 ± 2.7[b]	101.2 ± 0.5[a]	100.1 ± 0.5[a]	96.5 ± 0.1[a]
Spinosyn A	12.5 ± 2.6[d]	98.3 ± 0.3[b]	109.2 ± 0.7[a]	87.0 ± 0.5[c]
Spinosyn D	9.3 ± 3.2[c]	92.0 ± 0.1[b]	101.6 ± 0.4[a]	90.5 ± 0.3[b]
Tebuconazole	59.7 ± 1.6[b]	99.5 ± 0.7[a]	114.4 ± 0.6[a]	111.4 ± 0.4[a]
Tebufenozid	100.9 ± 0.5[b]	112.0 ± 0.4[a]	120.4 ± 1.5[a]	98.6 ± 0.4[b]
Temephos	25.4 ± 3.0[c]	105.9 ± 0.4[a]	100.7 ± 0.2[ab]	98.0 ± 0.6[b]
Tetraconazole	92.5 ± 0.6[c]	108.2 ± 1.0[a]	102.4 ± 0.5[ab]	93.4 ± 0.6[bc]

Table 2 Percentage recoveries (±SD) of 96 pesticides subjected to either QuEChERS clean-up methods or no clean-up *(Continued)*

Thiabendazol	$17.2 \pm 1.6^{(d)}$	$82.2 \pm 0.4^{(a)}$	$77.3 \pm 0.1^{(b)}$	$57.9 \pm 0.2^{(c)}$
Thiacloprid	$85.3 \pm 1.2^{(a)}$	$99.5 \pm 0.6^{(a)}$	$94.7 \pm 0.0^{(a)}$	$61.0 \pm 0.3^{(b)}$
Thiamethoxam	$95.6 \pm 0.6^{(a)}$	$101.2 \pm 0.1^{(a)}$	$99.7 \pm 0.2^{(a)}$	$59.7 \pm 0.1^{(b)}$
Thiodicarb	$43.3 \pm 2.4^{(c)}$	$101.4 \pm 0.6^{(a)}$	$103.5 \pm 0.5^{(a)}$	$76.8 \pm 0.3^{(b)}$
Triadimenol	$92.8 \pm 0.1^{(b)}$	$108.6 \pm 1.0^{(a)}$	$111.3 \pm 0.03^{(a)}$	$97.9 \pm 0.6^{(a)}$
Triadimefon	$117.7 \pm 0.3^{(a)}$	$107.4 \pm 0.8^{(a)}$	$115.7 \pm 0.1^{(a)}$	$110.6 \pm 0.7^{(a)}$
Tribenuron-methyl	$64.2 \pm 1.7^{(b)}$	$73.3 \pm 0.1^{(a)}$	$81.5 \pm 0.1^{(a)}$	$63.4 \pm 0.4^{(b)}$
Trifloxystrobin	$60.1 \pm 1.7^{(b)}$	$101.2 \pm 0.2^{(a)}$	$103.8 \pm 0.1^{(a)}$	$93.7 \pm 0.1^{(a)}$

*For each pesticide, mean recoveries with the same letter are not significantly different

MRM methods of targeted analytes as shown in Fig. 3. This figure illustrates an example of MRM chromatogram of the 96 pesticides targeted in this study that were extracted from spiked honey after PSA clean up. From this chromatogram, each colored peak represent a unique pesticide identified based on the MRM transition ions. A detailed summary indicating the identity of each peak shown in Fig. 3 and their corresponding retention times along with their molecular masses are provided in Table 1.

Validation of the selected method

The developed method was validated following the guidelines provided in the Guidance document on analytical quality control and validation procedures for pesticide residues analysis in food and feed (SANCO/12571/2013). To meet these guidelines, the method was validated in terms of recovery, linearity, LOQ, matrix effects, intra-day and inter-day precision. The mean recovery values used in this study were within the range of 70–120 %, with an associated repeatability, RSD <20 %, for all compounds within the scope of the method. Matrix-matched calibration standards were used to calculate recoveries as this helped in compensating for any matrix effects arising from matrix interferences or co-extractives that can change the ionization efficiency of an analyte causing signal suppression or enhancement leading to poor recoveries. This could have an adverse effect on the quality of the data and

can erroneously result in false positive or negative results. It is therefore imperative for any LC-MS/MS method to give acceptable quantitative results; matrix effects must be considered (Ferrer et al. 2011; Kittlaus et al. 2011).

Table 3 shows the list of pesticides validated and demonstrates the summarized recovery results along with the linearity of the validated analytes. This table illustrates recoveries obtained at LOR using PSA and no clean-up approach. Percent recovery values for these analytes were calculated using matrix-matched calibration curves. The LOR for the method was determined as the lowest spike level of the validation meeting these method performance acceptability criteria. Although the LOD and LOR varied depending on the pesticides in question, most compounds could be detected at 0.1 and quantified below 1 ng/g. Overall, the LOD and the LOR was set at 0.5 and 1 ng/g, respectively. From this study, approximately 10 % of the studied compounds had poor recoveries from either method but there was tremendous improvement on recoveries when both methods were combined. In this case, all pesticides, except for two (fluquinconazole –68 % and propamocarb - 63 %) had good recoveries which were well within the recommended limits provided in SANCO/12575/2013 document. It is worth noting that pesticides with good recoveries had good reproducibility (RSD <20 %) whereas those with poor recoveries were characterized by poor reproducibility. As a result, during

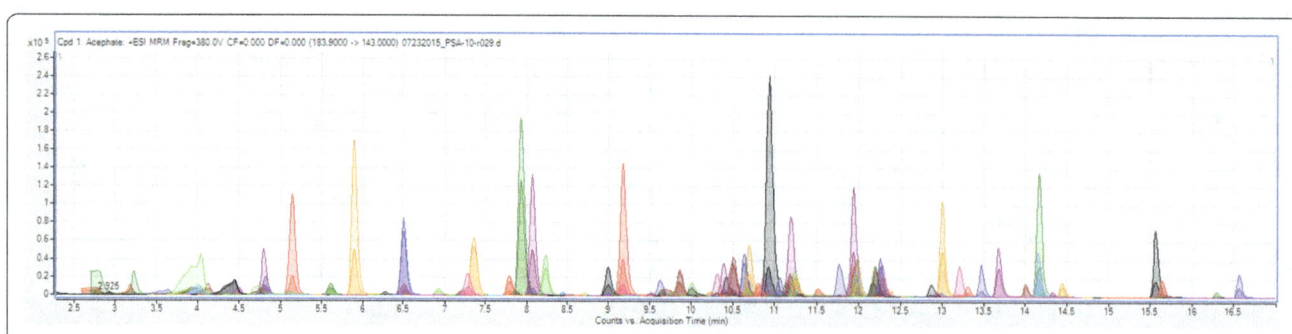

Fig. 3 Representative example of MRM chromatogram of 96 pesticides extracted from a spiked honey sample at 10 ng/g level and cleaned up using PSA only

Table 3 Extraction efficiencies of validated pesticides spiked at LOR, precision in terms of RSD ($n = 5$) and coefficients of determination for the investigated pesticides

| Compound name | % recovery at LOR (1 ng/g) | | | | |
| | No clean-up | | PSA | | |
	% recovery	% RSD, $n = 5$	% recovery	% RSD, $n = 5$	R^2
Acephate	70.1	2.6	97.4	4.7	0.9989
Acetamiprid	82.6	3.2	115.0	3.9	0.9982
Aldicarb fragment	73.0	0.9	102.8	1.5	0.9476
Amidosulfuron	87.1	1.2	54.2	24.3	0.9990
Azoxystrobin	84.5	4.5	90.8	4.9	0.9986
Benalaxyl	87.6	3.3	87.5	1.5	0.9989
Bifenazate	89.6	2.4	96.3	4.3	0.9919
Bifenthrin	91.3	6.5	99.0	5.0	0.9996
Bitertanol	85.7	0.4	89.9	0.1	0.9982
Bosclid (Nicobifen)	81.7	0.7	88.2	1.5	0.9977
Bromuconazole	117.3	11.5	102.3	9.3	0.9991
Bupirimate	87.6	1.7	95.6	4.1	0.9998
Buprofezin	81.9	9.0	86.1	4.3	0.9985
Carbaryl	83.8	1.8	118.3	1.3	0.9980
Carbofuran	67.8	3.1	115.8	5.1	0.9987
Chlorfenvinphos	95.0	5.7	103.0	1.1	1.0000
Chlorpyriphos	93.1	6.2	99.3	1.8	0.9996
Chlorpyriphos-methyl	82.9	16.9	92.7	8.6	0.9990
Clofentezin	93.8	1.8	94.0	1.9	0.9996
Coumaphos	70.3	4.7	75.9	3.3	0.9935
Cyazofamid	108.4	9.4	113.6	1.6	0.9992
Cymiazol	86.6	0.5	109.6	5.9	0.9987
Cyproconazole	84.8	0.9	89.2	4.1	0.9976
Cyprodinil	93.7	1.0	96.7	1.6	0.9995
DEET	68.5	3.3	96.1	4.0	0.9999
Diazinon	87.6	3.6	100.4	0.5	0.9999
Dichlorvos	72.8	15.4	101.4	10.9	0.9998
Diflubenzuron	95.6	0.4	99.5	3.9	0.9999
Dimethoate	67.5	2.9	108.7	0.3	0.9997
Diuron	55.9	0.7	98.4	4.7	0.9979
Epoxyconazol	100.4	8.4	100.7	0.4	1.0000
Ethion	74.9	7.1	82.8	8.9	0.9986
Ethoprophos	91.6	3.7	103.8	11.8	0.9998
Etofenprox	95.2	0.8	101.3	2.8	1.0000
Fenamiphos	100.2	8.9	103.8	2.0	0.9991
Fenazaquin	95.1	2.3	99.5	5.2	0.9999
Fenbuconazole	94.9	1.9	101.3	0.4	1.0000
Fenobucarb	83.0	5.1	96.7	0.3	0.9999
Fipronil	99.7	2.6	98.3	13.7	0.9982

Table 3 Extraction efficiencies of validated pesticides spiked at LOR, precision in terms of RSD ($n = 5$) and coefficients of determination for the investigated pesticides (Continued)

Flazasulfuron	87.1	3.7	25.7	62.6	0.9995
Fludioxonil	75.6	5.3	62.2	1.8	0.9981
Flufenacet	82.2	14.1	79.1	17.0	0.9926
Fluquinconazole	68.3	6.9	68.3	1.7	0.9981
Flusilazole	79.5	4.7	90.9	6.1	0.9991
Flutriafol	93.5	4.9	93.9	1.8	0.9999
Fosthiazate	73.2	4.1	109.4	1.6	0.9998
Hexaconazole	94.6	0.3	98.8	0.9	1.0000
Hexythiazox	89.7	2.7	98.2	0.8	0.9997
Imidacloprid	69.3	6.0	101.4	3.4	0.9996
Indoxacarb	93.2	1.1	96.8	1.9	0.9998
Ipconazole	89.8	2.4	95.7	4.9	0.9996
Iprovalicarb	90.7	13.6	100.6	9.9	0.9988
Isoxaflutole	91.4	0.5	76.0	5.7	0.9951
Kresoxim-methyl	108.8	10.7	113.5	13.2	0.9962
Linuron	76.3	7.1	74.0	3.0	0.9958
Lufenuron	91.9	15.8	90.2	9.0	0.9990
Malathion	78.8	8.3	86.3	1.0	0.9972
Metalaxyl	83.1	1.3	91.4	0.7	0.9995
Metconazole	78.2	11.0	81.0	8.2	0.9984
Methidathion	72.7	8.2	89.5	2.6	0.9993
Methomyl	96.1	0.4	114.6	3.5	0.9996
Metolachlor	111.6	1.1	115.3	3.7	0.9984
Metribuzin	66.3	14.3	115.4	2.8	0.9976
Metsulfuron-methyl	121.6	0.5	53.5	28.1	0.9994
Monocrotophos	77.9	0.4	112.5	2.7	0.9968
Nicosulfuron	90.1	2.1	10.5	39.9	0.9994
Novaluron	94.9	0.2	95.5	0.6	0.9999
Omethoate	78.3	5.9	81.3	15.8	0.9995
Oxamyl	72.9	15.3	112.4	2.5	0.9995
Picoxystrobin	93.1	2.6	101.4	9.4	0.9992
Pirimicarb	87.0	2.5	113.1	0.4	0.9992
Pirimiphos-methyl	99.3	1.0	112.2	4.4	0.9999
Prochloraz	79.9	0.4	80.1	1.2	0.9978
Procymidon	79.9	6.0	111.7	10.9	0.9956
Profenofos	98.5	3.7	109.2	11.2	0.9994
Promecarb	85.1	1.3	87.7	0.2	0.9993
Propamocarb	62.8	0.0	29.7	83.9	0.9996
Propargit	25.2	52.7	76.5	19.8	0.9983
Pyraclostrobin	80.0	10.4	88.1	5.6	0.9972
Pyridaben	84.3	1.5	87.2	1.2	0.9998
Pyriproxyfen	92.7	4.4	100.4	1.0	1.0000
Rotenone	97.7	1.9	113.0	1.7	0.9985

Table 3 Extraction efficiencies of validated pesticides spiked at LOR, precision in terms of RSD ($n = 5$) and coefficients of determination for the investigated pesticides *(Continued)*

Spinosyn A	92.0	0.8	95.9	0.8	0.9999
Spinosyn D	85.1	0.1	92.3	2.0	0.9997
Tebuconazole	92.1	4.0	97.1	0.1	0.9993
Tebufenozid	75.3	0.5	87.7	9.5	0.9958
Temephos	94.4	1.6	95.5	2.6	0.9999
Tetraconazole	83.1	6.1	108.6	3.1	0.9997
Thiabendazol	91.3	2.7	110.9	1.9	0.9946
Thiacloprid	67.7	0.4	108.2	1.8	0.9995
Thiamethoxam	55.4	5.5	100.0	1.6	1.0000
Thiodicarb	78.3	1.0	102.4	3.5	0.9996
Triadimenol	73.7	8.5	69.8	13.5	0.9903
Triadimefon	85.6	8.9	83.8	2.2	0.9948
Tribenuron-methyl	71.8	4.9	65.7	13.4	0.9997
Trifloxystrobin	94.7	3.1	100.6	0.6	0.9999

recovery studies, blank matrix was fortified at 10 times the LOR since it gave the best reproducibility for all studied compounds. The method linearity was evaluated by assessing the signal responses of the targeted analytes from matrix-matched calibration solutions prepared in blank extracts at seven concentration levels. The developed method was proven satisfactory with linear chromatographic response for the tested pesticides, ranging from 0.1 to 100 ng g^{-1}. Majority of the correlation coefficients (R^2) was higher or equal to 0.995, see Table 3.

Application of the method to real samples

As a natural product manufactured by bees, honey is considered to be free from any extraneous material. However, chemical residues have been reported in honey by several investigators. The presence of these residues in honey has prompted the need for setting up monitoring programs to determine the proper assessment of human exposure to pesticides (Choudhary and Sharma 2008). Unfortunately, there is no homogeneity on MRLs as different national regulations have established their own maximum concentrations of pesticide residues permitted in honey. In the absence of MRLs set for honey in the two African countries studied, the European Union set MRLs were employed and where no MRL existed, it was presumed at 10 which is the default MRL for pesticides with no specific value set as recommended in Regulation(EC)No 396/2005.

So far, there is little information that is currently available on chemical residues present in honey or hive products from most African countries (Muli et al. 2014; Eissa et al. 2014). Previous studies have shown that

whereas in North America honey bees are exposed to at least 7 pesticides per food visit, this is not the case in Africa (Mullin et al. 2010). Results from a recent study carried out in Kenya detected less than four pesticides for the whole study duration at very minimal concentrations in honey bees and their hive products (Muli et al. 2014). In the current study, a preliminary analysis of pesticide residues in 28 honey samples obtained from local farmers' markets and supermarkets from various regions in Kenya and Ethiopia during the period of November 2014 to July 2015 revealed the presence of 17 pesticide residues out of the 96 pesticides investigated. The concentrations for each detected pesticide were compared with the set MRL values. Table 4 indicates the summarized results obtained from the two countries. Our preliminary results show that, with the exception of malathion, an organophosphate that has multiple uses in Africa, no other pesticide was detected at a level higher than the set MRL levels. For most pesticides, the levels obtained were about 10-fold lower than the set MRL levels, with concentration levels at <100 ng/g. However, the maximum concentration detected for malathion was 0.092 mg/kg, a level that far exceeds its acceptable MRL of 0.05 mg/kg. Although this compound is quickly metabolized from the body and is known to be non-persistent in the environment, exposure to the levels detected (0.092 mg/kg) in this study over a long period could result in adverse health effects to both humans and honey bees. Thus, further investigation is required to determine its cumulative effects and whether there are any potential synergistic effects when other contaminants are present. Malathion is also considered to be highly toxic to honey bees with LD_{50} of 0.16 µg/bee (Allison 2011). It is worth noting that data from the present study does not reflect seasonality of pesticide present in honey samples obtained from the two countries. This would require in-depth systematic studies using large samples obtained directly from specific beekeeping sites over different seasons in the two countries. Follow up studies are underway to investigate how seasonality affects residues present in honey from various African countries.

Conclusion

A highly efficient approach for determining pesticide residues in honey with good recoveries was developed. This approach involved using a modified QuEChERS method along with or without any clean-up. The viability of this approach was demonstrated by using 96 pesticides. About 98 % of these pesticides investigated had recoveries that are well within the acceptable limits of 70–120 %. The methods were linear (>0.995) over the range tested (0.1–100 ng/g) with LOR for

Table 4 Detected pesticide residues in honey obtained from Kenya and Ethiopia

SampleID	ACTM	AF	CF	CAR	CHP	Cy	DEET	DDVP	DM	BPMC	HEX	Mal	Met	Metri	Rot	TBN	THIA
Kenya																	
Taita	<LOQ	N/D	N/D	N/D	<LOQ	<LOQ	0.708	N/D	N/D	N/D	<LOQ	56.9	N/D	49.4	N/D	<LOQ	<LOQ
VapA	<LOQ	1.37	<LOQ	N/D	<LOQ	N/D	N/D	N/D	N/D	N/D	N/D	92.3	1.81	N/D	N/D	N/D	<LOQ
Cab	<LOQ	N/D	N/D	N/D	<LOQ	1.59	<LOQ	N/D	N/D	<LOQ	<LOQ	N/D	1.95	14.1	N/D	N/D	N/D
Nak	N/D	<LOQ	N/D	<LOQ	N/D	N/D	<LOQ	N/D	N/D	N/D	N/D	N/D	N/D	N/D	N/D	N/D	N/D
Ken	<LOQ	<LOQ	N/D	N/D	N/D	N/D	<LOQ	N/D	N/D	N/D	N/D	N/D	N/D	N/D	N/D	N/D	N/D
Mwi	N/D	N/D	N/D	1.26	N/D	N/D	1.01	N/D	<LOQ	N/D	N/D	N/D	N/D	N/D	N/D	N/D	N/D
Kak	<L/OQ	<LOQ	N/D	N/D	N/D	<LOQ	<LOQ	N/D	N/D	N/D	N/D	N/D	N/D	<LOQ	N/D	N/D	N/D
ML	ND	N/D	N/D	<LOQ	N/D	N/D	<LOQ	N/D	N/D	N/D	N/D	N/D	N/D	34.0	N/D	N/D	N/D
HR	<LOQ	N/D	N/D	2.87	N/D	<LOQ	<LOQ	N/D	N/D	N/D	N/D	N/D	N/D	70.4	N/D	N/D	N/D
Gedi	N/D	N/D	N/D	N/D	N/D	N/D	N/D	N/D	N/D	N/D	N/D	N/D	N/D	N/D	N/D	N/D	N/D
K-B	N/D	N/D	N/D	<LOQ	N/D	N/D	1.37	2.58	N/D	N/D	N/D	N/D	N/D	N/D	N/D	N/D	N/D
K-M	N/D	N/D	N/D	N/D	N/D	N/D	<LOQ	<LOQ	<LOQ	N/D	N/D	N/D	N/D	<LOQ	N/D	N/D	N/D
K-N	<LOQ	N/D	N/D	<LOQ	N/D	N/D	<LOQ	N/D	N/D	N/D	N/D	N/D	N/D	N/D	N/D	N/D	N/D
VapB	<LOQ	<LOQ	N/D	N/D	<LOQ	N/D	1.09	N/D	N/D	N/D	1.66	76.7	5.29	N/D	N/D	<LOQ	<LOQ
Ethiopia																	
MB	N/D	N/D	N/D	N/D	N/D	N/D	N/D	N/D	N/D	N/D	N/D	N/D	N/D	9.52	N/D	N/D	N/D
Tol	<LOQ	N/D	N/D	N/D	<LOQ	<LOQ	<LOQ	N/D	N/D	<LOQ	N/D	60.5	4.77	11.2	N/D	<LOQ	N/D
Tig	<LOQ	<LOQ	N/D	N/D	<LOQ	<LOQ	<LOQ	N/D	N/D	N/D	<LOQ	15.3	N/D	N/D	N/D	N/D	N/D
SapV	<LOQ	<LOQ	<LOQ	N/D	<LOQ	<LOQ	<LOQ	N/D	N/D	<LOQ	<LOQ	45.1	1.25	14.2	N/D	<LOQ	<LOQ
E-1	N/D	N/D	N/D	N/D	N/D	N/D	N/D	N/D	N/D	N/D	N/D	N/D	N/D	2.60	N/D	N/D	N/D
E-2	N/D	N/D	N/D	N/D	N/D	N/D	N/D	1.16	<LOQ	N/D	N/D	N/D	N/D	44.2	N/D	N/D	N/D
E-3	<LOQ	N/D	N/D	N/D	N/D	N/D	N/D	N/D	N/D	N/D	N/D	N/D	N/D	N/D	N/D	N/D	N/D
E-4	ND	N/D	N/D	N/D	N/D	N/D	N/D	N/D	N/D	N/D	N/D	N/D	N/D	7.95	6.99	N/D	N/D
E-5	ND	N/D	N/D	N/D	N/D	N/D	N/D	N/D	N/D	N/D	N/D	N/D	N/D	N/D	N/D	N/D	N/D
E-M1	<LOQ	<LOQ	N/D	<LOQ	N/D	N/D	N/D	N/D	N/D	N/D	N/D	N/D	N/D	N/D	N/D	N/D	N/D
E-H	<LOQ	N/D	N/D	<LOQ	N/D	N/D	N/D	N/D	N/D	N/D	N/D	N/D	N/D	N/D	N/D	N/D	N/D
Tol2	N/D	N/D	1.10	N/D	N/D	N/D	N/D	3.46	<LOQ	N/D	N/D	N/D	N/D	N/D	N/D	N/D	N/D
Tig2	N/D	N/D	N/D	N/D	N/D	N/D	4.98	N/D	N/D	N/D	N/D	N/D	N/D	N/D	N/D	N/D	N/D
Sap S2	<LOQ	N/D	<LOQ	N/D	<LOQ	10.5	<LOQ	N/D	N/D	<LOQ	<LOQ	22.3	N/D	21.0	N/D	N/D	N/D
MRL	50	10	10	50	*10	*10	*10	*10	*10	*10	*10	50	50	100	10	50	10

*Set at default MRL value; *N/D* not detected, *<LOQ* below the quantification limits
Identified pesticide residues: *ACTM* Acetamiprid, *AF* Aldicarb fragment, *CF* Carbofuran, *CHP* Chlopyrifos, *Cy* Cymiazole, *DDVP* Dichlorvos, *DM* Dimethoate, *BPMC* Fenobucarb, *HEX* Hexaconazole, *Mal* Malathion, *Met* Metalaxyl, *Metri* Metribuzin, *Rot* Rotenone, *TBN* Tebuconazole, *THIA* Thiomethoxam

most pesticides at 1 ng/g or ppb. The applicability of the developed methods to real samples was tested by performing a preliminary study of commercial honey from Africa. A total of 17 pesticide residues were detected at levels 10-fold lower than their set MRL values except malathion which was detected at almost 2-fold higher than its set MRL. Overall, these results suggest that honey from these regions maybe safe for both bees and human consumption but further investigation is required to determine the cumulative effect of these pesticides. In-depth follow up studies using this method are underway to verify this observation in honey samples collected from different agro-ecological regions from various African countries.

Acknowledgements
The authors would like to thank *icipe* management for their support, Daisy Salifu of *icipe*, for her support in statistical analysis, colleagues from African Reference Laboratory Bee Health (ARLBH) at *icipe* for their support, Beatrice Njuguna of ARLBH for providing the honeybee samples for analysis. This work has been supported financially by the European Union grant number DCI-FOOD-2013/313-659.

Authors' contributions

JI designed the study and the experimental setting, performed the analytical work and wrote the manuscript. BT contributed in experimental design and critically revised the manuscript. SK edited and proofread the manuscript. All authors read and approved the final manuscript.

Competing interests

The authors declare that they have no competing interests.

References

Allison C. Malathion.toxipedia. 2011. http://www.toxipedia.org/display/toxipedia/Malathion. Accessed 13 July 2015.

Anastassiades M, Lehotay SJ, Stajnbaher D, Schenck FJ. Fast and easy multiresidue method employing acetonitrile extraction/partitioning and "dispersive solid phase extraction" for the determination pesticide residues in produce. J AOAC Int. 2003;86:412.

Barganska Z, Slebioda M, Namiesnik J. Pesticide residues levels in honey from apiaries located in Northern Poland. Food Control. 2013;31:196.

Blasco C, Fernandez M, Pena A, Lino C, Silveira MI, Font G, Pico Y. Assessment of pesticide residues in honey samples from Portugal and Spain. Agric Food Chem. 2003;51:8132–8.

Blasco C, Vazquez-Roig P, Onghena M, Masia A, Pico Y. Analysis of insecticides in honey by liquid chromatogrpahy-ion trap mass spectrometry: comparison of different extraction procedures. J Chromatogr A. 2011;1218:4892–901.

Bogdanov S. Contaminants of bee products. Apidologie. 2006;37:1–18.

Bogdanov S, Jurendic T, Sieber R, Gallmann P. Honey for nutrition and health. J Am Coll Nutr. 2008;27:677–89.

Brodschneider R, Crailsheim K. Nutrition and health in honey bees. Apidologie. 2010;41:278–94.

Campillo N, Pen˜alver R, Aguinaga N, Herna´ndez-Co´rdoba M. Solid-phase microextraction and gas chromatography with atomic emission detection for multiresidue determination of pesticides in honey. Anal Chim Acta. 2006;562:9–15.

Choudhary A, Sharma DC. Pesticide residues in honey samples from Himachal Pradesh (India). Bull Environ Contam Toxicol. 2008;80:417–22.

De Pinho GP, Neves AA, de Queiroz MELR, Silverio FO. Optimization of the liquid-liquid extraction method at low temperature purification (LLE-LTP) for pesticide residue analysis in honey samples by gas-chromatography. Food Control. 2010;21:1307–11.

Eissa F, El-Sawi S, Zidan NE. Determining pesticide residues in honey and their potential risk to consumers. Pol J Environ Stud. 2014;23:1573–80.

Erdog˘rul O˙. Levels of selected pesticides in honey samples from Kahramanmaras,Turkey. Food Control. 2007;18:866–71.

Fernández MA, Simal LJ. Simplified method for the determination of organochlorine pesticides in honey. Analyst. 1991;116:269–71.

Ferrer C, Lozano A, Agüera A, Girón AJ, Fernández-Alba AR. Overcoming matrix effects using the dilution approach in multiresidue methods for fruits and vegetables. J Chromatogr A. 2011;1218:7634.

Fontana AR, Camargo AB, Altamirano JC. Coacervative microextraction ultrasound-assisted back-extraction technique for determination of organophosphates pesticides in honey samples by gas chromatography-mass spectrometry. J Chromatogr A. 2010;1217:6334–41.

Garcia-Chao M, Agruna MJ, Calvete GF, Sakkas V, Llompart M, Dagnac T. Validation of an off line solid phase extraction liquid chromatography-tandem mass spectrometry method for the determination of systemic insecticide residues in honey and pollen samples collected in apiaries from NW Spain. Anal Chim Acta. 2010;672:107–33.

Gill RJ, Ramos-Rodriguez O, Raine NE. Combined pesticide exposure severely affects individual- and colony-level traits in bees. Nature. 2012;491:105–8.

Herrera A, Perez AC, Conchello P, Bayarri S, Lazaro R, Yague C, Arino A. Determination of pesticides and PCBs in honey by solid-phase extraction cleanup followed by gas chromatogram with electron-capture and nitrogen-phosphorus detection. Anal Bioanal Chem. 2005;381:695.

Irani M. Determination of pesticides residues in honey samples. Bull Environ Contam Toxicol. 2009;83:818–21.

Kittlaus S, Schimanke J, Kempe G, Speer K. Assessment of sample cleanup and matrix effects in the pesticide residue analysis of foods using postcolumn infusion in liquid chromatography-tandem mass spectrometry. J Chromatogr A. 2011;1218:8399.

Kruve A, Künnapas A, Herodes K, Leito I. Matrix effects in pesticide multi-residue analysis by liquid chromatography-mass spectrometry. J Chromatogr A. 2008;1187:58.

Kujawski MW, Namiesnik J. Challenges in preparing honey samples for chromatographic determination of contaminants and trace residues. TrAC Trends Anal Chem. 2008;27:785–93.

Kujawski MW, Namiesnik J. Levels of 13 multi-class pesticide residues in Polish honeys determined by LC-ESI-MS/MS. Food Control. 2011;22:914–9.

Kujawski MW, Barganska Z, Marciniak K, Miedzianowska E, Kujawski JK, Slebioda M, Namiesnik J. Determining pesticide contamination in honey by LC-ESI-MS/MS – comparison of pesticide recoveries of two liquid-liquid extraction based approaches. LWT Food Sci Technol. 2014;56:517–23.

Martel AC, Zeggane S, Auri`eres C, Drajnudel P, Faucon JP, Aubert M. Acaricide residues in honey and wax after treatment of honey bee colonies with Apivar or Asuntol50. Apidologie. 2007;38:534–44.

Muli E, Patch H, Frazier M, Frazier J, Torto B, Baumgarten T, Kilonzo J, Kimani JN, Mumoki F, Masiga D, Tumlinson J, Grozinger C. Evaluation of the distribution and impacts of parasites, pathogens, and pesticides on honey bee (Apis mellifera) populations in East Africa. PLoS One. 2014;9(4):e94459.

Mullin CA, Frazier M, Frazier JL, Ashcraft S, Simonds R, Vanengelsdorp D, Pettis JS. High levels of miticides and agrochemicals in North American apiaries: implications for honey bee health. PLoS One. 2010;5(3):e9754.

R Core Team. R: a language and environment for statistical computing. Vienna: R Foundation for Statistical Computing; 2014. URL http://www.R-project.org/.

Regulation (EC) No 396/2005 with annexes. European Commission: Food Safety/Plants/Pesticides/Maximum Residue Levels. http://ec.europa.eu/food/plant/protection/pesticides/community_legislation_en.htm. Accessed 9 July 2015.

Rissato SR, Galhiane MS, De Almeida MV, Gerenutti M, Apon BM. Multiresidue determination of pesticides in honey samples by gas chromatography-mass spectrometry and application in environmental contamination. Food Chem. 2007;101:1719–26.

SANCO/12571/2013. Guidance document on analytical quality control and validation procedures for pesticide residues analysis in food and feed. 2013. Available online: http://ec.europa.eu/food/plant/plant_protection_products/guidance_documents/docs/qualcontrol_en.pdf. Accessed 14 Apr 2015.

Vanengelsdorp D, Meixner MD. A historical review of managed honey bee populations in Europe and the United States and the factors that may affect them. J Invertebr Pathol. 2010;103:S80–95.

Wang J, Kliks MM, Jun S, Li QX. Residues of organochlorine pesticides in honeys from different geographical regions. Food Res Int. 2010;43:2329–34.

Weist L, Bulete A, Giroud B, Fratta C, Amic S, Lambert O, Pouliquen H, Arnaudguilhem C. Multiresidue analysis of 80 environmental contaminants in honeys, honeybees and pollens by one extraction procedure followed by liquid and gas chromatography coupled with mass spectrometric detection. J Chromatogr A. 2011;1218:5743.

4

Investigation of heavy metal contents in Cow milk samples from area of Dhaka, Bangladesh

Md Iftakharul Muhib[1], Muhammed Alamgir Zaman Chowdhury[1,2], Nusrat Jakarin Easha[1], Md Mostafizur Rahman[1,4]* (ID), Mashura Shammi[1,5], Zeenath Fardous[2], Mohammad Latiful Bari[3], M. Khabir Uddin[1], Masaaki Kurasaki[4] and Md Khorshed Alam[2]

Abstract

Background: Cow milk is considered as one of the responsible food sources contaminated with heavy metals. The objectives of the study were to assess the content of selected metals in cow milk and its associated human health risks in the food chain of Bangladesh. A total of 90 cow milk samples of Branded, Dairy and Domestically produced milk were collected randomly from different sources of Savar Upazila in Dhaka area. Cadmium (Cd), chromium (Cr), lead (Pb), manganese (Mn), copper (Cu) and iron (Fe) contents in collected milk samples were determined using Flame Atomic Absorption Spectrometry (FAAS). To ensure quality control, one of the best quality control parameters i.e. recovery test; from eight various sample digestion methods were used. The Hazard Quotient (HQ) and Carcinogenic Risk (CR) values were also calculated.

Results: From the results, it was found that, the orders of heavy metal content in brand, dairy and domestic cow milk were Cr > Fe > Cu > Mn > Cd > Pb, Cr > Fe > Mn > Cu > Cd > Pb and Fe > Cr > Mn > Cu > Cd > Pb, respectively. Among the six metals, only Cr showed to exceed the highest Estimated Daily Intake (EDI) rate (for brand cow milk: 0.413 mg/day, dairy farm cow milk: 0.243 mg/day, domestic cow milk: 0.352 mg/day),and the comparison percentages of calculated values per permeable values were as follows; 206.5 % for brand cow milk,121.5 % for dairy farm cow milk and 176.0 % for domestic cow milk. Hazard Quotients (HQ) values and Carcinogenic Risk (CR) values were found within the acceptable level.

Conclusion: Although, the metal content in sampled cow milks were within the safe limit, the potential human health risks cannot be neglected for the regular/long time consumption of heavy metal contained cow milk.

Keywords: Cow milk, Heavy metals, Hazard quotients (HQ), Carcinogenic risk (CR), Estimated daily intake (EDI)

Background

Milk has a positive influence on human health. It is considered as nearly complete food since they are good source of proteins, fats, vitamin supplements and major minerals (Enb et al. 2009; Qin et al. 2009; Yuzbas et al. 2009; Salah and Ahmed 2012; Seyed and Ebrahim 2012). There are about 38 micro and trace elements reported to be found in raw milk from different regions around the world (Dobrzański et al. 2005; Nwankwoala et al. 2002). These minerals content in raw cow milk may vary depending on several factors i.e. lactation period of cows, health conditions, seasonal variations, climatic conditions, annual feed composition and environmental contamination (Licata et al. 2004; Yahaya et al. 2010). The milk processing conditions may also have effective influence on the contents and retains of minerals in total composition of milk (Lante et al. 2006; Salah et al. 2013). All of these minerals including the trace elements in cow milk occurred as inorganic ions and remain with

* Correspondence: mmrahman@ees.hokudai.ac.jp
[1]Department of Environmental Sciences, Jahangirnagar University, Dhaka 1342, Bangladesh
[4]Faculty of Environmental Earth Science, Hokkaido University, Sapporo 060-0810, Japan
Full list of author information is available at the end of the article

proteins, peptides, carbohydrates and other molecules (Vegarud et al. 2000). Most of these trace elements have beneficial health importance. For example they act like enzymatic co-factors that can play vital roles in different physiological functions of human body and lack of these minerals may cause distribution and pathological problems mainly in vulnerable age (Enb et al. 2009). The essential elements become toxic when the concentration level exceeds 40 to 200 fold from their respective recommended threshold value (Rao 2005). Malhatet et al. (2012) found that the contamination in milk is considered as one of the main dangerous aspects within the last few years.

Increased environmental pollution has accelerated the problems of milk contamination and uncertainties about milk qualities (Farid and Baloch 2012). The worldwide milk contamination via environmental pollutants and xenobiotic compounds through cattle feeds like toxic metals, mycotoxin, dioxin and other pollutants are considered to have greater influence on public health (Seyed and Ebrahim 2012). Uptake of these contaminated milk acts like an additional source of heavy metal exposure (Ruqia et al. 2015). The main sources of metal contamination to humans are industrial or domestic effluents, combustion, bushfires, decomposition of chemical fertilizers, pesticides etc. (Degnon et al. 2012). Abdominal pain, hepatotoxicity, neurotoxicity, vomiting (Hussain et al. 2010), decreasing of intelligence quotient (IQ) level, Alzheimer's disease, behavioral disorders (Ahmad et al. 2011), tissue injury, irritation of lungs, cancer (Bushra et al. 2014) etc. could be generated due to over exposure of heavy metals. Besides heavy metals are non-biodegradable in nature and become accumulated in the food chains via bio-transformation, bio-accumulation and biomagnifications (Aslam et al. 2011). Complete elimination or prevention of chemical contaminants cannot be achieved from milk because the lipophilic contaminants will find its way into the persistent fat compounds from where heavy metals cannot be removed readily (Girma et al. 2014). Schematic diagram of heavy metals entering into food chain is given in the Fig. 1.

The heavy metal contamination of milk is less explored in less developed countries like Bangladesh (Islam et al. 2015; Shahriar et al. 2014). Islam et al. (2015) found that food chain around the nearby areas of Dhaka city in Bangladesh was contaminated by elements namely Cr, Ni, Cu, As, Cd and Pb through milk consumption in the study period of 2012–13. Besides the milk consumption rate in Bangladesh is very low (39.2 ml/day) while the recommended allowance is 250 ml/ day (Islam et al. 2015). According to previous survey, the annual milk production was 1.74 million tons during the year of 2001 and 2.28 million ton in 2007 (HIES 2011; BER 2007). Jamal and Fuad (2013),calculated that the milk production would be increased up to 4.55 million ton during the year of 2015–16. Moreover, with this increasing scenario in milk production, it is assumed that the consumer population of the country would face significant health threat in the long run from consuming contaminated milk and milk products. Thus the daily intake rate of heavy metal hazard quotients (HQ) and carcinogenic risk (CR) might be considered as exponentially increasing trend with the increasing rate of milk production.

The contamination of food stuffs due to metals and other toxins is one of the most important issues in developing countries. There are a lot of studies which have been conducted around the world associated with health risks for example; arsenic in cultivated rice in Srilanka (Channa et al. 2015), trace metal and alfatoxin in cassava flour in west Africa (Hayford et al. 2016),metals contaminated mushroom in Ethiopia (Medhanye et al. 2016), also health risk for contamination of foods and soils in China (Khan et al. 2008) and India (Sridhara Chary et al. 2008). However, it is observed that continuous long term exposures of consumers to heavy metal by consumption of cow milks get less emphasis in developing countries particularly in Bangladesh. Considering the aforementioned issues the study provides a significant importance in terms of public health hazard of Bangladesh. Therefore, the present study was designed to investigate concentration of selected heavy metals contaminating cow milks in Bangladesh particularly in city areas.

Methods
Study area and sample collection
The study was conducted in the period from December 2014 to October 2015. A total of 90 cow milk samples were collected from different areas of SavarUpazila, Dhaka, Bangladesh (Fig. 2). The milk samples were classified according to their collection sources as (i) the popular packaged cow milk was considered as Brand milk (33 samples), (ii) the dairy farm milk (30 samples) collected from the available dairy farms and (iii) the milk samples collected from the small household farmers as domestic cow milk (27 samples). All the samples were collected in a sterile glass bottle following standard methods and stored at 4 °C until analysis.

Instrumental analysis
Flame Atomic Absorption Spectroscopy (FAAS) (Model: AA-6300, Atomic Absorption Spectrophotometer, SHIMADZU, Japan) was used for heavy metal analysisforcadmium(Cd), chromium (Cr), copper (Cu), manganese (Mn), lead (Pb), and iron (Fe). Standard solution of each metal wasprepared at four different concentrations of 0.01, 0.1, 1.0, 5.0 ppm from Sigma-Aldrich (St. Louis,

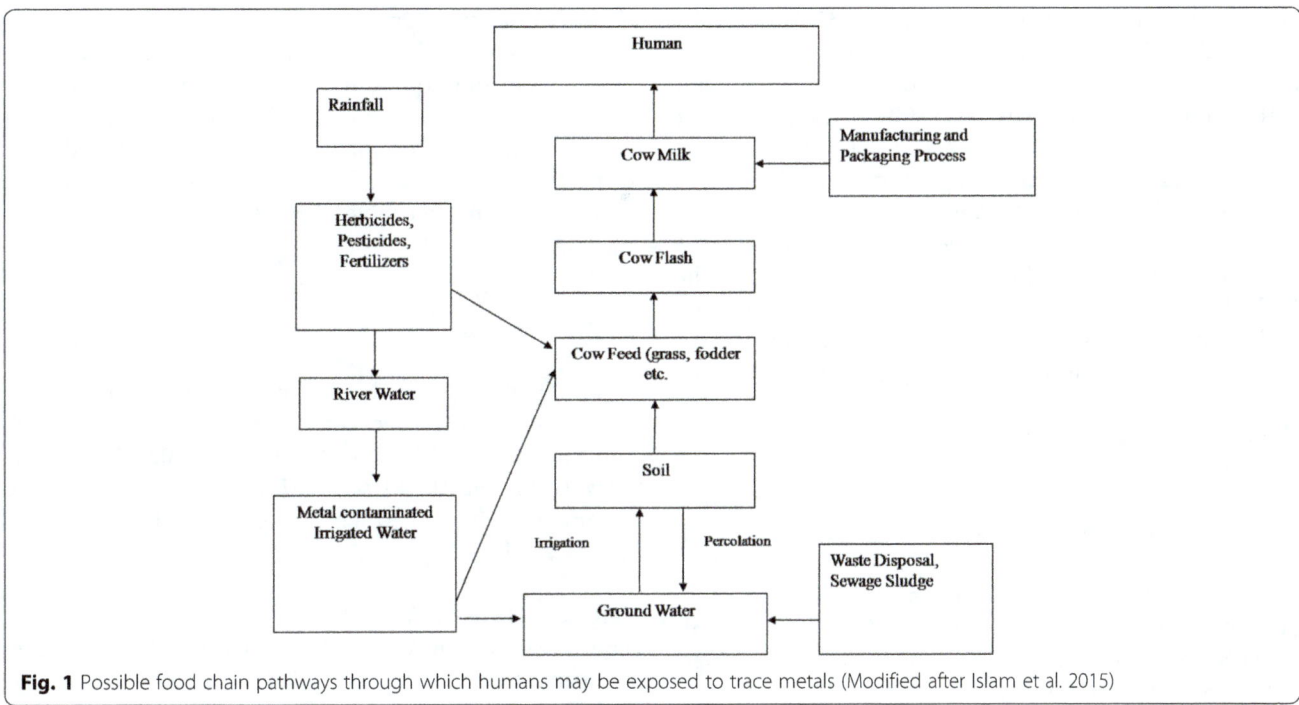

Fig. 1 Possible food chain pathways through which humans may be exposed to trace metals (Modified after Islam et al. 2015)

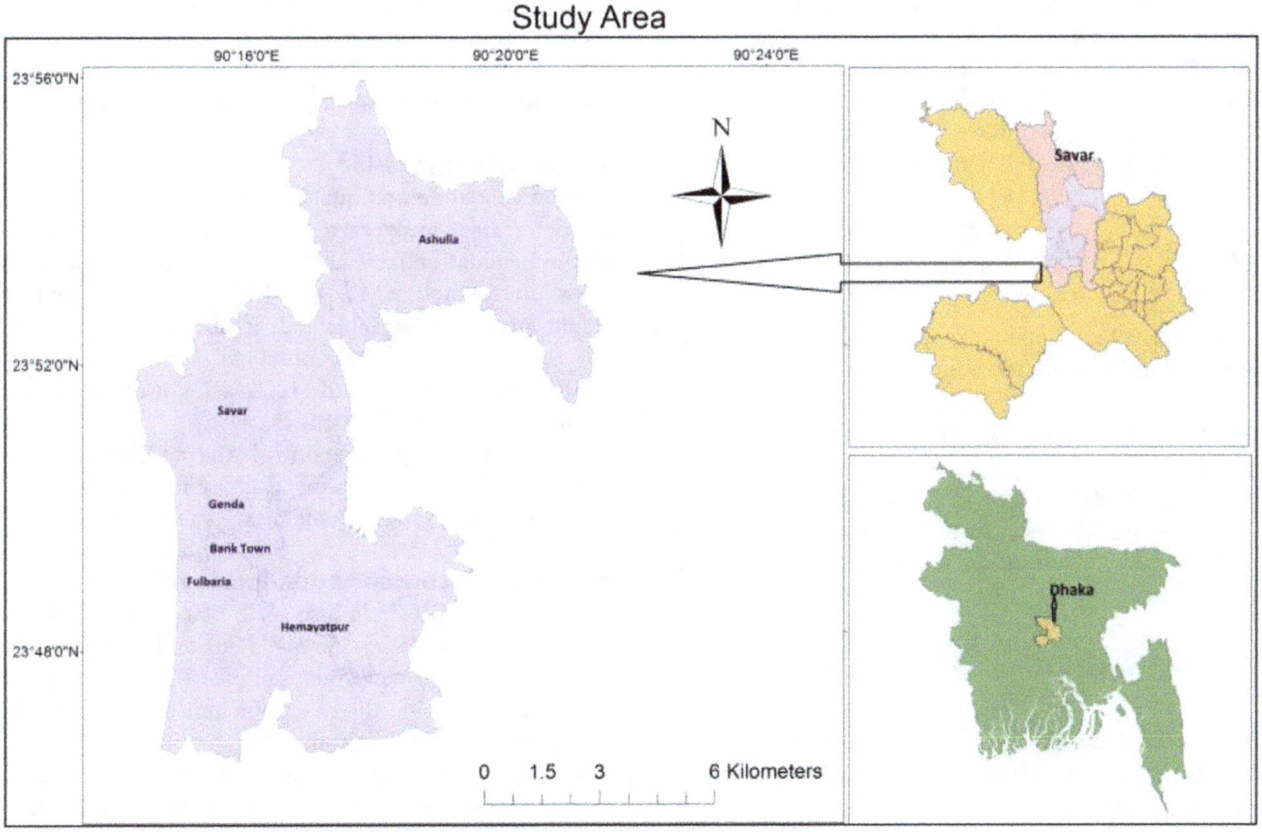

Study Area

Fig. 2 Location of the Sampling Areas in Dhaka, Bangladesh

USA). Spectral lines were set to 228.67, 357.65, 324.57, 279.43, 217.35 and 248.30 nm for Cd, Cr, Cu, Mn, Pb and Fe, respectively. The minimum detection limit (MDL) for Cd, Fe and Cu was 0.001 mg/kg; MDL for Mn and Pb was 0.002 mg/kg and for Cr MDL was 0.005 mg/kg. For ensuring the quality control we used the certified reference materials (CRM) for metal analysis and also performed the recovery test with the best digestion method for each metals (Table 2). The CRM for metal was purchased; Cd, Cr, Pb and Fe from Fluka, Sigma Aldrich (St. Louis, USA) and Mn and Cu from Kanto chemicals co. Inc. (Tokyo, Japan).

Data calculation

Recovery test

Eight various methods for milk samples digestion were selected from previous related works to perform recovery test. Acid mixing ratios$HNO_3 + HCLO_4$ (10 ml + 5 ml) were considered as M-1, $HNO_3 + H_2O_2$ (10 ml + 3 ml) as M-2, Sample + HNO_3 (1 gm + 5 ml) as M-3, $HNO_3 + HCLO_4$ (7 ml + 4 ml) as M-4, Supernatant Sample + HNO_3 (15 ml + 5 ml) as M-5, HNO_3 + HNO_3 $+HCLO_4 + H_2O_2$ (15 ml + 5 ml + 5 ml + drop wise) as M-6, HNO_3+ H_2O_2 (6 ml + 1 ml) as M-7 and $HNO_3 + HCl + HF$ (2 ml + 6 ml + 2 ml) were set as M-8 for digestion of selected cow milk (Seyed and Ebrahim 2012; Nnadozie et al. 2014; Elatrash and Atoweir 2014; Rubina et al. 2013; Dawd et al. 2012; Jolanta et al. 1996; Tassew et al. 2014; European Committee for Standardization 2002). All of the reagents were from Merck (Darmstadt,Germany). The recovery percentages were calculated by the following equation:

$$RecoveryPercentages = \frac{CE}{CM} \times 100 \qquad (1)$$

Where, CE = Experimental concentration (ppm) and CM = Spiked Concentration (ppm)

Estimated daily intake (EDI) of metals due to milk consumption

The estimated daily intake (EDI) of trace metals inmilk depends on metal concentrations (for dry weight basis), and daily milk consumption rate as well as the average body weight.

$$EDI\, for\, each\, metal\, (mg/kg) = (Ci \times 39.2)/60 \qquad (2)$$

Here, 39.2 mg/day = Daily milk consumption rate for Bangladesh (HIES 2011) and 60 kg = average body weight of an adult resident.

Ci = metal concentrations in milk (mg/l) (Islam et al. 2014).

Hazard quotients (HQs)

In the present study, the human health risks associated with the consumption of cow milk by the local community inhabitants were evaluated based on the hazard quotients (HQs). The method of estimating health risk using HQs was described in the USEPA Region III risk-based concentration table (USEPA 2000). The equation for HQ:

$$HQ = \frac{EDI}{RfD} \times 10^{-3} \qquad (3)$$

Here, EDI = estimated daily intake of metal (mg/day),
RfD = Oral Reference Dose (mg/kg/day). For Cr, Cd, Cu and Pb it is0.003, 0.001, 0.04 and 0.004, respectively (Islam et al. 2014; USEPA 2010). HQs indicate potential health risk when it is equal or higher than 1 (Islam et al. 2014).

Carcinogenic risk (CR)

The target carcinogenic risks (CR) were also calculated by using the equation provided in USEPA Region III Risk-Based Concentration Table (USEPA 2006):

$$CR = \frac{EFr \times ED \times EDI \times CSFo}{AT} \times 10^{-3} \qquad (4)$$

Here, EFr = exposure frequency (350 days/year), ED = exposure duration (30 years) (USEPA 2006). AT = averaging time for carcinogens (365 days/year × 70 years). CSFo stands for oral carcinogenic slope factor (USEPA 2010).

Results and discussion

Method validation and quality control

To determine recovery as one of the most important method validation parameters, eight various milk sample digestion methods (M-1 to M-8) were performed and results presented in Table 1. It is obvious that the highest recovery value was obtained for M-7 digestion method where HNO3 and H2O2 acids used were in the 6:1 ratio.(Tassew et al. 2014).

To ensure the quality control, the certified reference material (CRM) valuefor metal analysis with percentage of recovery for respective metalsarelisted in Table 2.

Heavy metal concentration incow milk

Concentrations of Cadmium (Cd), Chromium (Cr), Lead (Pb), Manganese (Mn), Copper (Cu) and Iron (Fe) were determined in 90 cow milk samples (brand, dairy farm and domestic) using the most efficient digestion method (M-7) and results are summarized in Table 3

Average concentrations of trace metals among the branded cow milk samples had shown a descending order of Cr>Fe>Cu>Mn>Cd>Pb (Fig. 3). On the other hand, the dairy farm cow milk samples had shown

Table 1 Metal recovery values for different milk sample digestion methods

Method Id	Spiked Concentration	Metal concentration (ppm)					
		Cr	Cd	Pb	Mn	Cu	Fe
M-1	10 (ppm)	5.5010±0.03	6.4110±0.026	6.421±0.001	4.8312±0.001	5.5128±0.040	5.5301±0.040
	Recovery percentage	55 %	64 %	64 %	48 %	55 %	55 %
M-2	10 (ppm)	7.5020±0.025	5.6133±0.001	7.5113±0.102	6.7014±0.030	7.2105±0.050	7.2311±0.003
	Recovery percentage	75 %	56 %	75 %	67 %	72 %	72 %
M-3	10 (ppm)	8.0010±0.075	8.1201±0.002	7.9012±0.120	7.9111±0.030	7.4127±0.001	8.1020±0.030
	Recovery percentage	80 %	81 %	79 %	79 %	74 %	81 %
M-4	10 (ppm)	8.6012±0.010	8.1101±0.002	7.7010±0.030	6.8013±0.013	5.028±0.040	9.2033±0.102
	Recovery percentage	86 %	81 %	77 %	68 %	50 %	92 %
M-5	10 (ppm)	4.5413±0.275	3.8014±0.001	3.6103±0.102	7.5103±0.030	4.1031±0.031	6.8102±0.050
	Recovery percentage	45 %	38 %	36 %	75 %	41 %	68 %
M-6	10 (ppm)	5.0322±0.085	5.1300±0.001	5.087±0.014	8.1078±0.130	4.1002±0.027	4.2341±0.050
	Recovery percentage	50 %	51 %	50 %	81 %	41 %	42 %
M-7	10 (ppm)	9.8621±0.002	9.8801±0.002	9.7805±0.006	10.1400±0.001	9.9320±0.008	9.7300±0.017
	Recovery percentage	98 %	98 %	97 %	101.4 %	99 %	97 %
M-8	10 (ppm)	7.8147±0.010	7.3101±0.220	7.7103±0.104	5.8310±0.027	5.1901±0.002	7.1713±0.050
	Recovery percentage	78 %	73 %	77 %	58 %	51 %	71 %

the descending order of Cr>Fe>Mn>Cu>Cd>Pb while the average concentrations of trace metals among the domestic cow milk samples had shown the descending order of Fe>Cr>Mn>Cu>Cd>Pb (Fig. 3). It is clear from the figure that chromium possessed the highest concentration of metal content for both branded milk (0.672±0.010) and dairy cow milk (0.373±.008), while iron had shown the highest concentration (0.631 ±0.101) for the domestic cow milk. Lead had shown the least concentration for all types of sampled milk including 0.033±0.006 ppm for branded cow milk, 0.015±0.002 ppm for dairy farm cow milk and 0.012 ±0.001 ppm for domestic cow milk, respectively. Heavy metal contaminations in milk samples are found different countries all over the world in both brand milk and non-brand milk. A comparative scenario among previous studies around the world is illustrated in Table 4

Concentration of Cd was found extremely higher in one report from Pakistan (Mohammed et al. 2013) for both branded and non-brand milk samples (USEPA 2010) compared to the present study. In case of chromium both previous study from Bangladesh (Islam et al. 2015) and the present study had shown higher concentration compared to the other countries (USEPA 2006; Zodape et al. 2012). Similar results of higher concentration were also reported from branded milk of Indian study (Islam et al. 2014). Concentration of manganese was not reported previously from any type of cow milk. Lead samples had been reported higher in India (Islam et al. 2014) and Egypt (USEPA 2010) followed by Pakistan (Mohammed et al. 2013), Palestine (Abdul et al. 2012) and Nigeria (Ali et al. 2011) compared to the other reported countries (Seyed and Ebrahim 2012; Elatrash and Atoweir 2014; Khalil and Seliem 2013) including present study and previous study from Bangladesh (Islam et al. 2015). Concentration of cupper and iron had been found lower compared to the previous reports (Table 4).

Table 2 Metal concentration and recovery values for CRM milk samples digested by M-7 method

Metal	CRM value (mg/l)	Measured con.(mg/l)	Recovery (%) with M-7[a]	Minimum detection limit
Cd	1000 mg/l ± 4 mg/l	9.8801±0.002	98.8	0.001 mg/kg
Cr	1000 mg/l ± 4 mg/l	9.8621±0.002	98.6	0.005 mg/kg
Pb	1000 mg/l ± 4 mg/l	9.7805±0.006	97.8	0.002 mg/kg
Fe	1000 mg/l ± 4 mg/l	9.7300±0.017	97.3	0.001 mg/kg
Mn	1005 mg/l	10.1400±0.001	101.4	0.002 mg/kg
Cu	1001 mg/l	9.9320±0.008	99.3	0.001 mg/kg

[a]M-7: HNO_3+ H_2O_2 (6 ml + 1 ml) for digestion of milk

Table 3 Concentration of Cd, Cr, Pb, Mn, Cu and Fe in Milk samples

Metal	Brand Cow Milk (ppm)			Dairy Cow Milk (ppm)			Domestic Cow Milk (ppm)		
	Min	Max	Mean ± SD	Min	Max	Mean ± SD	Min	Max	Mean ± SD
Cd	BDL[a]	0.075	0.053±0.022	BDL	0.073	0.024±.009	BDL	0.081	0.047±0.026
Cr	0.165	1.099	0.672±0.010	BDL	1.233	0.373±.008	0.081	1.533	0.539±0.013
Pb	BDL	0.200	0.033±0.006	BDL	0.200	0.015±0.002	BDL	0.204	0.012±0.001
Mn	0.032	0.167	0.092±0.02	0.069	0.173	0.126±0.02	0.042	0.198	0.130±0.023
Cu	0.042	1.778	0.163±0.031	0.008	0.224	0.064±0.013	0.040	0.184	0.127±0.029
Fe	0.250	0.861	0.486±0.077	0.196	0.624	0.333±0.054	0.355	0.949	0.631±0.101

[a]*BDL* bellow detection limit

Health risk assessment

The estimated daily intake (EDI) of metals from cow milk consumption had been investigated for selected metals. EDI and Permissible Values (PV) for metals studied, together with the contribution of EDI to PV (%), for adult consumers of cow milk (brand, dairy farm and domestic) are listed in Table 5.

To evaluate the daily intake, mean concentrations of metals in each cow milk category were multiplied by the milk consumption rate and divided by the body weight

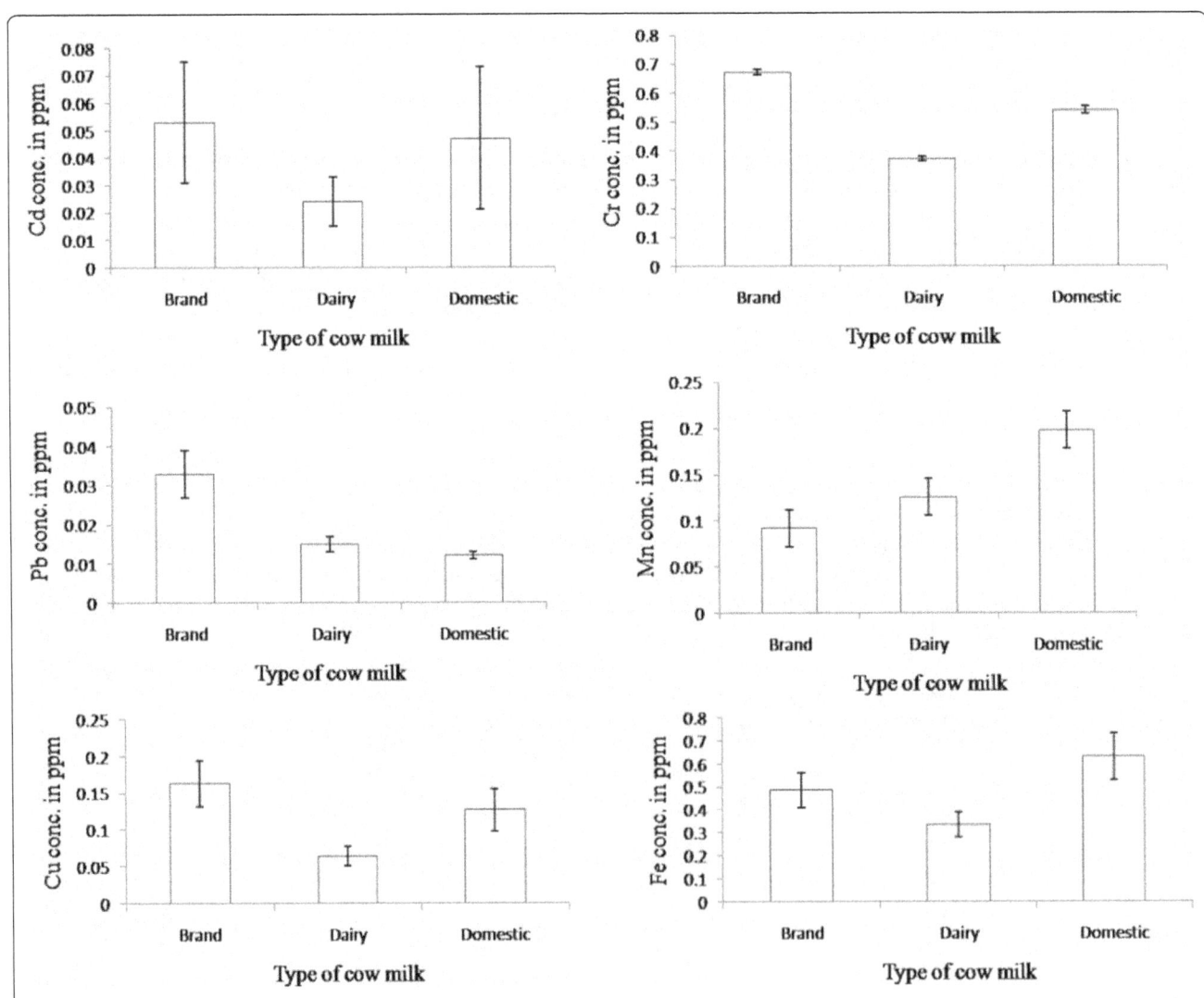

Fig. 3 Bar diagram of metal (Cd, Cr, Pb, Mn, Cu and Fe) concentrations versus type of cow milk, error bars indicating mean ± SD (n (number of samples) = 33, 30 and 27 for brand, dairy and domestic, respectively)

Table 4 Heavy metal concentrations (ppm) in different countries around the world

Country	Cd	Cr	Pb	Mn	Cu	Fe	Milk Type	References
Nigeria	-	-	0.63 ± 0.24	-	0.59±0.01–0.56±0.02	-	Non-Brand	Ali et al. 2011
Iran	-	-	0.013±0.006	-	-	-	Non-Brand	Seyed and Ebrahim 2012
Palestine	0.054	-	0.93	-	0.66	12.91	Brand	Abdul et al. 2012
	0.036	-	0.20	-	0.62	8.23	Non-Brand	
Philippines	0.003 to 0.01	0.0008 to 0.001	-	-	-	-	Brand	Solidum et al. 2012
Egypt	0.288	-	4.404	-	2.836	16.38	Non-Brand	Farag et al. 2012
India	-	0.175 to 0.013	5.904 to 0.139	-	37.290 to 0.039	-	Brand	Zodape et al. 2012
Pakistan	1.97±0.40	-	0.68±0.15	-	-	-	Non-Brand	Mohammed et al. 2013
	4.06±1.9	-	3.32±1.66				Brand	
Saudi Arabia	-	-	0.01 to 0.02	-	0.16 to 0.42	1.13	Brand	Khalil and Seliem 2013
Libia	0.001	-	0.003	-	-	-	Non-Brand	Elatrash and Atoweir 2014
Ethiopia	-	0.064± 0.010	-	-	0.206±0.024	-	Non-Brand	Alem et al. 2015
Bangladesh	0.029±0.026	1.6± 0.41	0.20±0.23	-	2.3±1.2	-	Non-Brand	Islam et al. 2015
Bangladesh	0.053±0.022	0.672±0.010	0.033±0.006	0.092±0.02	0.163±0.031	0.486±0.077	Brand	Present study
Bangladesh	0.024±.009	0.373±.008	0.015±0.002	0.126±0.02	0.064±0.013	0.064±0.013	Dairy Cow Milk	Present study
Bangladesh	0.047±0.026	0.539±0.013	0.012±0.001	0.130±0.023	0.127±0.029	0.631±0.101	Domestic Cow Milk	Present study

of the adult resident consumer. Metal specific EDIs revealed that EDI of Cr (for brand cow milk: 0.413 mg/day, for dairy farm cow milk: 0.243 mg/day and for domestic cow milk 0.352 mg/day) exceeded the permissible value (0.2 mg/day) and possess the highest concentration percentages to permissible value. EDIs of Cd, Pb, Mn, Cu and Fe were found below the permissible limits and also agreed with previous reports (Islam et al. 2015; Ademola 2014; Salah et al. 2012; Anita et al. 2010). Based on these data, this can be concluded that Cr was the major components contributing to the potential health risk via the consumption of all milk samples collected from Savar Upazila.

Hazard quotients (HQ) and carcinogenic risk (CR)

By definition, risk assessment is the evaluation process of the potential health effects from doses to human of one contaminant received through one or more exposure pathways. So, the potential health effects from doses to humans can be evaluated from risk assessment. By evaluating the hazard quotients (HQ), non-carcinogenic risks from consumption of foodstuffs by the adult inhabitants can be assessed. Based on the HQ we evaluated the non-carcinogenic risks due to consumption of cow milk for the adult resident and the estimated HQ values of metals are given in Table 6. From the results (Table 6) all the metals showed the HQ value below the threshold value of 1 suggested that there are no obvious health

Table 5 PV for metals studied in cow milk samples, mean EDI values and their contributions to PV for adult consumers

Metal	Permissible Value (PV) (mg/day)	References	Brand Milk		Dairy Farm Milk		Domestic Milk	
			EDI (mg/day)	Contribution to PV	EDI (mg/day)	Contribution to PV	EDI (mg/day)	Contribution to PV
Cd	0.046	(JECFA 2003)	0.043	73.91 %	0.016	34.78 %	0.030	65.21 %
Cr	0.2	(Oliver 1997)	0.413	206.5 %	0.243	121.5 %	0.352	176.0 %
Pb	0.21	[JECFA 2003]	0.021	10 %	0.10	4.76 %	0.003	1.42 %
Mn	5	(Ogabiela et al. 2011)	0.068	1.36 %	0.082	1.64 %	0.082	1.64 %
Cu	30	(JECFA 2003)	0.106	0.35 %	0.045	0.15 %	0.082	0.27 %
Fe	40	(FAO/WHO 2002)	0.317	0.79 %	0.215	0.53 %	0.412	1.03 %

Table 6 Non-carcinogenic human healthrisk of trace metalsdue to consumption of cow milk in area of Dhaka city, Bangladesh

Metals	Hazard quotients (HQs)			Carcinogenic Risk (CR)	
	Brand Milk	Dairy Milk	Domestic Milk	Milk type	Pb
Cd	0.043	0.016	0.030	Brand milk	7.33×10^{-7}
Cr	0.137	0.081	0.117		
Pb	0.005	0.025	0.0007	Dairy milk	3.5×10^{-7}
Mn	0.004	0.005	0.005		
Cu	0.002	0.001	0.002	Domestic milk	1.0×10^{-7}
Fe	0.0004	0.0003	0.0005		
Total ΣHQ	1.9×10^{-1}	1.28×10^{-1}	1.55×10^{-1}		

risks related to these metals associated with the consumption of cow milk in the study area. This finding agrees with Anita et al. 2010 and Islam et al. 2015. However the HQ value for each metal due to consumption of cow milk in the study area decreased in the order of: for brand milk; Cr>Cd>Pb>Mn>Cu>Fe, for dairy milk; Cr>Pb>Cd>Mn>Cu>Fe and for domestic cow milk; Cr>Cd>Mn>Cu>Fe>Pb. The data in the Table 6 also show the cumulative HQ (HQs) did not exceed the suggested threshold valueof 1 but the HQs value had decreased in the order of brand milk>domestic milk>dairy milk. This had revealedthat the brand milk hadhigher vulnerability to reach at the threshold for human health risks than the domestic and dairy cow milk. The carcinogenic risk (CR) of Pb due to consumption of cow milk by adult inhabitants in the study area was assessed using the target carcinogenic risk (CR). The result from the Table 6 showed the CR of Pb (Brand; 7.33×10^{-7}, Dairy; 3.5×10^{-7} and Domestic; 1.0×10^{-7}) due to consumption of cow milk was below 10^{-6} and considered as negligible. Due to the unavailability of carcinogenic slope factor values in USEPA 2010, most of the studied metals were not considered for direct CR assessment except for Pb. Therefore, the potential of CR for the inhabitants of the study area is within the safe limit, but the cumulative hazard quotients are nearing the threshold. Therefore, the non-carcinogenic health risk of the inhabitants due to consumption of cow milk should not be neglected.

Conclusion

To evaluate the safety of cow milk samples from Dhaka city area, selected heavy metal contents were analyzed by the most validated methods. Among the metals analyzed, Cr concentration along with their daily intake rate was found to be dominant percentages for both branded and non-branded cow milk samples. The current cumulative risks of studied metals due to consumption of cow milk remained below unity (HQ<1), indicating that

people would not experience significant risk due to cow milk consumption. The studied direct carcinogenic risk of Pb is also below the recommended level (CR<10^{-6}). But the cumulative HQs value is nearing the to the threshold, meaning due to regular consumption of cow milk along with its potential risk of contamination could lead to human health risks in the near future. It can be recommended that proper monitoring of cattle feed quality as well as the techniques of milk processing should be carefully considered for the public health safety in Bangladesh.

Abbreviations
AT: Averaging time for carcinogens; BDL: Below detection limit; CR: Carcinogenic risk; CRM: Certified reference materials; CSFo: stands for oral carcinogenic slope factor; ED: Exposure duration; EDI: Estimated daily intake; EFr: Exposure frequency; FAAS: Flame atomic absorption spectrometry; HQ: Hazard quotient; MDL: Minimum detection limit; PV: Permissible value; RfD: Oral reference dose

Author's contribution
MIM contributed to study design, sampling, instrumental analysis and writing; MAZC, NJE, ZF, MLB and MKA contributed to carry out instrumental analysis and sampling; MMR, MS, MKU, MK contributed to study design, manuscript preparing, data analysis, paper review. All authors read and approved the final manuscript.

Competing interests
This is an original manuscript that has not been submitted elsewhere for publication. All authors have read the manuscript and agreed that the work is ready for submission to the journal with no conflict of interests. The author Md. Mostafizur Rahman will represent for all correspondence.

Author details
[1]Department of Environmental Sciences, Jahangirnagar University, Dhaka 1342, Bangladesh. [2]Agrochemicals and Environmental Research Division, Institute of Food & Radiation Biology, Atomic Energy Research Establishment, G.P.O. Box 3787, Savar 1349, Bangladesh. [3]Food Analysis Research Laboratory, Center for Advanced Research in Sciences, University of Dhaka, Dhaka 1000, Bangladesh. [4]Faculty of Environmental Earth Science, Hokkaido University, Sapporo 060-0810, Japan. [5]Department of Environmental Pollution and Process Control, Xinjiang Institute of Ecology and Geography, ChineseAcademy of Sciences, Urumqi 830011, Xinjiang, People's Republic of China.

References
Abdul KA, Swaileh KM, Hussein RM, Matani M. Levels of metals (Cd, Pb, Cu and Fe) in cow's milk, dairy products and hen's eggs from the West Bank, Palestine. Int Food Res J. 2012;19(3):1089–94.

Ademola AK. Assessments of Natural Radioactivity and Heavy Metals in Commonly Consumed Milk in Oke-Ogun Area, Nigeria and Estimation of Health Risk Hazard to the Population. J Environ Anal Toxicol. 2014;4:253.

Ahmad N, Rahimb M, Mas H. Toxocological Impact Assessement of heavy metals in human blood and milk samples collected in district Shangla. Pakistan: SciInt (Lahore); 2011.

Alem G, Tesfahun K, Kassa B. Quantitative Determination of the Level of Selected Heavy Metals in the Cows' Milk from the Dairy Farm of the Haramaya University, Eastern Ethiopia. Int J Chem Nat Sci. 2015;3(1):240–8.

Ali JA, Bukar DE, Jimoh N, Hauwa NT, Yusuf N, Umar ZT. Determination of copper, zinc, lead and some biochemical parameters in fresh cow milk from different locations in Niger State, Nigeria. Afr J Food Sci. 2011;5(3):156–60.

Anita S, Rajesh KS, Madhoolika A, Fiona MM. Health risk assessment of heavy metals via dietary intake of foodstuffs from the wastewater irrigated site of a dry tropical area of India. Food Chem Toxicol. 2010;48:611–9.

Aslam B, Javed I, Hussain KF, Ur-Rahman Z. Uptake of Heavy Metal Residues from Sewerage Sludge in the Milk of Goat and Cattle during Summer Season. Pak Vet J. 2011;31(1):75–7.

BER. Bangladesh Economic Review, Ministry Of Finance. Dhaka: The Government of Bangladesh; 2007.

Bushra I, Saatea A, Samina S, Riaz K. Assessment of Toxic Metals in Dairy Milk and Animal Feed in Peshawar, Pakistan. British BiotechnoJ. 2014; 4(8):883–93.

Channa J, Priyani P, Saranga F, Mala A, Sarath G, Sisira S. Presence of arsenic in Sri Lankan rice. Int J Food Contaminat. 2015;2:1.

Dawd AG, Gezmu TB, Haki GD. Essential and toxic metals in cow's whole milk from selected sub-cities in Addis Ababa, Ethiopia. Online Int J Food Sci. 2012;1(1):12–9.

Degnon RG, Dahouenon-Ahoussi E, Adjou ES, Soumanou MM, Dolganova NV, Sohounhloue DCK. Heavy metal contamination of the Nokoué Lake (southern Benin) and the dynamic of their distribution in organs of some fish's species (Mugilcephalus L. and Tilapia guineensis). J Anim Sci Adv. 2012;2(7):589–95.

Dobrzañski Z, Kolacz R, Górecka H, Chojnacka K, Bartkowiak A. The content of microelements and trace elements in raw milk from cows in the Silesian region. Polish J Environ Stud. 2005;14(5):685–9.

Elatrash S, Atoweir N. Determination of lead and cadmium in raw cow's milk by graphite furnace atomic absorption spectroscopy. Int J Chem Sci. 2014;12(1):92–100.

Enb A, AbouDonia MA, Abd-Rabou NS, Abou-Arab AAK, El-Senaity MH. Chemical Composition of Raw Milk and Heavy Metals Behavior During Processing of Milk Products. Global Veterinaria. 2009;3(3):268–75.

European Committee for Standardization. Characterization of waste-Microwave assisted digestion wihhydrofluoric (HF), nitric (HNO₃) andhydrochloric (HCl) acid mixture for subsequent determination of elements. 2002. EN 13656.

FAO/WHO. Codex Alimentarius-general standards for contaminants and toxins in food. Schedule 1 Maximum and guideline levels for contaminants and toxins in food, Joint FAO/WHO food standards programme. Rotterdam: Codex committee; 2002. Reference CX/FAC 02/16.

Farag M, Mohammed H, Ayman S, Abd EF. Contamination of Cows Milk by Heavy Metal in Egypt. Bull Environ Contam Toxicol. 2012;88:611–3.

Farid S, Baloch MK. Heavy metal ions in milk samples collected from animals feed with city effluent irrigated fodder. Greener J Physical Sciences. 2012;2(2):36–43.

Girma K, Tilahun Z, Haimanot D. Review on Milk Safety with Emphasis on Its Public Health. World J Dairy Food Sci. 2014;9(2):166–83.

Hayford O, Paa TA, Nanam TD. Variations in trace metal and aflatoxin content during processing of High Quality Cassava Flour (HQCF). Int J Food Contaminat. 2016;3:1.

HIES (Household Income and Expenditure Survey). Preliminary report on household income and expenditure survey-2010. Dhaka, Bangladesh: Bangladesh Bureau of Statistics, Statistics Division, Ministry of Planning; 2011.

Hussain Z, Nazir A, Shafique U, Salman M. Comparative study for the determination ofmetals in milk samples using Flame- AAS and EDTA complexometric titration. J Sci Res. 2010;1:55–76.

Islam MS, Ahmed MK, Al-mamun MH, Masunaga S. Trace metals in soil and vegetables and associated health risk assessment. Environ Monit Assess. 2014;186:8727–39.

Islam MS, Kawser MA, Habibullah MAM, Shigeki M. Assessment of trace metals in foodstuffs grown around the vicinity of industries in Bangladesh. J Food Compos Anal. 2015;42:8–15.

Jamal HM, Fuad HM. Forecasting of Milk, Meat and Egg Production in Bangladesh. Res J Animal, Vet Fishery Sci. 2013;1(9):7–13.

JECFA. Summary and conclusions of the 61st meeting of the joint FAO/WHO Expert committee on food additives (JECFA). Rome, Italy: JECFA/61/SC; 2003.

Jolanta BB, Ewa S, Wiestaw Z. Determination of Major and Trace Elements in Powdered Milk by Inductively Coupled Plasma Atomic Emission Spectrometry. J Chem Anal. 1996;41:625.

Khalil HM, Seliem AF. Determination of Heavy Metals (Pb, Cd) and some Trace Elements in Milk and Milk Products Collected from Najran Region in K.S.A. Life Sci J. 2013;10(2):648–52.

Khan S, Cao Q, Zheng YM, Huang YZ, Zhu YG. Health risks of heavy metals in contaminated soils and food crops irrigated with wastewater in Beijing, China. Environ Pollut. 2008;152:686–92.

Lante A, Lomolino G, Cagnin M, Spettoli P. Content and characterization of minerals in milk in crescenza and squaquerone Italian fresh cheeses by ICP-OES. Food Control. 2006;17:229–33.

Licata P, Trombetta D, Cristani M, GiofreF MD, Calo M, Naccari F. Levels of "toxic" and "essential" metals in samples of bovine milk from various dairy farms in Calabria, Italy. Environ Int. 2004;30:1–6.

Malhat F, Hagag M, Saber A, Fayz AE. Contamination of cow's milk by heavy metal in Egypt. Bull Environ Contam Toxicol. 2012;88(4):611–3.

Medhanye G, Negussie M, Abi MT. Levels of essential and non-essential metals in edible mushrooms cultivated in Haramaya, Ethiopia. Int J Food Contaminat. 2016;3:2.

Mohammed AGA, Abubakar Musa KE, WaleedAboshora WZ. Evaluation of some physicochemical parameters of three commercial milk products. Pak J Food Sci. 2013;23(2):62–5.

Nnadozie CU, Birnin-Yauri UA, Muhammad C. Assessment of Some Diary Products Sold in Sokoto Metropolis, Nigeria. Int J Adv Res Chem Sci. 2014;1(10):31–7.

Nwankwoala A, Odueyungbo S, Nyavor K, Egiebor N. Levels of 26 elements in infant formula from USA, UK and Nigeria by microwave digestion and ICP-OES. Food Chem. 2002;77(4):439–47.

Ogabiela EE, Udiba UU, Adesina OB, Hammuel C, Ade-Ajayi FA, Yebpella GG, Mmereole UJ, Abdullahi M. Assessment of metal levels in fresh milk from cows grazed around Challawa Industrial Estate of Kano, Nigeria. J Basic Appl Sci Res. 2011;1(7):533–8.

Oliver MA. Soil and human health: A Review. Eur J Soil Sci. 1997;48:573–92.

Qin LQ, Wang XP, Li W, Tong WJ TX. The minerals and heavy metals in cow's milk from China and Japan. J Health Sci. 2009;55(2):300–5.

Rao AN. Trace element estimation: methods and clinical context.Online. J Health Allied Sci. 2005;4(1):1–9.

Rubina P, Abbas B, Darakhshan A, Shahid SS, Qamar-ul H. Elucidation of physico-chemical characteristics and mycoflora of bovine milk available in selected area of Karachi, Pakistan. J Appl Sci Environ Manag. 2013; 17(2):259–65.

Ruqia N, Muslim K, Hameed UR, Zubia M, Muhammad M, Rumana S, Naila G, Faryal S, Irum P, Fathma S, Muhammad Z, Noor UA, Nelofer J. Elemental Assessment of Various Milk Packs Collected From KPK, Pakistan. Am-Eurasian J Toxicol Sci. 2015;7(3):157–61.

Salah F, Ahmed AEA. Assessment of Toxic Heavy Metals in Some Dairy Products and the Effect of Storage on its Distribution. J Ameri Sci. 2012;8(8):665–70.

Salah AEA, Esmat AI, Rania MKM. Prevalence of Some Trace and Toxic Elements in Raw and Sterilized Cow's Milk. J Am Sci. 2012;8(9):753–61.

Salah FA, Esmat IA, Mohamed AB. Heavy metals residues and trace elements in milk powder marketed in DakahliaGovernorate. Int Food ResJ. 2013;20(4):1807–12.

Seyed MD, Ebrahim R. Determination of Lead Residue in Raw Cow Milk from Different Regions of Iran by Flameless Atomic Absorption Spectrometry. Am-Eurasian J Toxicol Sci. 2012;4(1):16–9.

Shahriar SMS, Akther S, Akter F, Morshed S, AlamMK SI, Halim MA, Hassan MM. Concentration of Copper and Lead in Market Milk Concentration of Copper and Lead in Market Milk. Int Letters Chem, Phys Astro. 2014;27:56–63.

Solidum JN, Burgos SG, dela Cruz KM, Padilla R. A Quantitative Analysis on Cadmium and Chromium Contamination in Powdered Children's Milk. Metro Manila, Philippines: International Conference on Environment and Bioscience; 2012.

Sridhara Chary N, Kamala CT, Samuel Suman Raj D. Assessing risk of heavy metals from consuming food grown on sewage irrigated soils and food chain transfer. Ecotoxicol Environ Saf. 2008;69:513–24.

Tassew B, Ahmed H, Vegi MR. Determination of Concentrations of Selected Heavy Metals in Cow's Milk. J Health Sci. 2014;4(5):105–12.

USEPA. Risk-based Concentration Table US Environmental Protection Agency Washington. DC/Philadelphia: PA; 2000.

USEPA. USEPA Region III Risk-Based Concentration Table: Technical Background Information. Washington: Unites States Environmental Protection Agency; 2006.

USEPA (2010) Risk Based Concentration Table. Available from: http://www.epa.gov/reg3hwmd/risk/human/index.htm.

Vegarud GE, Landsrud T, Svaning C. Mineral-binding milk proteins and peptides; occurrence, biochemical and technological characteristics. Br J Nutr. 2000;84:91–8.

Yahaya MI, Ezo GC, Musa YF, Muhamad SY. Analysis of heavy metals concentration in roadside soils in Yauri, Nigeria. Afr J Pure Appl Chem. 2010;4(3):022–30.

Yuzbas I, Sezgin NE, Yldrm Z, Yldrm M. Changes In Pb, Cd, Fe, Cu and Zn Levels during the Production of Kasar Cheese. J Food Qual. 2009;32:73–83.

Zodape GV, Dhawan VL, Wagh RR. Determination of Metals in Cow Milk Collected From Mumbai City, India. Srilanka: Eco Revolution Colombo; 2012. p. 270–4.

5

Microbiota of frozen Vietnamese catfish (*Pangasius hypophthalmus*) marketed in Belgium

Anh Ngoc Tong Thi[1,2], Simbarashe Samapundo[1], Frank Devlieghere[1*] and Marc Heyndrickx[3,4]

Abstract

Background: Vietnamese catfish (*Pangasius hypophthalmus*) is highly appreciated in many European countries, the U.S., Canada, Japan etc. This paper presents an overview of the microbiota of frozen Vietnamese catfish products marketed in Belgium. Samples of *Pangasius* steaks, portions and fillets from six brands were collected from supermarkets located in Ghent, Belgium.

Results: The total psychrotrophic and mesophilic aerobic counts of the samples evaluated from each brand did not differ significantly (p > 0.05) and ranged from 3.8-5.2 log CFU/g and 3.8-4.8 log CFU/g, respectively. Lactic acid bacteria counts varied from 2.2 to 4.1 log CFU/g while the counts of presumptive Enterobacteriaceae ranged from 1.6 to 3.8 log CFU/g. A total of 132 isolates were collected from the plates used to enumerate the microbial parameters mentioned above. Fourteen different genera and 18 different species were identified by means of 16S rRNA gene sequencing. The most prevalent genera of *Lactococcus* (31.2 %), *Staphylococcus* (11.7 %), *Serratia* (10.4 %), *Acinetobacter* (9.1 %), *Enterococcus* (7.8 %) and *Pseudomonas* spp. (6.5 %) were identified by means of 16S rRNA gene sequencing.

Conclusion: The results obtained provide an overview of the dominant microbiota on frozen *Pangasius* which is useful for the development of appropriate preservation techniques for thawed *Pangasius* products.

Keywords: Microbiota, Catfish, *Pangasius*, 16S rRNA gene sequence

Background

Vietnamese catfish (*Pangasius hypophthalmus*) products, both farmed and freshwater, are appreciated in many European countries, the U.S., Canada, Japan etc. *Pangasius* has become an affordable 'white fish' substitute for cod and other white fleshed fish species in the West (Phan et al. 2009). Although high quantities of Vietnamese *Pangasius* products are exported to Western countries in frozen form (e.g. 158,000 tons in 2011) (VASEP 2014), fresh *Pangasius* is preferred. Hence, the trading of thawed products as 'fresh' fish is common in Western countries. However, once thawed the fish fillets deteriorate primarily through microbiological spoilage (ICMSF 2005).

The initial microbiological flora on freshly harvested, properly handled pond reared fish products typically consists of a diverse mixture of *Acinetobacter, Aeromonas, Citrobacter, Enterobacter, Escherichia, Flavobacterium, Micrococcus, Moraxella, Pseudomonas, Staphylococcus, Streptococcus* and *Vibrio* spp. (ICMSF 2005). Obligate anaerobic lactic acid bacteria such as *Carnobacterium, Lactobacillus, Enterococcus* and *Vagococcus* are also commonly recovered from the guts of freshwater fish (Austin 2002; ICMSF 2005). More recently, isolates collected from the intestines and gills of *Pangasius* fish were identified by API strips as Enterobacteriaceae (49.1 % of the isolates), pseudomonads (35.2 %) and Vibrionaceae (15.7 %) (Sarter et al. 2007). Spoilage-related microbiota on *Pangasius* products has been well documented (Noseda et al. 2012; Tong Thi et al. 2013). The dominant microbiota on thawed *Pangasius* fillets stored in air and vacuum at the end of the shelf life (7 and 10 days, respectively) generally consists of *Serratia* and *Pseudomonas* spp. while lactic acid bacteria (e.g. *Carnobacterium maltaromaticum* and *Carnobacterium divergens*) and *Brochothrix thermosphacta* have been reported to be dominant on *Pangasius* products which are

---

* Correspondence: frank.devlieghere@ugent.be
[1]Department of Food Safety and Food Quality, Laboratory of Food Microbiology and Food Preservation, Food2Know, Ghent University, Coupure Links 653, Ghent 9000, Belgium
Full list of author information is available at the end of the article

packaged under modified atmospheres (Noseda et al. 2012). *Serratia* and *Pseudomonas* spp. have also been isolated from imported frozen Vietnamese *Pangasius* products retailed in Denmark (Noor Uddin et al. 2013). Tong Thi et al. (2013) reported the high prevalence of *Aeromonas, Acinetobacter, Lactococcus* and *Enterococcus* spp. on fresh *Pangasius* fillets during processing at two companies in Vietnam. It was also determined in the same study that the microbial diversity on the products depended on the location, source of water, suppliers (fish farms) and production capacity (Tong Thi et al. 2013). The microbiota on frozen products could influence the shelf life of thawed products. Therefore, identification of the spoilage microorganisms on frozen *Pangasius* products will provide an overview of the spoilage microbiota on thawed products which would allow processors to select appropriate preservation methods for thawed products.

The major objective of this study was to determine the microbiota of frozen Vietnamese *Pangasius* products sold in Belgium by means of a combination of culture-dependent techniques and 16S rRNA gene sequencing.

Methods

Six different brands of frozen fish products sold as Vietnamese *Pangasius* in various retail outlets in Ghent (Belgium) were available at the time of sampling and were evaluated in this study. Four brands were in the form of fillets (*ca.* 200–220 g/fillet), one brand was in the form of steaks (*ca.* 70–100 g/steak) and another in the form of portions (*ca.* 70–80 g/piece). Three packages of each brand were purchased at the same time and kept at −20 °C until the microbiological and chemical analyses were performed. As sub-lethal injury of bacterial cells may occur during frozen storage, the samples were initially thawed over a 24 h period in a refrigerator at 4.0 ± 0.7 °C in order to enhance the recovery of the contaminating bacteria before the analyses were performed.

Physico-chemical characteristics: drip loss, water content, water activity, pH and salt content

The drip (thawing) loss, water content, water activity (a_w), pH and salt content of all the *Pangasius* products were determined as follows. The drip loss was determined as the difference (%) between the weight of the packaged *Pangasius* products after thawing with and without the exudates. The weight of the packages was determined after thawing before the exudates were removed by decanting after which the weight of the package was measured again. Thereafter a 150–200 g composite sample from each package was homogenised for 1 min in a commercial blender (Braun 600 W, Spain). The a_w and pH of the homogenates were measured in duplicate by means of a_w-kryometer (NAGY, Gaeufelden, Germany) and a SevenEasy pH meter (Mettler Toledo

GmbH, Schwerzenbach, Swizerland), respectively. The water content of each sample was determined in duplicate gravimetrically by drying a 5 g aliquot of homogenate in aluminium dishes containing sea sand to avoid spattering for 12 h at 105 °C. The salt was extracted by boiling a 5 g homogenate in distilled water for 10 min. The chloride content in the extract was determined by titration with silver nitrate (Merck, Darmstadt, Germany) using a 5 % (*w/v*) chromate indicator (Merck, Darmstadt, Germany) according to the Mohr method (ISO 9297:1989).

Microbiological analyses

The fish samples for microbiological analyses were prepared separately from the fish samples for physico-chemical analyses. A 150–200 g composite sample from each package was prepared for microbial analysis. A 25 g sample was aseptically transferred to sterile stomacher bags by means of sterile scalpels and tweezers. Primary decimal dilutions were prepared by adding 225 ml of sterile physiological peptone saline (PPS, 0.85 g NaCl and 1 g neutralized bacteriological peptone (Oxoid, Basingstoke, U.K.) per L) to each of the 25 g samples. The mixtures were then homogenized in a stomacher for 1 min. Further decimal dilutions were prepared in PPS. The total psychrotrophic and mesophilic aerobic counts were determined by pour plating the decimal dilutions on Plate Count Agar (PCA, Oxoid, Basingstoke, U.K.) followed by incubation for 72 ± 4 h at 22 °C and 30 °C, respectively. The counts of presumptive Enterobacteriaceae were determined by pour plating (with an additional over layer) the decimal dilutions on Violet Red Bile Glucose agar (VRBGA, Oxoid, Basingstoke, U.K.). The VRBGA plates were incubated for 24 h at 37 °C after which all colonies were counted. Psychrotrophic lactic acid bacteria (LAB) were determined by pour plating (with an additional over layer) the decimal dilutions on de Man Rogosa Sharpe agar (MRS, Oxoid, Basingstoke, U.K) followed by incubation for 72 ± 4 h at 22 °C.

Isolation and identification of dominant microbiota
Sample preparation

From the three samples evaluated of each brand, 20–30 isolates were selected for identification taking into account as many different morphologies (e.g. color, size, and shape) as possible. These originated from the PCA, VRBGA, and MRS plates used for enumeration. A total of 132 isolates were purified by successive 4 × 4 streak plating (and microscopic analysis). The DNA was extracted from these isolates as described below.

DNA-extraction

DNA extraction was performed according to the protocol of Flamm et al. (1984) with minor modifications. In brief, lysostaphine (0.5 mg/ml; Sigma) and mutanolysine-lysozyme solution (1 U/ml mutanolysine, Sigma; 2.5 mg/ml lysozyme, Roche) dissolved in HPLC water and TE-buffer [0.05 M Tris, (Invitrogen); 0.02 M EDTA (Merck), pH 8], were added to a pellet of each isolate grown on Tryptone Soya Agar (Oxoid, Basingstoke, U.K.) at 30 °C for 24 h. The quality and quantity of DNA templates were tested beforehand by means of a spectrophotometer (Nanodrop, Isogen).

rep-PCR

All isolates were grouped into clusters on the basis of the similarity of their fingerprints obtained with $(GTG)_5$-PCR, which is a rep-PCR technique. The microbial DNA was used as a template in the PCR-reaction. Reactions were carried out in 25 µl volume containing microbial DNA (50 ng/µl), 1x RedGoldstar buffer (75 mM Tris-HCl; Eurogentec) and a final concentration of 3.4 mM of $(GTG)_5$ primer (Eurogentec), 1.5 mM Mg_2Cl (Applied Biosystems), 1 U RedGoldStar DNA polymerase (Eurogentec) and 0.2 mM of each deoxynucleotide triphosphate (GE Healthcare Europe GmbH). Amplification was done in a Geneamp PCR 9700 Thermocycler (Applied Biosystems) using the amplification conditions as follows: initial denaturation at 95 °C for 7 min, 30 cycles of 1 min at 94 °C, 1 min at 40 °C, 8 min at 65 °C and a final 16 min extension at 65 °C (Versalovic et al. 1994). PCR products were size separated in a 1.5 % Seakem LE agarose gel (Lonza) in 1xTBE buffer (0.1 M Tris, 0.1 M Boric acid, 2 mM EDTA) at 120 V for 4 h. The $(GTG)_5$ profiles were visualized under UV light after staining with ethidium bromide for 30 min. and a digital image was captured using the G:BOX camera (Syngene). The resulting fingerprints were compared using the Bionumerics version 6.5 software package (Applied Maths, Sint-Martens-Latem, Belgium) using the EZ load 100 bp PCR Molecular Ruler (Biorad) as normalization reference. The similarity between the fingerprints was calculated using the Pearson correlation (1 % optimization and 1 % position tolerance). The fingerprints were grouped according to their similarity by use of UPGMA (unweighted pair group method with arithmetic averages algorithm).

Identification of the microbial isolates by sequence analysis

A 1500 bp fragment of the 16S rRNA gene was amplified by PCR using forward 16 F27 and reserve 16R1522 primers (Brosius et al. 1978). Amplification was performed as follows: initial denaturation at 94 °C for 1 min, 25 cycles at 94 °C for 15 s, 60 °C for 15 s and 72 °C for 30 s followed by an elongation step at 72 °C for 8 min. All PCR products were purified for sequencing with a High Pure PCR product purification kit (Roche) according to manufacturer's protocol and stored at –20 °C. The quality and quantity of the purified PCR products were verified on a 1.5 %

agarose gel. The sequence reactions were then performed at Macrogen (Seoul, Korea), using a template of 30–50 ng PCR product DNA and 0.2 µM of primer 16 F27 (16S forward primer). The partial 16S rDNA sequences (around 900 bp) were compared with validly published prokaryotic names in the EzTaxon server (http://www.eztaxon.org/; (Chun et al. 2007) to determine the closest phylogenetic relatives of the strains and calculate levels of 16S rDNA gene sequence similarity. A minimum of 98.5 % of similarity (unless otherwise indicated) with a EZTaxon entry was used to identify the isolates to the genus level and to the tentative species level. All isolates were additionally characterized by Gram staining, oxidase and catalase test.

Statistical analysis

Results of the physico-chemical characteristics and the microbiological analysis (log CFU/g) were reported as mean value ± standard deviation of triplicates per product (brand). Differences in the mean counts (log CFU/g) of the sampled products were statistically assessed using one way Analysis of Variance (ANOVA) in SPSS version 20 (IBM Inc., Chicago, Ill., USA) when a Shapiro-Wilk test indicated that the means were normally distributed. If a Levene test confirmed heteroscedasticity, a Tamhane's T2 test was used. A non-parametric Kruskal-Wallis H-type test was performed in case the data showed non-normality. Thereafter, comparison of the paired means was done using the Mann-Whitney U test ($\alpha = 0.05$).

Results

Physico-chemical characteristics of frozen *Pangasius* marketed in Belgium

The results of physico-chemical characteristics performed on the samples are shown in Table 1. The mean water content of the thawed *Pangasius* fillets ranged from 79.3 to 87.7 %. Fillets from brand 3 had significantly higher water content than the fillets from the other brands. The *Pangasius* steaks evaluated in this study had significantly lower water content (74.0 %) than the fillets and portions ($p < 0.05$). The mean drip (thaw) losses of the fillets ranged from 7.5 to 16.8 %. The drip losses of fillets from brand 4 (mean = 7.5 %) were significantly the lowest ($p < 0.05$) of the four brands of filleted *Pangasius* products evaluated. The portions had the smallest drip losses ($p < 0.05$) of any of the products evaluated; these being on average *ca.* 3 and 6 times lower than the fillets of brand 4 and 1, respectively. No correlation occurred between the water content and drip losses. The mean a_w values of the fillets ranged from 0.990 to 0.995, with fillets from brand 3 (a_w 0.990) having significantly lower a_w ($p < 0.05$) than those of fillets from brand 1 to 2. The steaks and portions had a_w values (both a mean of 0.994) which were in the same range as fillets.

Table 1 Physico-chemical characteristics of Vietnamese *Pangasius* products marketed in Belgium

Product type	Water content (g/100 g wet fish)	Drip loss (%)	a_w	pH	Salt content (%)
Fillets (brand 1)	$79.3 \pm 1.2^{a*}$	$16.8 \pm 0.2^{e*}$	$0.9950 \pm 0.0001^{a*}$	$6.5 \pm 0.0^{a*}$	$0.12 \pm 0.0^{a*}$
Fillets (brand 2)	80.5 ± 1.5^{a}	10.5 ± 1.7^{d}	0.9947 ± 0.0002^{a}	6.7 ± 0.3^{a}	0.28 ± 0.2^{abcd}
Fillets (brand 3)	87.7 ± 0.9^{c}	11.9 ± 3.9^{d}	0.9896 ± 0.0007^{c}	8.2 ± 0.2^{c}	0.93 ± 0.2^{b}
Fillets (brand 4)	80.0 ± 0.3^{a}	7.5 ± 0.1^{c}	0.9947 ± 0.0003^{abc}	6.5 ± 0.1^{a}	0.23 ± 0.1^{c}
Steaks (brand 5)	74.0 ± 2.1^{b}	$12.6 \pm 1.8^{bd*}$	0.9939 ± 0.0008^{b}	6.2 ± 0.1^{b}	0.49 ± 0.1^{d}
Portions (brand 6)	80.0 ± 1.4^{a}	2.6 ± 1.4^{a}	0.9944 ± 0.0002^{ab}	6.5 ± 0.1^{a}	0.22 ± 0.1^{ac}

*Data are expressed as mean value ± standard deviation of three replicates. Means with a different superscript letter in the same column indicate where statistically ($p \leq 0.05$) differences occurred between products that were evaluated in this study

The mean pH values of the fillets ranged from 6.5 to 8.2. The pH of the *Pangasius* fillets from brand 3 (mean = 8.2) were significantly higher ($p < 0.05$) than those of the fillets from the other three brands evaluated. The mean pH values of the portions did not differ significantly ($p > 0.05$) from that of fillets from brands 1, 2 and 4, whilst the *Pangasius* steaks (mean pH value = 6.2) had significantly lower pH values ($p < 0.05$) than those of the fillets and portions. The salt content (based on the chloride content) of the fillets ranged from 0.12 to 0.93 %. As for the pH, the NaCl content of the *Pangasius* fillets from brand 3 (mean = 0.93 %) were significantly higher ($p < 0.05$) than those of the fillets from the other three brands evaluated. The portions had a similar NaCl content to fillets of brands 1, 2 and 4 whilst the steaks had a mean NaCl content (0.49 %) which was significantly higher ($p < 0.05$) than those of the portions and fillets of brands 1 and 4, but significantly smaller ($p < 0.05$) than that of fillets from brand 3.

Microbiota of frozen *Pangasius* fish

The microbial quality of frozen *Pangasius* products marketed in Belgium is shown in Table 2. The total psychrotrophic aerobic counts (TPC) ranged from 3.8 to 5.2 log CFU/g, whilst the total mesophilic aerobic counts (TMC) ranged from 3.8 to 4.8 CFU/g. The TPC and TMC for each brand of fish did not differ significantly from each other ($p > 0.05$). With regard to the fillets, it can be seen that the TPC and TMC on fillets from brand 2 were both

significantly lower ($p < 0.05$) than those on fillets from the other three brands. The counts of lactic acid bacteria (LAB) varied greatly between products, with significantly lower ($p < 0.05$) LAB occurring on the fillets from brand 2 than those found on the fillets from the other three brands. However, LAB counts from brand 2 did not differ significantly ($p > 0.05$) from these counts on the steaks and portions sampled. The counts of presumptive Enterobacteriaceae were highest on the fillets from brand 1 (3.8 ± 0.2 log CFU/g) while the lowest counts were found on the portions from brand 6 (1.6 ± 0.6 log CFU/g).

Identification of the isolates collected from different products

A total of 132 isolates were collected from the plates used to enumerate the aerobic counts, presumptive Enterobacteriaceae and lactic acid bacteria on the frozen *Pangasius* products evaluated in this study. These isolates were clustered based on their rep-PCR fingerprints. Each cluster consisted of at least four isolates with a similarity level of at least 65 %. From this cluster analysis, two representative isolates of each cluster were selected for further analysis by 16S rRNA gene sequencing and thereafter the tentative identification was extrapolated for the entire group of isolates in each cluster.

The identification of 76 selected isolates (of which 33 were Gram negative and 43 were Gram positive) included 14 different genera and 18 different species

Table 2 Microbiota of Vietnamese *Pangasius* products marketed in Belgium

Product type	Total psychrotrophic aerobic counts (TPC)	Total mesophilic aerobic counts (TMC)	Psychrotrophic lactic acid bacteria	Presumptive Enterobacteriaceae
Fillets (brand 1)	$4.7 \pm 0.3^{bd*}$	4.8 ± 0.4^{a}	4.0 ± 0.7^{a}	3.8 ± 0.2^{a}
Fillets (brand 2)	3.8 ± 0.1^{a}	3.8 ± 0.0^{b}	2.2 ± 0.5^{bc}	2.9 ± 0.0^{b}
Fillets (brand 3)	5.2 ± 0.2^{b}	4.5 ± 0.3^{ac}	4.1 ± 0.1^{a}	3.0 ± 0.4^{bd}
Fillets (brand 4)	5.1 ± 0.4^{bd}	4.6 ± 0.2^{a}	4.0 ± 0.1^{a}	2.9 ± 0.2^{b}
Steaks (brand 5)	4.6 ± 0.2^{cd}	4.8 ± 0.1^{a}	2.7 ± 0.1^{c}	2.5 ± 0.2^{d}
Portions (brand 6)	4.4 ± 0.1^{c}	4.3 ± 0.1^{c}	2.3 ± 0.1^{b}	1.6 ± 0.6^{c}

*Data are expressed as mean value ± standard deviation (log CFU/g) of three replicates. Means with a different superscript letter in the same column indicate where statistically ($p \leq 0.05$) differences occurred between products that were evaluated in this study

(Table 3 and Additional file 1). On the basis of the total number of isolates identified, *Acinetobacter, Serratia, Staphylococcus* and *Lactococcus* spp. were highly frequent, representing 10.5, 7.9, 11.8, and 31.6 % of the isolates, respectively. *Lactococcus* spp. were isolated from five of the six brands evaluated, the only exception was fillets from brand 3. *Enterococcus, Stenotrophomonas, Chryseobacterium* and *Empedobacter* spp. were found only on fillets from brand 3. In addition, *Serratia* spp. were identified on the portions and steaks while *Enterobacter* and *Morganella* spp. were identified only on fillets from brand 1 to 4, respectively. *Staphylococcus* spp. was identified on the portions and fillets (from brands 1 to 2). The other microbiota identified on *Pangasius* sampled included *Klebsiella* (brand 1 and 5), *Pseudomonas* (brand 1 and 4), *Arthrobacter* (brand 2), and *Macrococcus* spp. (brand 1).

Table 3 Genera and species isolated from different *Pangasius* products sold in Belgium

Isolate[a]	Fillets (brand 1)	Fillets (brand 2)	Fillets (brand 3)	Fillets (brand 4)	Steaks (brand 5)	Portions (brand 6)	Total isolates	Prevalence (%)
Acinetobacter spp.	1[b]		3		4		8	10.5
Acinetobacter johnsonii					2			
Acinetobacter beijerinckii	1		3					
Acinetobacter haemolyticus					2			
Pseudomonas spp.	2			3			5	6.6
Pseudomonas mosselii				3				
Pseudomonas beteli	2							
Stenotrophomonas spp.			1				1	1.3
Stenotrophomonas maltophilia			1					
Serratia spp.					1	5	6	7.9
Serratia nematodiphila					1	5		
Enterobacter spp.	4						4	5.3
Enterobacter hormaechei	4							
Klebsiella spp.	1				2		3	3.9
Klebsiella pneumoniae	1				2			
Morganella spp.				2			2	2.6
Morganella morganii				2				
Chryseobacterium spp.			2				2	2.6
Chryseobacterium indologenes			2					
Arthrobacter spp.		2					2	2.6
Arthrobacter protophormiae		2						
Lactococcus spp.	2	3		8	7	4	24	31.6
Lactococcus garvieae		3		8	7	4		
Lactococcus lactis	2							
Enterococcus spp.			6				6	7.9
Enterococcus casseliflavus			6					
Macrococcus spp.	2						2	2.6
Macrococcus caseolyticus	2							
Staphylococcus spp.	2	1				6	9	11.8
Staphylococcus sciuri	2	1				6		
Empedobacter spp.			2				2	2.6
Empedobacter brevis			2					
Total of strain abundance	14	6	14	13	14	15	76	100
Number of species	7	3	5	3	5	3		

[a]Identification results on genus and species level; species identifications are only tentative
[b]The frequency of identified isolates based on rep-clustering and partial 16S rRNA gene sequence analysis with cut-off value of 98.5 % similarity with type strains of validly published prokaryotic names in EZTaxon database. The percentage of total number of isolates of each genus is listed in the last column

Discussion

Physico-chemical characteristics

The results of the physico-chemical characteristics analyses of the frozen Vietnamese *Pangasius* products marketed in Belgium generally confirmed the findings of previous studies on *Pangasius* products. In agreement with our findings, Usydus et al. (2011) determined that the water content of frozen Vietnamese *Pangasius* products marketed in Poland was 84.7 ± 0.3 %. Karl et al. (2010) reported that frozen Vietnamese *Pangasius* products marketed in Germany had lower values of water content which ranged from 78.1 to 83.3 %, value of drip losses between 12.5 and 24.6 % and pH values between 6.3 and 7.6. Orban et al. (2008) reported that frozen Vietnamese *Pangasius* products marketed in Italy had values of water content which ranged from 80.1 to 85.0 %, and pH values between 7.56 and 7.96. In the same study, a high sodium content (0.222–0.594 %) was determined in the *Pangasius* products. This was assumed to be a result of the fish being possibly treated with water-binding additives of polyphosphate before freezing (Orban et al. 2008). The same conclusion was also derived by Karl et al. (2010) for *Pangasius* products marked in Germany by means of differential scanning calorimetry which showed a decreased thermal stability in the protein domains of the fish. The use of phosphate in both fish and meat can increase water retention and reduce thaw loss as a result of an increase in the pH and ionic strength and binding of phosphate to the protein (Thorarinsdottir et al. 2001; Kaufmann et al. 2005; Gonçalves et al. 2008). Brand 3 fillets were most likely treated with water-binding additives as they had the highest water content (87.7 ± 0.9 %), lowest water activity (0.9896 ± 0.0007), highest pH (8.2 ± 0.2) and a very high salt content (0.93 ± 0.2 %). It is necessary to investigate further the composition of additives used during processing and their impacts on the quality and safety in general and the microbiota of *Pangasius* products in particular.

Microbiota of frozen *Pangasius* products marketed in Belgium

The TPC (3.8 to 5.2 log CFU/g) on the *Pangasius* products were not significantly different ($p > 0.05$) from the TMC (3.8 to 4.8 log CFU/g). These values are below the acceptance limit for frozen *Pangasius* fillets established by Vietnamese Science & Technology Ministry (TCVN 2010) and for fresh fish developed by the Laboratory of Food Microbiology and Food Preservation (Ghent University) (Uyttendaele et al. 2010). Moreover, the TMC cannot always give a realistic estimation of the microbial contamination levels in food, especially foods stored at chilling or freezing temperatures. Some previous studies have reported that the TMC are not sufficient for enumerating the microbial counts of a

packaged food product stored at chilling or freezing temperature (Broekaert et al. 2011; Pothakos et al. 2012). However, either TPC or TMC can use to enumerate the total microbia counts for frozen *Pangasius* fillets (orginated from the tropical areas of Vietnam). This is consistent with previous results observed on frozen *Pangasius* originating from Vietnam that were processed for export to Belgium and other European countries (Tong Thi et al. 2013; Noseda et al. 2013; Noseda et al. 2012). The lactic acid bacteria (LAB) counts varied greatly between the products. The highest LAB counts, 4.1 ± 0.1 log CFU/g, were observed on the fillets from brand 3. These counts were in agreement with those obtained by Noseda et al. (2012), who found 3.9 ± 0.1 log CFU/g of LAB on thawed frozen Vietnamese *Pangasius* intended for a study regarding the effect of modified atmosphere packaging. However, these and our counts were generally higher than those observed on frozen *Pangasius* fish sampled after production in Vietnam (1.5 ± 0.3 to 3.4 ± 0.4 log CFU/g) (Tong Thi et al. 2013). It is possible that the mesophilic LAB on the frozen *Pangasius* fish sampled after production in Vietnam were less dominant than the psychrotrophic LAB on the *Pangasius* fillets evaluated in this study and by Noseda et al. (2012). Of the LAB identified in the samples, *Lactococcus* spp. were the most prevalent (31.6 %). These results were in agreement with the findings of our previous study on *Pangasius* fish during processing (Tong Thi et al. 2013). *Lactococcus* and *Enterococcus* spp. have previously been isolated from lightly preserved salmon products such as cold smoked, salted and dried salmon (Françoise 2010). Both *L. lactis* and *L. garvieae* are also associated with fresh and marine water in tropical areas (Michel et al. 2007). In addition to their association with fish farm environments, they are also sometimes isolated from human and other mammalian clinical cases (Michel et al. 2007).

Enterococcus spp. can tolerate high salts concentrations (Harwood et al. 2000), which can explain that they were only isolated from brand 3 fillets. With regard to the presumptive Enterobacteriaceae, various isolates of *Enterobacter*, *Klebsiella*, *Morganella* and *Serratia* spp. were identified. The incidence of these isolates appeared to be dependent on how the frozen *Pangasius* was processed as *Serratia* was isolated from steaks and portioned *Pangasius* products, whilst *Enterobacter* and *Klebsiella* were isolated from fillets (brand 1) and *Morganella* from fillets (brand 4). Differences between the types of Enterobacteriaceae contaminating frozen *Pangasius* products in Vietnam have also been found on the basis of the size of the processing plant. Frozen *Pangasius* fillets processed in a large plant were determined to be contaminated by *Serratia* spp. whereas those processed in a small scale plant were contaminated by *Enterobacter*, *Klebsiella*, and *Morganella* spp. (Tong Thi et al. 2013). Kim et al. (2003) also pointed to the importance of sanitation in the fish processing plant to prevent cross-contamination with *Enterobacter*, *Klebsiella*, and *Morganella*. Moreover, species belonging to the Enterobacteriaceae family are in general frequently isolated from intestines of tropical and farmed fish (ICMSF 2005; Apun et al. 1999) and Vietnamese *Pangasius* (Sarter et al. 2007; Tong Thi et al. 2013). Contamination of the flesh by these species is likely to occur during gutting (for *Pangasius* steaks) and manually filleting (for fillets and portions) (Tong Thi et al. 2013; Noseda et al. 2013; Tong Thi et al. 2014). Based on our previous studies, it is suggested that producers have to pay a lot of attention during gutting and filleting such as sanitation of knives, gloves, cutting boards, etc. in order to control the microbial quality. In other words, both gutting and filleting steps may be suggested as critical control points during processing *Pangasius* products. Similarly, based on our findings of Tong Thi et al. (2013) combined with the results obtained from this study, we suggest that *Enterobacter*, *Klebsiella*, *Morganella* and *Serratia* spp. may be good indicators of the microbial quality for *Pangasius* fish. The growth of these species can be prevented by applying low temperature (i.e. 0 to 4 °C) and modified atmosphere packaging techniques such as vacuum or packaging in modified atmospheres (Baylis 2006). Furthermore, the counts of presumptive Enterobacteriaceae were highly variable among the products, ranging from 1.6 ± 0.6 to 3.8 ± 0.2 log CFU/g. It has to be mentioned that the approach used in this study can only give an idea of the presumptive Enterobacteriaceae. *Acinetobacter* and *Pseudomonas* spp. were also identified for isolates growing not only on non-selective PCA but also on selective VRBGA plates.

The ability of *Acinetobacter* spp. to grow on both non-selective (PCA) and selective media (VRBGA), may have contributed to the high prevalence of *Acinetobacter* spp. (10.5 %) on the frozen *Pangasius* fish evaluated in this study. In addition, *Acinetobacter* and *Pseudomonas* spp. have been isolated from the intestines of fish (Hovda et al. 2007; Merrifield et al. 2009; Ringø et al. 2006) and therefore may potentially contaminate the fish during processing. This is supported by our previous findings where *Acinetobacter* and *Pseudomonas* spp. were detected on *Pangasius* samples collected at the filleting and trimming steps of processing in Vietnamese companies (Tong Thi et al. 2013). Moreover, *Pseudomonas* spp. have also been determined to be the dominant spoilage bacteria on thawed *Pangasius* stored in air at 4 °C (Noseda et al. 2012). The spoilage capacity of the isolates identified in this study should be further evaluated to provide better insights into their spoilage mechanisms.

The prevalence of *Staphylococcus sciuri* (11.8 %) on frozen *Pangasius* was relatively high. *Staphylococcus sciuri* has found in humans (Stepanović et al. 2005) and also in gut (Boari et al. 2008), therefore they could have been transferred to the products *via* human contact during handling and processing. *S. sciuri* has been identified in various food products of animal origins such as meat and meat products; milk and milk products and fish and fish products (García et al. 2002; Papamanoli et al. 2002). *S. sciuri* is the most frequently reported histamine-forming bacterium in cod, escolar steaks, swordfish fillets, cold smoked rainbow trout and whole and filleted catfish (Hwang et al. 2012; Chang et al. 2008; Ramos and Lyon 2000).

Empedobacter, *Macrococcus*, *Arthrobacter*, *Chryseobacterium* and *Stenotrophomonas* spp. were less prevalent in the frozen *Pangasius* products evaluated in this study. *Stenotrophomonas maltophilia*, an important opportunistic pathogen, has been also isolated from channel catfish in China (Geng et al. 2010). *Chryseobacterium indologenes* has also been found on *Pangasius* fish sampled during processing in Vietnam (Tong Thi et al. 2013) and frozen *Pangasius* exported to Denmark (Noor Uddin et al. 2013). *Chryseobacterium indologenes* has been isolated from diseased yellow perch (Pridgeon et al. 2013) whilst *Chryseobacterium* spp. are known to be widely distributed in the environment and soil (Benmalek et al. 2010), and fresh water (Kim et al. 2008; Park et al. 2008).

Conclusion

A high prevalence of Pseudomonas (6.5 %), Enterococcus (7.8 %), Acinetobacter (9.1 %), Serratia (10.4 %), Staphylococcus (11.7 %) and Lactococcus spp. (31.2 %) was determined on thawed Vietnamese Pangasius products marketed in Belgium. These results are crucial as currently very little is known about the microbiota of thawed Pangasius products marketed in the West as 'fresh' Pangasius products. This knowledge is important with regards to the development of suitable preservation techniques such as vacuum and modified atmosphere packaging to inhibit the microorganisms contaminating thawed Pangasius fish.

Acknowledgments

Grants from Research Foundation Flanders (FWO/project number: GA02012N) and Ministry of Education and Training of Viet Nam (MOET) are acknowledged. We are grateful to Pieter Siau, Ann Vanhee and Ann Dirckx for their practical guidance and assistance.

Authors' contributions

ANTT contributed to the sampling and analysis. FD and MH contributed to the design of the study. ANTT, FD and MH contributed to the interpretation of data and drafting the manuscript. SS contributed to critical revision of the paper. All authors read and approved the final manuscript.

Competing interests

The authors declare that they have no competing interests.

Author details

[1]Department of Food Safety and Food Quality, Laboratory of Food Microbiology and Food Preservation, Food2Know, Ghent University, Coupure Links 653, Ghent 9000, Belgium. [2]Department of Food Technology, Faculty of Agriculture and Applied Biology, Can Tho University, 3-2 Street, Can Tho City, Viet Nam. [3]Institute for Agricultural and Fisheries Research (ILVO), Technology and Food Science Unit, Brusselsesteenweg 370, Melle 9090, Belgium. [4]Department of Pathology, Bacteriology and Avian Diseases, Faculty of Veterinary Medicine, Ghent University, Salisburylaan 133, Merelbeke 9820, Belgium.

References

Apun K, Yusof AM, Jugang K. Distribution of bacteria in tropical freshwater fish and ponds. Int J Environ Health Res. 1999;9(4):285–92.

Austin B. The bacterial microflora of fish. Sci World J. 2002;2(3):558–72.

Baylis CL. Food Spoilage Microorganisms. In: Blackburn CW, editor. Food Science, Technology And Nutrition. New York: Woodhead Publishing; 2006. p. 624–67.

Benmalek Y, Cayol J-L, Bouanane NA, Hacene H, Fauque G, Fardeau M-L. *Chryseobacterium solincola* sp. nov., isolated from soil. Int J Syst Evol Microbiol. 2010;60(8):1876–80.

Boari CA, Pereira GI, Valeriano C, Silva BC, Morais VM, Figueiredo HCP, Piccoli RH. Bacterial ecology of tilapia fresh fillets and some factors that can influence their microbial quality. Food Sci Tech (Campinas). 2008;28(4):863–7.

Broekaert K, Heyndrickx M, Herman L, Devlieghere F, Vlaemynck G. Seafood quality analysis: molecular identification of dominant microbiota after ice storage on several general growth media. Food Microbiol. 2011;28(6):1162–9.

Brosius J, Palmer ML, Kennedy PJ, Noller HF. Complete nucleotide sequence of a 16S ribosomal RNA gene from *Escherichia coli*. Proc Natl Acad Sci. 1978;75:4801–5.

Chang S-C, Kung H-F, Chen H-C, Lin C-S, Tsai Y-H. Determination of histamine and bacterial isolation in swordfish fillets (*Xiphias gladius*) implicated in a food borne poisoning. Food Control. 2008;19(1):16–21.

Chun J, Lee J-H, Jung Y, Kim M, Kim S, Kim BK, Lim Y-W. EzTaxon: a web-based tool for the identification of prokaryotes based on 16S ribosomal RNA gene sequences. Int J Syst Evol Microbiol. 2007;57(10):2259–61.

Flamm RK, Hinrichs DJ, Thomashow MF. Introduction of pAMb1 into *Listeria monocytogenes* by conjugation and homology between native *L. monocytogenes* plasmids. Infect Immun. 1984;44:157–61.

Françoise L. Occurrence and role of lactic acid bacteria in seafood products. Food Microbiol. 2010;27(6):698–709.

Garcia M, Rodriguez M, Bernardo A, Tornadijo M, Carballo J. Study of enterococci and micrococci isolated throughout manufacture and ripening of San Simón cheese. Food Microbiol. 2002;19(1):23–33.

Geng Y, Wang K, Chen D, Huang X, He M, Yin Z. *Stenotrophomonas maltophilia*, an emerging opportunist pathogen for cultured channel catfish, *Ictalurus punctatus*, in China. Aquaculture. 2010;308(3–4):132–5.

Gonçalves AA, Rech BT, Rodrigues P, Pucci D. Quality evaluation of frozen seafood (*Genypterus brasiliensis*, *Prionotus punctatus*, *Pleoticus muelleri* and *Perna perna*) previously treated with phosphates. Pan-Am J Aquat Sci. 2008;3(3):248–58.

Harwood VJ, Whitlock J, Withington V. Classification of antibiotic resistance patterns of indicator bacteria by discriminant analysis: use in predicting the source of fecal contamination in subtropical waters. Appl Environ Microbiol. 2000;66(9):3698–704.

Hovda MB, Lunestad BT, Fontanillas R, Rosnes JT. Molecular characterisation of the intestinal microbiota of farmed Atlantic salmon (*Salmo salar* L.). Aquaculture. 2007;272(1):581–8.

Hwang C-C, Lin C-M, Huang C-Y, Huang Y-L, Kang F-C, Hwang D-F, Tsai Y-H. Chemical characterisation, biogenic amines contents, and identification of fish species in cod and escolar steaks, and salted escolar roe products. Food Control. 2012;25(1):415–20.

ICMSF. Micro-organisms in Food 6: Microbiological ecology of food commodities. New York: Kluwer Academic/Plenum Publishers; 2005. p. 174–249.

Karl H, Lehmann I, Rehbein H, Schubring R. Composition and quality attributes of conventionally and organically farmed *Pangasius* fillets (*Pangasius hypophthalmus*) on the German market. Int J Food Sci Tech. 2010;45(1):56–66.

Kaufmann A, Maden K, Leisser W, Matera M, Gude T. Analysis of polyphosphates in fish and shrimps tissues by two different ion chromatography methods: Implications on false-negative and-positive findings. Food Addit Contam. 2005;22(11):1073–82.

Kim KK, Lee KC, Oh H-M, Lee J-S. *Chryseobacterium aquaticum* sp. nov., isolated from a water reservoir. Int J Syst Evol Microbiol. 2008;58(3):533–7.

Kim SH, An H, Wei CI, Visessanguan W, Benjakul S, Morrissey MT, Su YC, Pitta TP. Molecular Detection of a Histamine Former, Morganella morganii, in Albacore, Mackerel, Sardine, and a Processing Plant. J Food Sci. 2003;68(2):453–7.

Merrifield DL, Burnard D, Bradley G, Davies SJ, Baker RTM. Microbial community diversity associated with the intestinal mucosa of farmed rainbow trout (*Oncoryhnchus mykiss Walbaum*). Aqua Res. 2009;40(9):1064–72.

Michel C, Pelletier C, Boussaha M, Douet DG, Lautraite A, Tailliez P. Diversity of lactic acid bacteria associated with fish and the fish farm environment, established by amplified rRNA gene restriction analysis. Appl Environ Microbiol. 2007;73(9):2947–55.

Noor Uddin GM, Larsen MH, Guardabassi L, Dalsgaard A. Bacterial flora and antimicrobial resistance in raw frozen cultured seafood imported to Denmark. J Food Prot. 2013;76(3):490–9.

Noseda B, Islam MT, Eriksson M, Heyndrickx M, De Reu K, Van Langenhove H, Devlieghere F. Microbiological spoilage of vacuum and modified atmosphere packaged Vietnamese *Pangasius hypophthalmus* fillets. Food Microbiol. 2012; 30(2):408–19.

Noseda B, Tong Thi AN, Rosseel L, Devlieghere F, Jacxsens L. Dynamics of microbiological quality and safety of Vietnamese *Pangasianodon hypophthalmus* during processing. Aquac Int. 2013;21(3):709–27.

Orban E, Nevigato T, Lena GD, Masci M, Casini I, Gambelli L, Caproni R. New trends in the seafood market. Sutchi catfish (*Pangasius hypophthalmus*) fillets from Vietnam: Nutritional quality and safety aspects. Food Chem. 2008;110(2):383–9.

Papamanoli E, Kotzekidou P, Tzanetakis N, Litopoulou-Tzanetaki E. Characterization of Micrococcaceae isolated from dry fermented sausage. Food Microbiol. 2002;19(5):441–9.

Park SC, Kim MS, Baik KS, Kim EM, Rhee MS, Seong CN. *Chryseobacterium aquifrigidense* sp. nov., isolated from a water-cooling system. Int J Syst Evol Microbiol. 2008;58(3):607–11.

Phan LT, Bui TM, Nguyen TT, Gooley GJ, Ingram BA, Nguyen HV, Nguyen PT, De Silva SS. Current status of farming practices of striped catfish, *Pangasianodon hypophthalmus* in the Mekong Delta, Vietnam. Aquaculture. 2009;296(3):227–36.

Pothakos V, Samapundo S, Devlieghere F. Total mesophilic counts underestimate in many cases the contamination levels of psychrotrophic lactic acid bacteria (LAB) in chilled-stored food products at the end of their shelf-life. Food Microbiol. 2012;32(2):437–43.

Pridgeon J, Klesius P, Garcia J. Identification and virulence of *Chryseobacterium indologenes* isolated from diseased yellow perch (*Perca flavescens*). J Appl Microbiol. 2013;114(3):636–43.

Ramos M, Lyon WJ. Reduction of endogenous bacteria associated with catfish fillets using the Grovac process. J Food Prot. 2000;63(9):1231–9.

Ringø E, Sperstad S, Myklebust R, Refstie S, Krogdahl Å. Characterisation of the microbiota associated with intestine of Atlantic cod (*Gadus morhua* L.): The effect of fish meal, standard soybean meal and a bioprocessed soybean meal. Aquaculture. 2006;261(3):829–41.

Sarter S, Kha Nguyen HN, Hung LT, Lazard J, Montet D. Antibiotic resistance in Gram-negative bacteria isolated from farmed catfish. Food Control. 2007; 18(11):1391–6.

Stepanović S, Dakić I, Martel A, Vaneechoutte M, Morrison D, Shittu A, Ježek P, Decostere A, Devriese LA, Haesebrouck F. A comparative evaluation of phenotypic and molecular methods in the identification of members of the *Staphylococcus sciuri* group. Syst Appl Microbiol. 2005;28(4):353–7.

TCVN. Officially legal criteria for frozen Tra fish (*Pangasius hypophthalmus*) fillet established by Vietnamese Science & Technology Ministry. Reference number: TCVN 8338: 2010. 2010.

Thorarinsdottir K, Arason S, Bogason SG, Kristbergsson K. Effects of phosphate on yield, quality, and water-holding capacity in the processing of salted cod (*Gadus morhua*). J Food Sci. 2001;66(6):821–6.

Tong Thi AN, Jacxsens L, Noseda B, Samapundo S, Nguyen BL, Heyndrickx M, Devlieghere F. Evaluation of the microbiological safety and quality of Vietnamese *Pangasius hypophthalmus* during processing by a microbial assessment scheme in combination with a self-assessment questionnaire. Fish Sci. 2014;80(5):1117–28.

Tong Thi AN, Noseda B, Samapundo S, Nguyen BL, Broekaert K, Rasschaert G, Heyndrickx M, Devlieghere F. Microbial ecology of Vietnamese Tra fish

(*Pangasius hypophthalmus*) fillets during processing. Int J Food Microbiol. 2013;167(2):144–52.

Usydus Z, Szlinder-Richert J, Adamczyk M, Szatkowska U. Marine and farmed fish in the Polish market: Comparison of the nutritional value. Food Chem. 2011;126(1):78–84.

Uyttendaele M, Jacxsens L, De Loy-Hendrickx A, Devlieghere F, Debevere J. Microbiological guide values and legal criteria. Ghent: Ghent University; 2010.

VASEP. Viet Nam Association of Seafood Exporters and Producers; 2014. http://www.pangasius-vietnam.com/378/Daily-News-p/About-Pangasius.htm. Accessed on 15 Aug 2014.

Versalovic J, Schneider M, De Bruijn FJ, Lupski JR. Genomic fingerprinting of bacteria using repetitive sequence-based polymerase chain reaction. Methods Mol Cell Biol. 1994;5(1):25–40.

6

Screening of caffeine, preservatives and antioxidants in dairy products available in Bangladesh using an RP-HPLC method

Md. Shahadat Hossain, Md. Samiul Islam, Subrata Bhadra and Abu Shara Shamsur Rouf[*]

Abstract

Background: The aim of this current study was to investigate the presence and to determine the contents of caffeine, preservatives and antioxidants in dairy products available in Bangladesh, one of the alarming countries in terms of food security. As high as 41 marketed dairy products, of both locally and internationally manufactured, were collected from local markets of Dhaka, Bangladesh in mid–2013 and then analyzed using an RP–HPLC method to determine the contents of caffeine, benzoic acid, propylparaben, butylparaben, butylated hydroxyanisole and butylated hydroxytoluene. The evaluation was performed using a C_{18} column (150 mm × 4.6 mm i.d., 5 μm particle size) with a gradient flow rate of acetonitrile and diluted sulfuric acid (0.002 M) as mobile phase from ratio 15:85 to 80:20 (%v/v) at a flow rate of 2.0 ml/min at 265 nm.

Results: The retention times of CF, BA, PP, BP, BHA and BHT were found about 3.6, 11.6, 13.0, 13.3, 13.6 and 16. 7 min, respectively. Results revealed that BA was found in 17.1 % of the products within the concentration range of 11 ~ 2067 mg/L; among these, 42.9 % products exceeded the allowable limit of 300 mg/L set by JECFA. Moreover, 14.6 % products showed positive response to CF at a concentration of 10 ~ 18 mg/L, which was well below the tolerance limit (<200 mg/L) set by FDA. However, none of these investigated dairy products were found to contain any detectable amount of PP, BP, BHA or BHT.

Conclusion: Presence of excess amount of BA in dairy products, which are one of the most favorite healthy food items for all generations, can easily jeopardize public health sector. Moreover, CF was found in some CF–free products, which is also an alarming issue to be considered seriously to prevent the future occurrence of it.

Keywords: Dairy products, HPLC, Caffeine, Preservatives, Antioxidants, Food security

Background

People, especially children, consume substantial amount (30 to 150 kg/capita/year) of fresh milk and dairy products every day (FAO 2013a). Milk is a valuable nutritious food with a short shelf–life and requires careful handling as it is highly perishable being an excellent medium for the growth of microorganisms–particularly bacterial pathogens–that can cause spoilage and diseases amongst the consumers. Milk processing like mechanical refrigeration, pasteurization etc. allows the preservation of milk for days, weeks or even months and helps to reduce food–borne illness (FAO 2013b). Processing of dairy products gives small–scale dairy producers higher cash incomes than selling raw milk, and offers better opportunities to reach regional and urban markets and helps to deal with seasonal fluctuations in milk supply. The transformation of raw milk into processed milk and products can benefit entire communities by generating off–farm jobs in milk collection, transportation, processing and marketing (FAO 2013b).

Bangladesh is nearly self–sufficient in food production, but unfortunately food security remains an elusive goal nowadays. Shortage of storage facilities, lack of academic education, frequent natural disasters, poverty, and lack of strict monitoring systems sometimes lead the crooked producers and businessmen not to apply the standard food processing procedures; rather frequently they add uncontrolled amount of different food preservatives and/ or antioxidants in the food products, thus make the

* Correspondence: rouf321@yahoo.com
Institutional address: Department of Pharmaceutical Technology, Faculty of Pharmacy, University of Dhaka, Dhaka 1000, Bangladesh

product adulterated. It is hard to believe that the percentage of these sorts of debased food products are increasing at an alarming rate and consumed by the common people across the country. Therefore, food–borne illness, intoxication and malnutrition are the common scenario in Bangladesh–more severely in the underdeveloped rural areas (Ali 2013; Rahman and Kabir 2012; Hossain et al. 2008). Numerous reports on adulteration of foods and food products are often published in the mainstream national newspapers as well as broadcasted in the television or radio that also inspired us to do in–depth research in this field. Even the National Food Safety Laboratory (NFSL) of our country found the presence of two harmful and banned pesticides–Aldrin and DDT in their investigated raw milk and milk products in 2014 (NFSL 2014). Beside this, in 2012, Chanda et al. found that all fifty milk samples collected from Barisal district of Bangladesh were adulterated with water and more interestingly, 10 % of those samples were adulterated with toxic formalin and 20 % with sodium bicarbonate. These types of abominable activities are increasing day by day due to lack of government and public awareness, greediness of businessmen, lack of regulations and standards, and finally lack of implementation of existing laws (Rahman and Kabir 2012).

Usually coliforms, lactic acid bacteria, yeasts, and molds are responsible for the spoilage of milk, but psychrotrophic organisms play the most dominant role. In addition, various bacteria of public health concern such as *Salmonella* spp., *Listeria monoytogenes, Camphylobacter jejuni, Yersinia enterocolitica*, pathogenic strains of *Eshcherichia coli* and enterotoxigenic strains of *Staphylococcus aureus* may also be found in milk and dairy products (Cempírková and Mikulová 2009; Varga 2007). These microorganisms and their metabolic end products are greatly harmful for human body. So, suitable preservation techniques are to be applied to prevent microbial and chemical spoilage of dairy products. The most important means of controlling bacteria, yeasts and molds are application of heat, cold, chemicals, radiation etc. Usually food industries incorporate two or more preservation techniques to preserve their products (Potter and Hotchkiss 1999). Mechanical cooling, thermization, pasteurization and ultra–heat treatment are regularly applied to dairy products (Guizani 2007). However, according to WHO/FAO authorized "codex general standard for food additives" guideline published in 1995, our all investigated dairy products have no permission to apply chemical preservation technique using any preservatives and antioxidants, except flavored yogurt in which only benzoic acid (BA) (Fig. 1) can be used in a concentration of less than 300 mg/L (Codex 2015). There is no specific rule on the use of caffeine (CF) in dairy products (Nutri pro 2013), but it may be present in chocolate milk prepared with cocoa powder considering the fact that its concentration should be well below the maximum daily allowable level of it. Excess intake of CF causes headache, insomnia, agitation, tremor, hypertonicity, systemic hypokalemia, tinnitus, tachyarrhythmia, delirium, hyperventilation, hyperthermia, polyuria, dehydration, seizures, and coma (Simmons and Kidner 1998; Davies et al. 2012). On the other hand, BA may cause asthma, urticaria, metabolic acidosis, convulsions etc. if exceeds 5 mg/kg b.w. and it may also form benzene by reacting with ascorbic acid (usually present in yogurt) in the presence of transition–metal catalyst (Qi et al. 2008; Roye et al. 2009). Decline in serum testosterone level in male persons and hypersensitivity reaction may occur if the intake amount of propylparaben (PP) and butylparaben (BP) exceed 10 mg/kg b.w. (WHO 2010; Roye et al. 2009). Moreover, butylated

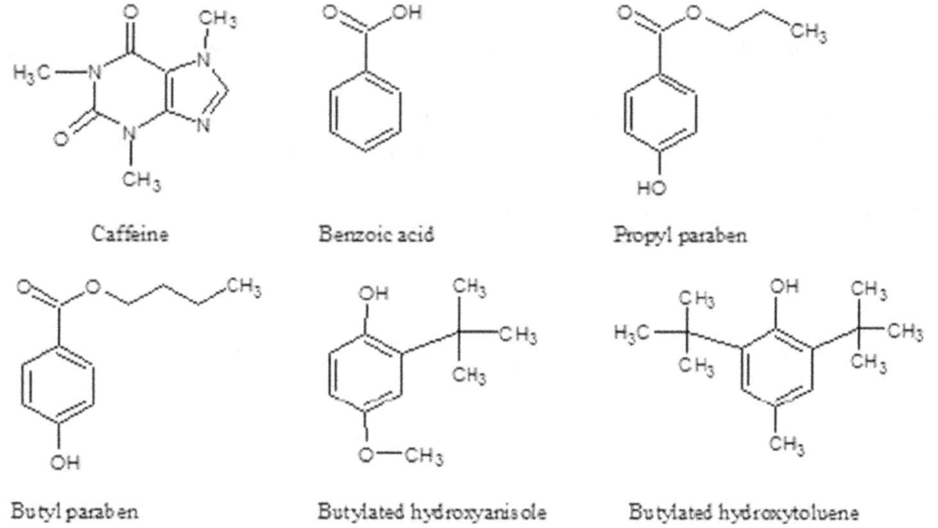

Fig. 1 Structural formula of the six additives

Table 1 The types and sources of collected dairy products

Parameters	Non-heat treated	Heat treated	
		UHT[a]-processed	Pasteurized
Products quantity	16	16	9
Source of origin	Local: 16	Local: 12	Local: 9
	Imported: Nil[b]	Imported: 4	Imported: Nil[b]

[a]UHT stands for Ultra Heat Treatment
[b]Nil means zero product of that type

hydroxyanisole (BHA) may cause carcinogenic effect when the daily ingestion crosses the limit of 500 mcg/kg b.w. (Report on Carcinogens 2014), whereas severe nausea and vomiting may occur following the ingestion of butylated hydroxytoluene (BHT), whose ADI is 125 mcg/kg b.w. (Roye *et al.* 2009). So, it would be a better approach to collect milk by maintaining proper cleanliness and sanitation, and then applying cooling and/or thermization technique (Guizani 2007).

CF can be quantitatively analyzed using gas chromatography–mass spectrometry (Kerrigan and Lindsey 2005; Ayala et al. 2009), HPLC (Sather and Vernig 2011; Srdjenovic et al. 2008), surfactant–mediated matrix–assisted laser desorption/ionization time–of–flight mass spectrometry (Grant and Helleur 2008), micellar electrokinetic capillary chromatography (Injac 2008), ultraviolet visible absorption spectrometry and thin layer chromatography (Sather and Vernig 2011). The determination of BA and parabens can be carried out by HPLC (Saad et al. 2005; Mroueh et al. 2008; Petronela and Elena 2009; Khosrokhavar et al. 2010; Tawalbeh et al. 2014), capillary electrophoresis (Tang and Wu 2007). The analysis of BHA and BHT can be done with liquid chromatography–mass spectrometry (Lee et al. 2006), non–isothermal pressurized–differential scanning calorimetry (Dunn 2005), gas chromatography (Yang et al. 2002), and square–wave voltammetry (Medeiros et al. 2010).

Studying inquisitively and applying many of these practically, a suitable and precise RP–HPLC method was developed for the simultaneous determination of CF, BA, PP, BP, BHA and BHT. As a part of our continuous interest in the researches of adulteration studies on food, soft drinks, cosmetics and dairy products, we herein report a validated method and the result of our study to explore the presence or absence of CF, BA, PP, BP, BHA and BHT in dairy products, and their compliance with

Table 2 Composition of mobile phase of the RP-HPLC method

Time (min)	Diluted sulfuric acid (0.002 M) (%)	Acetonitrile (%)
0.01–10.00	85	15
10.01–11.00	40	60
11.01–13.00	30	70
13.01–16.00	20	80
16.01–18.00	85	15

Table 3 System suitability of the proposed method

Standards	Peak area (Mean ± %RSD)[a]	Retention time (Mean ± %RSD)[a]	Tailing factor (Mean ± %RSD)[a]
CF	1865839 ± 0.71	3.6 ± 0.04	0.866 ± 0.44
BA	986908 ± 0.50	11.6 ± 0.05	1.075 ± 0.37
PP	994208 ± 0.35	13.0 ± 0.02	1.248 ± 0.33
BP	858343 ± 0.43	13.3 ± 0.04	1.229 ± 0.20
BHA	234037 ± 0.94	13.6 ± 0.03	1.224 ± 0.22
BHT	177374 ± 0.59	16.7 ± 0.03	1.017 ± 0.49

CF caffeine, *BA* benzoic acid, *PP* propylparaben, *BP* butylparaben, *BHA* butylated hydroxyanisole, *BHT* butylated hydroxytoluene
[a]All determinants are in sextuplicate

the existing rules and regulations listed by the regulatory bodies. To the best of our knowledge, this is the first study for the assay of such additives in dairy products available in Bangladesh.

Methods
Chemicals and reagents
All six working standards of CF, BA, PP, BP, BHA and BHT were gifted by Eskayef Bangladesh Limited, Gazipur, Bangladesh. HPLC grade sulfuric acid (Sigma Aldrich, Germany), HPLC grade acetonitrile (RCI Labscan, Thailand) and analytical grade trichloroacetic acid (BDH Chemicals, England) were used for analytical purposes. Milli–Q water was used to prepare the mobile phase.

Sample collection and storage
Forty one marketed dairy products were collected from departmental stores, supermarkets and dairy farms in Dhaka, Bangladesh during mid–2013. Among them, 37 products were local and 4 products were imported. There were 10 full cream milk, 7 locally collected milk (bought from local dairy farms), 6 chocolate milk, 5 mango milk, 8 buttermilk, 2 skimmed milk, 1 banana milk, 1 flavored yogurt and 1 lassi i.e. a sweet or savory Indian drink made from yogurt or buttermilk base with water. The different types of collected dairy products are listed in Table 1. In the product label claim it was observed that only two pasteurized chocolate milk and four UHT–processed

Table 4 Linearity study of the proposed method

Standards	Concentration range (mg/L)	Regression equation (Y = mX + c)	Correlation coefficient (R²)
CF	2–200	Y = 23019X–42400	0.996
BA	50–500	Y = 14211X–43609	0.992
PP	2–100	Y = 10605X–11766	0.992
BP	2–100	Y = 9294X–84770	0.997
BHA	50–300	Y = 1720X +10364	0.996
BHT	50–300	Y = 1855X–46598	0.998

CF caffeine, *BA* benzoic acid, *PP* propylparaben, *BP* butylparaben, *BHA* butylated hydroxyanisole, *BHT* butylated hydroxytoluene

Table 5 LOD and LOQ of the standards in the proposed method

Standards	LOD (μg/L)	LOQ (μg/L)
CF	2.66	8.06
BA	2.70	8.18
PP	0.21	0.64
BP	0.20	0.60
BHA	2.13	6.45
BHT	1.42	4.30

CF caffeine, *BA* benzoic acid, *PP* propylparaben, *BP* butylparaben, *BHA* butylated hydroxyanisole, *BHT* butylated hydroxytoluene

Table 6 Results of accuracy determination

Standards	Amount used (mg/L) eq. to (80 %–120 %) respectively	Amount recovered (mg/L)	% Recovery (Mean ± %RSD)[a]
CF	60	59.66	99.44 ± 0.025
	67.5	67.36	99.79 ± 0.04
	75	74.62	99.48 ± 0.045
	82.5	82.23	99.56 ± 0.047
	90	89.69	99.64 ± 0.02
BA	200	198.59	99.30 ± 0.04
	225	223.65	99.41 ± 0.03
	250	247.62	99.05 ± 0.02
	275	273.22	99.33 ± 0.02
	300	297.61	99.20 ± 0.02
PP	20	19.23	96.03 ± 0.4
	22.5	22.03	97.92 ± 0.25
	25	24.23	96.77 ± 0.14
	27.5	27.08	98.4 ± 0.22
	30	29.25	97.33 ± 0.21
BP	20	19.27	96.18 ± 0.16
	22.5	21.99	97.78 ± 0.12
	25	24.28	96.96 ± 0.19
	27.5	27.03	98.28 ± 0.2
	30	29.26	97.31 ± 0.22
BHA	160	158.31	98.9 ± 0.045
	180	178.33	99.06 ± 0.036
	200	197.58	98.77 ± 0.087
	220	218.42	99.25 ± 0.06
	240	238.42	99.34 ± 0.015
BHT	160	158.72	99.16 ± 0.04
	180	178.22	99.0 ± 0.04
	200	198.75	99.34 ± 0.09
	220	218.22	99.17 ± 0.025
	240	238.10	99.2 ± 0.02

CF caffeine, *BA* benzoic acid, *PP* propylparaben, *BP* butylparaben, *BHA* butylated hydroxyanisole, *BHT* butylated hydroxytoluene
[a]All determinants are in triplicate

chocolate milk products declared the presence of cocoa powder, which is a common source of CF but most of the manufacturers did not claim about the presence of BA, PP, BP, BHA and BHT. These samples were stored in a refrigerator below 4 °C till further use.

Equipment
Kubota–2100 centrifuge machine (Kubota, Japan) and Cyberscan 500 pH meter (Eutech, Singapore) were used to prepare the samples. Cole–Parmer Filtration machine (USA) was used to filter the mobile phase. To analyze samples a binary HPLC system (Shimadzu, Japan) equipped with a vacuum degasser (Model: DGU–20A3) and a UV/Vis detector (Model: SPD–20A) was used.

Chromatographic conditions
The separation was achieved using a 5 μm particle sized Capcell Pak octadecylsilyl (ODS) column (150 mm × 4.6 mm i.d.) (Shiseido, Japan) with the mobile phase consisting of acetonitrile and diluted sulfuric acid (0.002 M) at a binary gradient as shown in Table 2. The mobile phase flow rate was 2.0 mL/min and the injection volume was 20 μL. The column oven temperature was set at 30 °C and the analysis was performed at 265 nm wavelength.

Preparation of solutions
Preparation of standard solution
A standard solution was prepared by adding working standards of CF (75 mg/L), BA (250 mg/L), PP (25 mg/L), BP (25 mg/L), BHA (200 mg/L) and BHT (200 mg/L) and was used to analyze the investigated sample preparations.

Preparation of sample solution
Accurately measured 40 mL of marketed product was taken into a centrifuge tube with 4 mL of trichloroacetic acid, used to precipitate fat and protein present in milk and milk products and then it was centrifuged for 10 min at 3000 rpm. The supernatant was then filtered using a Whatmann 41 filter paper to obtain the final

Table 7 Precision of the proposed method

Standards	Spike level[a] (%)	Intra–day (Mean ± %RSD)[b]	Inter–day (Mean ± %RSD)[b]
CF	100	1865839 ± 0.71	1855736 ± 0.63
BA	100	986908 ± 0.50	982929 ± 0.56
PP	100	994208 ± 0.35	985310 ± 0.46
BP	100	858343 ± 0.43	852766 ± 0.76
BHA	100	216226 ± 0.80	213500 ± 0.86
BHT	100	234037 ± 0.94	214671 ± 0.69

CF caffeine, *BA* benzoic acid, *PP* propylparaben, *BP* butylparaben, *BHA* butylated hydroxyanisole, *BHT* butylated hydroxytoluene
[a]Spike level indicates the corresponding nominal concentration of each standard
[b]All determinants are in sextuplicate

Table 8 Ruggedness study of the proposed method

Standards	Analyst 1		Analyst 2	
	Amount found (mg/L)	% Recovery (Mean ± % RSD)[a]	Amount found (mg/L)	% Recovery (Mean ± % RSD)[a]
CF	74.61	99.48 ± 0.045	74.67	99.51 ± 0.05
BA	247.62	99.05 ± 0.02	247.57	99.03 ± 0.06
PP	24.19	96.77 ± 0.41	24.21	96.81 ± 0.54
BP	24.23	96.96 ± 0.19	24.16	96.15 ± 0.63
BHA	197.76	98.77 ± 0.09	197.56	98.13 ± 0.07
BHT	198.16	98.88 ± 0.03	197.99	98.12 ± 0.09

CF caffeine, *BA* benzoic acid, *PP* propylparaben, *BP* butylparaben, *BHA* butylated hydroxyanisole, *BHT* butylated hydroxytoluene
[a]All determinants are in sextuplicate

sample (King 1962). And the blank consisted of acetonitrile, diluted sulfuric acid (0.002 M) and trichloroacetic acid.

Validation of the test procedure
Method validation study was performed based on the current pharmaceutical regulatory guidelines i.e., ICH Q2 (R1). A number of parameters such as precision, accuracy, specificity, linearity, ruggedness and robustness were investigated for this purpose.

System suitability
For the evaluation of system suitability, the repeatability, retention time and tailing factor of six replicates of working standards of CF (75 mg/L), BA (250 mg/L), PP (25 mg/L), BP (25 mg/L), BHA (200 mg/L) and BHT (200 mg/L) were used and percentage relative standard deviation (%RSD) values were calculated in each case.

Linearity
The linearity was checked by analyzing different concentrations of CF (2–200 mg/L), BA (50–500 mg/L), PP (2–100 mg/L), BP (2–100 mg/L), BHA (50–300 mg/L) and BHT (50–300 mg/L). Calibration curves were made using MS Excel 2007 for each standard component. The regression line was calculated as $Y = mX + c$, where X

was the concentration of standard and Y was the response (peak area expressed as AU).

Limits of detection (LOD) and Limits of quantitation (LOQ)
LOD and LOQ were calculated according to ICH Q2 (R1) recommendations in accordance with the 3.3 s/m and 10s/m criteria respectively, where 's' is the standard deviation of the peak area and 'm' is the slope of the calibration curve determined from linearity investigation.

Accuracy (Recovery Test)
Recovery test was done by analyzing a sample of known concentration of standard solutions. Then percent recoveries (mean ± %RSD of three replicates) were calculated.

Precision
Repeatability (intra–day precision) and intermediate precision (inter–day precision) of the method were determined by using the solution of standards i.e. CF (75 mg/L), BA (250 mg/L), PP (25 mg/L), BP (25 mg/L), BHA (200 mg/L) and BHT (200 mg/L) and the solutions were analyzed in six replicates in the same day (intra–day precision) and daily for six times over a period of three days (inter–day precision).

Table 9 Robustness study of the proposed method

Standards	Standards (mg/L)	Amount detected (mg/L) (mean ± %RSD) [a]					
		Change in flow rate			Change in pH		
		1.9 mL/min	2.0 mL/min	2.1 mL/min	2.2	2.3	2.4
CF	75	74.69 ± 0.08	74.62 ± 0.05	74.59 ± 0.02	74.63 ± 0.02	74.62 ± 0.05	74.63 ± 0.04
BA	250	247.64 ± 0.01	247.61 ± 0.02	247.55 ± 0.01	247.68 ± 0.01	247.61 ± 0.02	247.59 ± 0.01
PP	25	24.24 ± 0.04	24.23 ± 0.04	24.18 ± 0.04	24.23 ± 0.04	24.23 ± 0.04	24.23 ± 0.04
BP	25	24.25 ± 0.06	24.28 ± 0.04	24.23 ± 0.09	24.27 ± 0.06	24.28 ± 0.04	24.25 ± 0.10
BHA	200	197.61 ± 0.01	197.58 ± 0.02	197.52 ± 0.01	197.58 ± 0.01	197.58 ± 0.02	197.58 ± 0.02
BHT	200	198.7 ± 0.01	198.75 ± 0.03	198.67 ± 0.01	198.71 ± 0.01	198.75 ± 0.03	198.75 ± 0.02

CF caffeine, *BA* benzoic acid, *PP* propylparaben, *BP* butylparaben, *BHA* butylated hydroxyanisole, *BHT* butylated hydroxytoluene
[a]All determinants are in triplicate

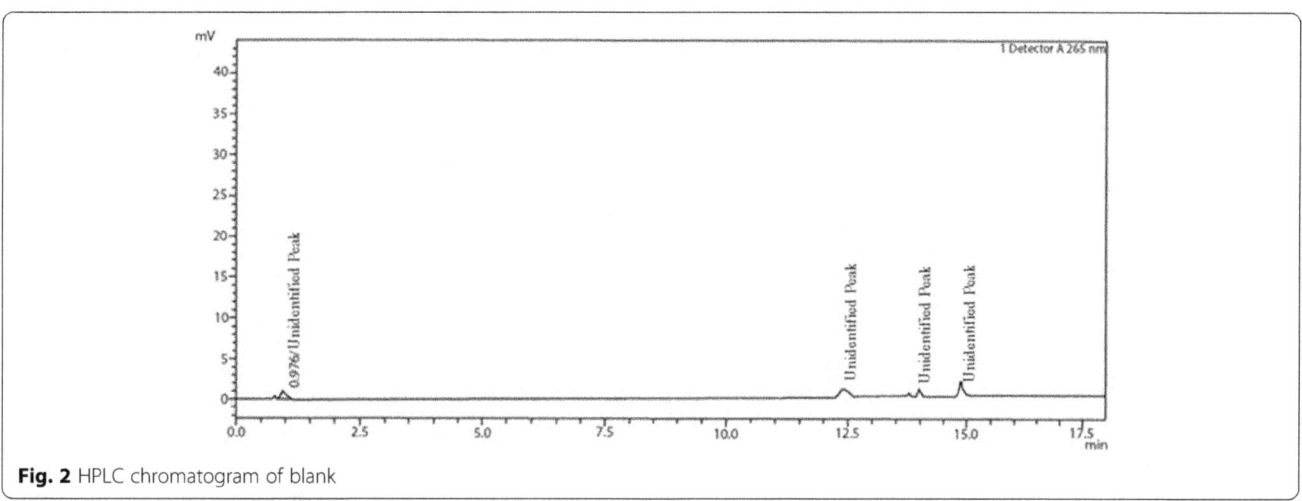

Fig. 2 HPLC chromatogram of blank

Ruggedness

Ruggedness of the method was determined by analyzing six assay sample solutions of standard CF (75 mg/L), BA (250 mg/L), PP (25 mg/L), BP (25 mg/L), BHA (200 mg/L) and BHT (200 mg/L) by two analysts in the same laboratory to check the reproducibility of the result. The percentage recovery and %RSD were calculated in both cases.

Robustness

To determine the robustness of the method, flow rate of the mobile phase and the pH of the diluted sulfuric acid were slightly changed. The %RSD of robustness testing under these conditions was calculated in all cases.

Results and discussion
Method validation
System suitability

The results (Mean ± %RSD of six replicates of the standards) of the chromatographic parameters are shown in Table 3, which indicate the good performance of the system.

Linearity

The relevant information is shown on Table 4. The correlation coefficient values of all six investigated ingredients are shown well within (0.99–1.00), which indicate the linearity of the proposed method.

Limits of detection (LOD) and Limits of quantitation (LOQ)

The LOD and LOQ values of CF, BA, PP, BP, BHA and BHT are shown in Table 5, indicating that the proposed method is well sensitive.

Accuracy (Recovery Test)

The overall recoveries of the standards are summarized in Table 6. Almost all results showed very good recoveries at five different concentration levels.

Precision

The results obtained are presented in Table 7. Intra–and inter–day results were found very close to those of the

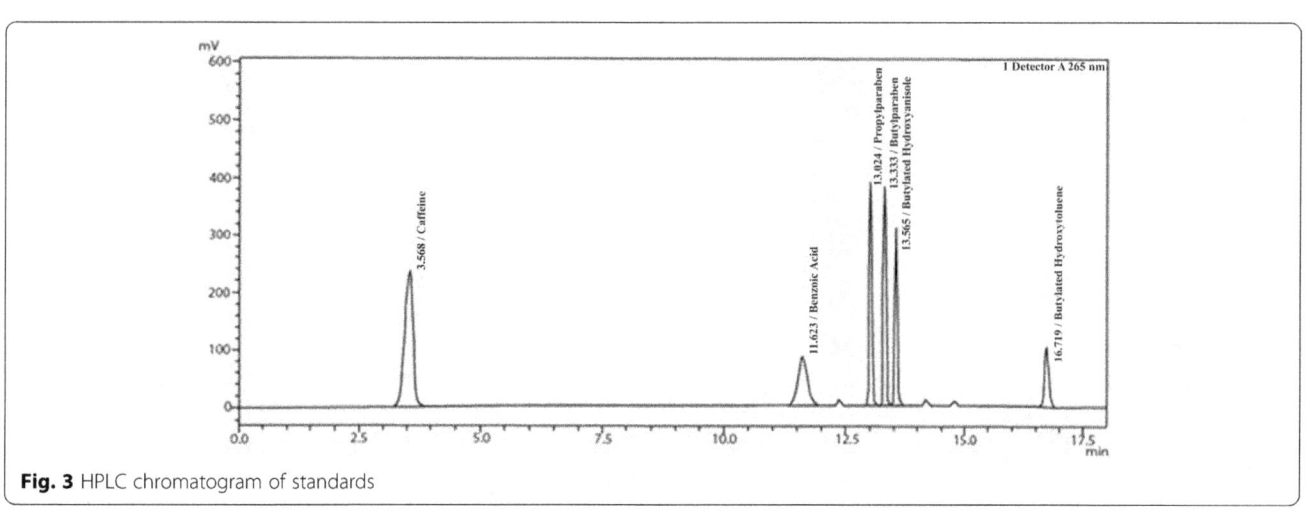

Fig. 3 HPLC chromatogram of standards

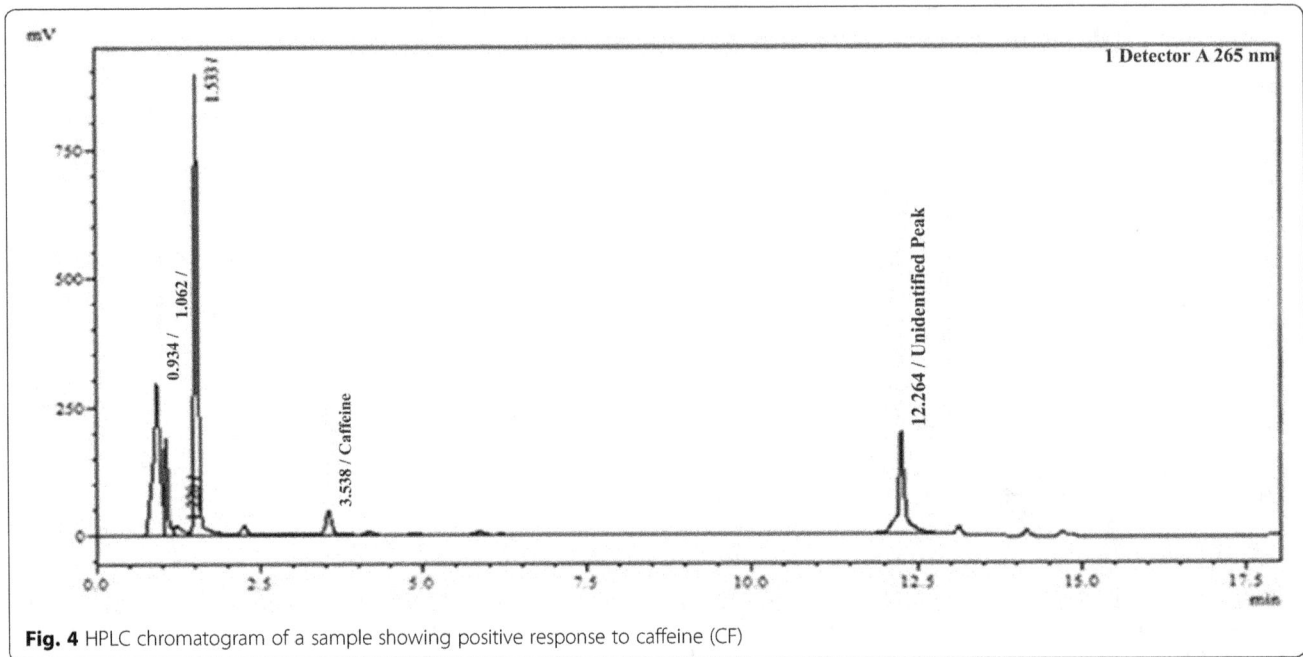

Fig. 4 HPLC chromatogram of a sample showing positive response to caffeine (CF)

corresponding freshly prepared nominal concentration, but inter–day results are slightly more deviated from the norms.

Ruggedness

The results (mean of %recovery ± %RSD of six assay samples) of the ruggedness test are presented in Table 8.

Robustness

The pH of the 0.002 M diluted sulfuric acid was reproducibly found about 2.3. The robustness of the method has been checked changing the pH of diluted sulfuric acid as well as flow rate of mobile phase. The results (mean ± %RSD of three assay samples) of the robustness study are given in Table 9.

Quantitation of CF, BA, PP, BP, BHA and BHT in the dairy products

A precise, robust and rugged RP–HPLC method for simultaneous analysis of CF, BA, PP, BP, BHA and BHT was developed to investigate 41 marketed dairy products within a shortest possible time. Our designed analytical

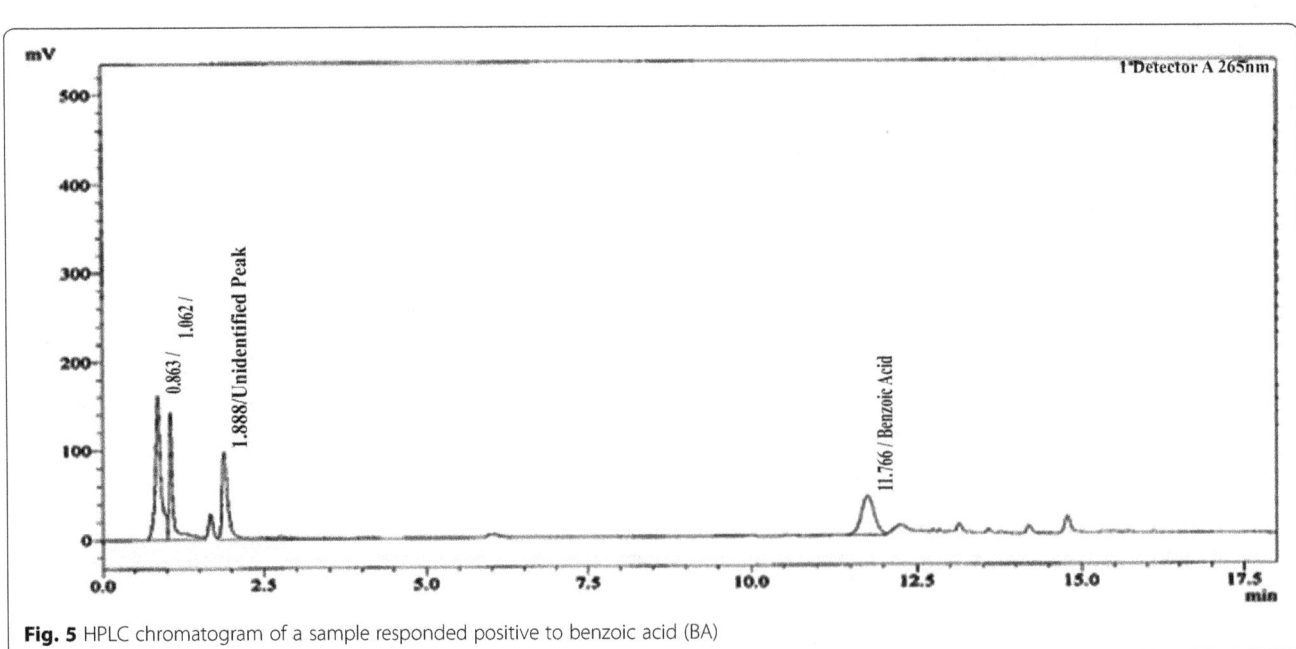

Fig. 5 HPLC chromatogram of a sample responded positive to benzoic acid (BA)

Table 10 The concentrations of CF and BA in dairy products

Estimation of CF		Estimation of BA	
CF positive dairy products	Concentration of caffeine present (mg/L) (mean ± %RSD)[a]	BA positive dairy products	Concentration of BA present (mg/L) (mean ± %RSD)[a]
C1	12.28 ± 0.66	M1	65.22 ± 0.09
C2	14.4 ± 0.40	M5	42.28 ± 0.16
C3	17.76 ± 0.34	B1	55.30 ± 0.12
C4	10.93 ± 0.33	Y1	11.68 ± 0.39
C5	13.57 ± 0.28	G1	1875.65 ± 0.15
C6	14.79 ± 0.26	G2	1074.85 ± 0.16
–	–	P1	2066.39 ± 0.15

CF caffeine, *BA* benzoic acid, *PP* propylparaben, *BP* butylparaben, *BHA* butylated hydroxyanisole, *BHT* butylated hydroxytoluene

[a]All determinants are in triplicate

method revealed their retention times of about 3.56, 11.62, 13.02, 13.33, 13.56 and 16.72 min, respectively, which is so far the best approach to analyze such a large number of dairy products. Figs. 2, 3, 4 and 5 represent the chromatograms of blank, six standards, a CF–positive product and a BA–positive product sequentially. Among the 41 dairy products, none was labeled to contain CF; rather 6 chocolate milks (composed of 14.6 % of milk products) were labeled to contain cocoa powder instead. Only these 6 products showed positive response towards CF in the concentration range of 10 ~ 18 mg/L. This range is well below the maximum allowable limit of CF, which is 200 mg/L. The resultant concentrations of CF in these 6 milk products are individually presented in Table 10.

Dairy products are available in Bangladesh as pasteurized form, UHT form and some are not heat treated. Two products out of 9 pasteurized dairy products and 4 products out of 16 UHT treated products contained CF and all of 16 not heat treated products did not contain CF. Independent of this type of treatment, every chocolate milk products whether they are heat treated or not as well as pasteurized or UHT treated were found CF positive. So, our study could not establish any relationship between heat treatment of the milk and the presence of CF. The presence of CF in pasteurized, UHT treated and not heat treated products are shown in Fig. 6.

There were 37 domestic milk and milk products and 4 products were imported from Australia, New Zealand, Ukraine and South Africa. Six out of 37 domestic dairy products were found to contain CF. The imported dairy products did not contain CF. Moreover, none of the imported products was chocolate milk products. There is no imported chocolate milk product available in Bangladesh. So, caffeine content could not be correlated with the products' origin i.e. whether they are imported or domestic. Interestingly, the presence of CF in the chocolate milks observed in this study is similar to the average content of 11 mg/Kg collected by Craig and Nguyen (1984). However, though the CF content found is low, the CF present in milk products may contribute to the exceeding of tolerance limit along with other CF rich products like tea, coffee, energy drinks etc.

Again, 7 out of 41 analyzed dairy products were found BA positive. The concentration of BA in these products was between 11 ~ 2067 mg/L. Among these products, 2 were mango milk, 1each from banana milk, flavored yogurt, lassi, and 2 from buttermilk. The comparison is shown in Fig. 7. Beside this, all the pasteurized milk and milk products were BA negative, 5 among 16 UHT milk and milk products were found to contain BA whereas 2 out of 16 not heat–treated products showed positive response to BA which is shown in Fig. 8. Seven dairy

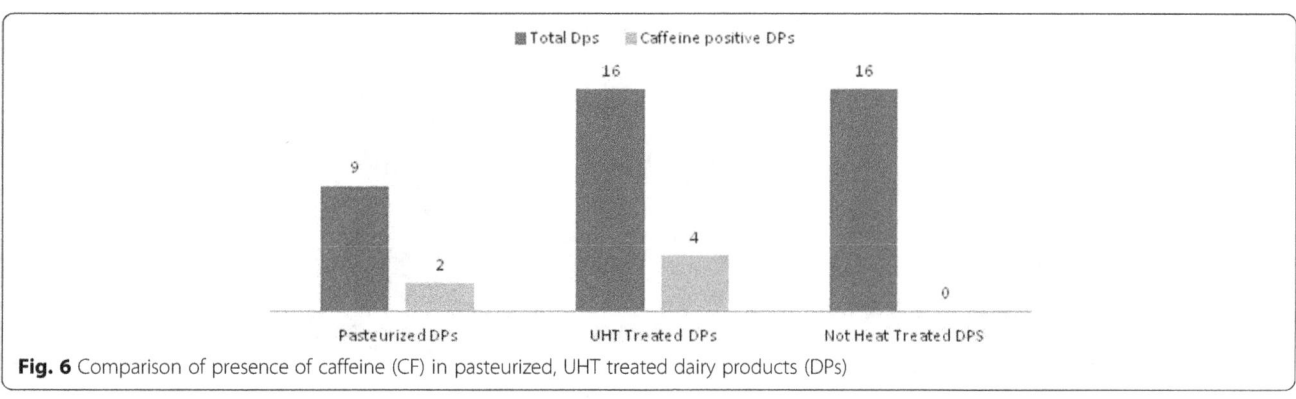

Fig. 6 Comparison of presence of caffeine (CF) in pasteurized, UHT treated dairy products (DPs)

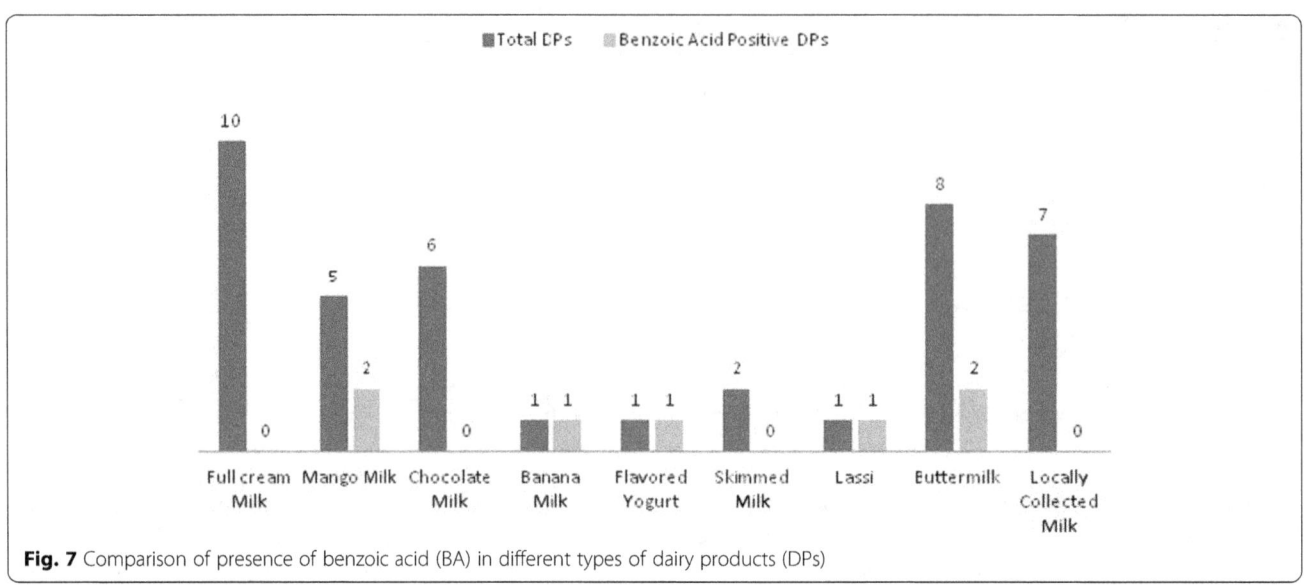

Fig. 7 Comparison of presence of benzoic acid (BA) in different types of dairy products (DPs)

products among 37 Bangladeshi milk products were found to contain BA. None of the imported milk products was found to contain BA.

According to the Joint FAO/WHO Expert Committee on Food Additives (JECFA), BA is only permitted to use in flavored yogurt up to the maximum use level of 300 mg/L. Among 7 BA positive dairy products, only 1 product was flavored yogurt. Therefore, BA was approved to be used in only 1 product and the concentration of this product was about 11.68 mg/L, which was well below the maximum use level set by JECFA. Six other BA positive products are not permitted to use BA. Moreover, 3 of them crossed the maximum permitted level of 300 mg/L, even though BA is not permitted to be used in these products. Apart from these 3 products, 4 other BA positive products contained relatively very low amount of the additive which may be originated due to feeding practices, contamination, storage condition, and veterinary drugs. It is also well known that hippuric acid, a natural component of milk, may be converted to benzoic acid by the fermentation of lactic acid bacteria.

Further study needs to be carried out to confirm these sources.

However, the mean concentrations of benzoic acid observed by Qi et al. (2008) in pasteurized and UHT milk products in China were 3.6 ± 3.3 and 2.4 ± 2.2 mg/kg, ranging between 0.51–8.8 and 0.62–13 mg/Kg, respectively, where, the UHT milk products was plain milk. In our study, none of the analyzed pasteurized and non-flavored UHT milk products was found BA positive. Beside this, in Jordan, Mihyar et al. (1998) found BA in labaneh (concentrated yogurt) ranging from not detected to 2000 mg/Kg, whereas, Tfouni and Toledo (2002) did not find BA in yogurt in Brazil while our study detected BA in the flavored yogurt. The concentration of BA found in the product was well below the maximum use level.

But, every analyzed milk products showed negative response to PP, BP, BHA and BHT. However, Pattono et al. (2009) revealed the presence of BHT in conventional milk and organic milk ranging from not detected to 1.45 mg/L and not detected to 1.67 mg/L, respectively, in Italy,

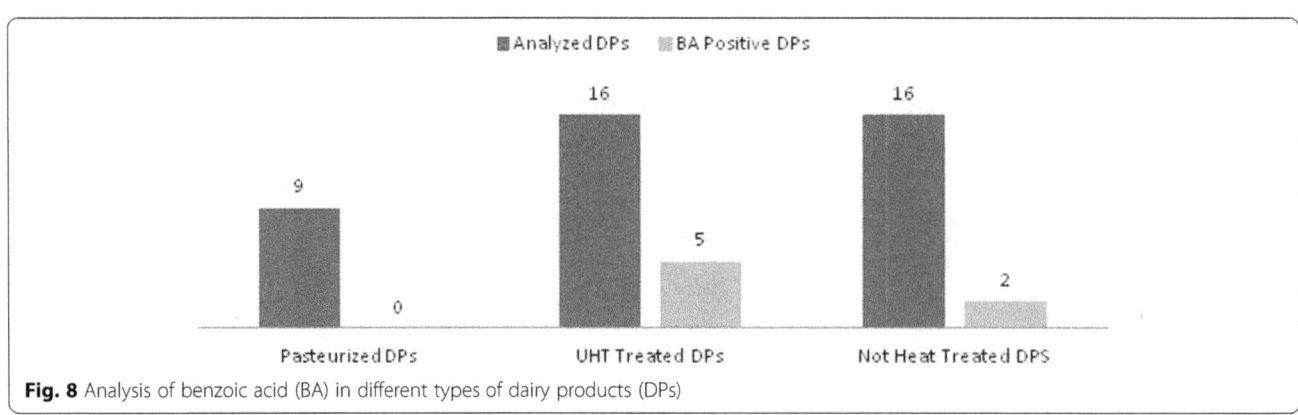

Fig. 8 Analysis of benzoic acid (BA) in different types of dairy products (DPs)

though, they did not find BHA in their samples. Though these 4 additives were not found within detectable range in our samples, continuous monitoring is necessary to maintain the situation.

Finally, since this is the first report from Bangladesh about using these six additives in dairy products, more studies should be carried out on a regular basis by researchers and respective regulatory bodies to protect common people from the uncontrolled consumption of these potentially harmful additives with highly nutritious dairy products.

Conclusion

The aim of this study was to explore the six additives in dairy products available in Bangladesh by developing a simple RP–HPLC method. The proposed time–programmed method can do it within 20 min, including reequilibration. Our screening revealed the presence of CF and BA in several dairy products. The non–permitted use of BA in dairy products and the presence of CF in CF–free products can be injurious to human health. So, we expect that the respective policy makers of Bangladesh will consider this outcome seriously and devise rules on them as quickly as possible.

Abbreviations
ADI, Acceptable Daily Intake; b.w., Body–weight; BA, Benzoic Acid; BHA, Butylated hydroxyanisole; BHT, Butylated hydroxytoluene; BP, Butylparaben; CF, Caffeine; FDA, Food and Drug Administration; JECFA, Joint FAO/WHO Expert Committee on Food Additives; PP, Propylparaben; RP–HPLC, Reversed Phase High Performance Liquid Chromatography; WHO, World Health Organization.

Acknowledgement
This analytical research has been funded by Higher Education Quality Enhancement Project (HEQEP), Window II, Round III, CP 3245, University Grants Commission of Bangladesh.

Authors' contributions
SB and ASSR did the experimental design of this study. MSH and MSI performed the laboratory analyses. MSH, MSI, and SB interpreted the data. MSH wrote the manuscript. MSH, SB, and ASSR edited and proofread the manuscript. All authors read and approved the final manuscript.

Competing interests
The authors declare that they have no competing interests.

References
Ali ANAA. Application of responsive regulation in the food safety regulations of Bangladesh. J South Asian Stud. 2013;01(01):01–9.
Ayala J, Simons K, Kerrigan S. Quantitative determination of caffeine and alcohol in energy drinks and the potential to produce positive transdermal alcohol concentrations in human subjects. J Anal Toxicol. 2009;33(January/February):27–33.
Cempírková R, Mikulová M. Incidence of psychrotrophic lipolytic bacteria in cow's raw milk. Czech J Anim Sci. 2009;54(2):65–73.
Chanda T, Debnath GK, Hossain ME, Islam MA, Begum MK. Adulteration of raw milk in the rural areas of Barisal district of Bangladesh. Bangladesh J Anim Sci. 2012;41(2):112–5.
Craig WJ, Nguyen TT. Caffeine and theobromine levels in cocoa and carob products. J Food Sci. 1984;49(1):302–3.
Davies S, Lee T, Ramsey J, Dargan PI, Wood DM. Risk of caffeine toxicity associated with the use of 'legal highs' (novel psychoactive substances). Eur J Clin Pharm. 2012;68(4):435–9.
Dunn RO. Effect of antioxidants on the oxidative stability of methyl soyate (biodiesel). Fuel Processing Technol. 2005;86(10):1071–85.
FAO. Dairy production and products: Milk and milk products. 2013a. http://www.fao.org/agriculture/dairy-gateway/milk-and-milk-products. Accessed 25 Dec 2014.
FAO. Dairy production and products: Milk processing. 2013b. http://www.fao.org/agriculture/dairy-gateway/milk-processing. Accessed 25 Dec 2014.
Codex general standard for food additives 192–1995. Food and Agriculture Organization of the United Nations (FAO) and World Health Organization (WHO). 2015.
Grant DC, Helleur RJ. Simultaneous analysis of vitamins and caffeine in energy drinks by surfactant–mediated matrix–assisted laser desoprtion/ionization. Anal Bioanal Chem. 2008;391:2811–8.
Guizani N. Postharvest handling of milk. In: Rahman MS, editor. Handbook of food preservation. 2nd ed. New York: Taylor & Francis Group; 2007.
Hossain MM, Heinonen V, Islam KMZ. Consumption of foods and foodstuffs processed with hazardous chemicals: a case study of Bangladesh. Int J Consumer Stud. 2008;32:588–95.
Injac R, Srdjenovic B, Prijatelj M, Boskovic M, Rajic KK, Strukelj B. Determination of caffeine and associated compounds in food, beverages, natural products, pharmaceuticals, and cosmetics by micellar electrokinetic capillary chromatography. J Chromatogr Sci. 2008;46(February):137–43.
Kerrigan S, Lindsey T. Fatal caffeine overdose: two case reports. Forensic Sci Int. 2005;153(1):67–9.
Khosrokhavar R, Sadeghzadeh N, Amini M, Ghazi-Khansari M, Hajiaghaee R, Sh EM. Simultaneous determination of preservatives (sodium benzoate and potassium sorbate) in soft drinks and herbal extracts using high performance liquid chromatography (HPLC). J Med Plants. 2010;9(35):80–7.
King RL. Oxidation of milk fat globule membrane material. I. Thiobarbituric acid reaction as a measure of oxidized flavor in milk and model systems. J Dairy Sci. 1962;45(10):1165–71.
Lee MR, Lin CY, Li ZG, Tsai TF. Simultaneous analysis of antioxidants and preservatives in cosmetics by supercritical fluid extraction combined with liquid chromatography–mass spectrometry. J Chromatogr A. 2006;1120(1–2):244–51.
Medeiros RA, Rocha-Filho RC, Fatibello-Filho O. Simultaneous voltammetric determination of phenolic antioxidants in food using a boron-doped diamond electrode. Food Chem. 2010;123(3):886–91.
Mihyar GF, Yousif AK, Yamani MI. Determination of benzoic and sorbic acids in labaneh by high-performance liquid chromatography. J Food Composit Anal. 1998;12:53–61.
Mroueh M, Issa D, Khawand J, Haraty B, Malek A, Kassaify Z, Toufeili I. Levels of benzoic and sorbic acid preservatives in commercially produced yoghurt in Lebanon. J Food Agri Environ. 2008;6(1):62–6.
[NFSL] The National Food Safety Laboratory. UN-FAO-backed lab finds chemicals in Bangladeshi food. 2014. http://www.bdfoodsafety.org/state-halts-raw-milk-sales-from-idaho-dairy. Accessed 10 April 2016.
Nutri pro. Caffeine. Nestle professional. 2013. http://www.nestleprofessional.com/australia/en/Insights/Nutrition-Fact-Sheets/Pages/Caffeine. Accessed 16 May 2013.
Pattono D, Battaglini LM, Barberio A, Castelli LD, Valiani A, Varisco G, Scatassa ML, Davit P, Pazzi M, Civera T. Presence of synthetic antioxidants in organic and conventional milk. Food Chem. 2009;115:285–9.
Petronela CENE, Elena DIACU. High-performance liquid chromatography method for the determination of benzoic acid in beverages. U P B Sci Bull B. 2009;71(4):81–8.
Potter NN, Hotchkiss JH. Food Science. In: Potter NN, editor. Food Deterioration and its Control. 5th ed. New York: Springer; 1999.
Qi P, Hong H, Liang X, Liu D. Assessment of benzoic acid levels in milk in China. Food Control. 2008;20:414–8.
Rahman MM, Kabir SML. Developing awareness profiling force and activities linking safety and quality of foods of animal origin in Bangladesh. Scient J Review. 2012;1(3):84–104.
Report on carcinogens. BHA. National Toxicology Program, Department of Health and Human Services. 2014. https://ntp.niehs.nih.gov/ntp/roc/content/profiles/butylatedhydroxyanisole.pdf. Accessed 16 May 2013.
Roye RC, Sheskey PJ, Quinn ME. Handbook of pharmaceutical excipients. In: Roye RC, Sheskey PJ, Quinn ME, editors. Pharmaceutical Press, London and American Pharmacists Association, Washington. 6th ed. 2009.

Saad B, Bari MF, Saleh MI, Ahmad K, Khairuddin M, Talib M. Simultaneous determination of preservatives (benzoic acid, sorbic acid, methylparaben and propylparaben) in foodstuffs using high-performance liquid chromatography. J Chromatogr A. 2005;1073:393–7.

Sather K, Vernig T. Determination of caffeine and vitamin B6 in energy drinks by high-performance liquid chromatography (HPLC). Concordia College J Anal Chem. 2011;2:84–91.

Simmons CRF, Kidner N. Caffeine toxicity in a bodybuilder. Emerg Med J. 1998;15:196–7. doi:10.1136/emj.15.3.196.

Srdjenovic B, Milic VD, Grujic N, Injac R, Lepojevic Z. Simultaneous HPLC determination of caffeine, theobromine, and theophylline in food, drinks, and herbal products. J Chromatogr Sci. 2008;46(February):144–9.

Tang Y, Wu M. The simultaneous separation and determination of five organic acids in food by capillary electrophoresis. Food Chem. 2007;103(1):243–8.

Tawalbeh Y, Ajo R, Al-Udatt M, Gammoh S, Maghaydah S, Al-Qudah Y, Al-Sunnaq A, Al-Natour F. Investigation of the antimicrobial preservatives in the dairy product (labneh). Food Sci Qual Management. 2014;31:117–21.

Tfouni SAV, Toledo MCF. Determination of benzoic and sorbic acids in Brazilian food. Food Control. 2002;13:117–23.

Varga L. Microbiological quality of commercial dairy products. 2007. Formatex, http://www.formatex.org/microbio/papers.htm. Accessed 26 July 2016.

[WHO] World Health Organization. Propylparaben. 2010. http://apps.who.int/food-additives-contaminants-jecfa-database/chemical.aspx?chemID=330. Accessed 16 May.

Yang MH, Lin HJ, Choong YM. A rapid gas chromatographic method for direct determination of BHA, BHT and TBHQ in edible oils and fats. Food Res Int. 2002;35(7):627–33.

7

Presence of organochlorine pesticides in xoconostle (*Opuntia joconostle*) in the central region of Mexico

José J. Pérez[1], Rutilio Ortiz[1*], María L. Ramírez[1], Javier Olivares[2], Daniel Ruíz[2] and David Montiel[2]

Abstract

Background: Mexico has many natural resources for use in the food industry. In recent years Xoconostle, which is considered as a traditional food, has gained importance due to key its components such as fiber, antioxidants, and fermentative bacteria found in both the peel and fruit, which could benefit human and animal health. Information on the presence of organic contaminants such as pesticides in Xoconostle is important in protecting its reputation as a healthy food. The objective of this research was to measure the concentration of pesticides in the peel of Xoconostle in four 50 m^2 plots in the same production area. The samples were taken in the state of Hidalgo, which is one major region of production. Fruits were collected from upper and lower levels in five plots. Organochlorine pesticides in peel samples were extracted by Soxhlet extraction, purified with chromatographic columns, and finally analyzed by gas chromatography with electron capture detector (ECD).

Results: The results showed a predominance of heptachlor, HCH derivatives and DDT isomers but there is no permissible limit for these compounds in this fruit.

Conclusions: The presence of organochlorines suggests the use of these compounds in control of pests in crops and livestock, in spite of the substitution of these compounds, principally by organophosphorus pesticides. Regulation is necessary to ensure the safety of the product for consumption and trade.

Keywords: Traditional food, Organochlorine compounds, Contamination

Background

Mexico has 93 species of cactus, some of which are commercially important for food and health uses. *Opuntia joconostle* (known as Xoconostle) is one such example. In recent years, Xoconostle uses have been growing due to its health benefits, as well as other applications such as a condiment in Mexican cuisine, and in the preparation of candies, jellies and beverages. The Xoconostle may remain in the processing plant for several months without deteriorating, and it can be kept for several weeks in a dry and cool environment without losing its flavor or moisture (Osorio-Esquivel et al. 2011). Xoconostle has been used as a so-called natural treatment for diabetes, hypertension, obesity, and respiratory ailments. There is

evidence that its consumption causes a reduction in cholesterol levels, a gradual decrease of seric glucose levels, and an increase in seric insulin levels (Morales et al. 2012, Osorio-Esquivel et al. 2011). This fruit has been consumed since pre-Hispanic times, especially in semi-arid regions in the central area of Mexico (Saenz et al. 2006; Morales et al. 2012).

Cactus fruits have attracted the attention of national and international researchers due to their commercial value. For example, in some regions the cactus has been employed as an ornamental plant due to its colorful flowers, and as an agricultural crop due to its fruits and the ability of its roots to help improve soil structure and to work in association with nitrogen fixing bacteria/mycorrhiza to help absorption of phosphorous (Saenz et al. 2006). The cactus has a high adaptability to extreme conditions of drought, cold, and thin soils poor in nutriments (Guzmán et al. 2010). Xoconostle peel is useful as food for livestock and alcoholic beverages by fermentation, and in the central region of

* Correspondence: guppyabanico@gmail.com
[1]Departamento de Producción Agrícola y Animal, Laboratorio de Análisis Instrumental, Universidad Autónoma Metropolitana, Unidad Xochimilco, Calzada del Hueso No 1100, Colonia Villa Quietud, Delegación Coyoacán C.P. 04960, D.F., México
Full list of author information is available at the end of the article

Mexico it is considered a good option to provide income for the population, and food for animals (Guzmán et al. 2010). It has been exported to others countries for similar uses in conservation of soil and as an alternative source of nutrition for livestock and human populations (Morales et al. 2012).

The increasing use of the Xoconostle suggests the need to investigate the possibility of contamination of the crop (Saenz et al., 2006). Farmers apply pesticides such as organochlorine and organophosphorous compounds to control pests. Some are prohibited while others including some organochlorine compounds (mainly heptachlor and DDT) may still be used in some countries. Pesticides have played a key role in providing reliable supplies of agricultural produce at prices affordable to consumers, and ensuring high profits to farmers (Wang 2013). Many investigations have shown the occurrence of organochlorine compounds in soil, water, forage, foods (animal and vegetable) in production systems. Levels of contamination are often found to be broadly related to anthropogenic activities (contamination sources) such as cities, commercial plantations/livestock, tourism and health campaigns to malaria (Narayan et al. 2012).

Contamination by pesticides is influenced by physicchemical characteristics and environmental variables (temperature, oxygen, water, light, microorganisms, organic matter and others). Inevitably pesticides are present in the environment and can be found in agricultural products, food for livestock, wood and derivatives. Despite their popularity and excessive use, exposure of the general population to pesticides originating from pesticide residues in food, air and drinking water generally involves low doses and is chronic (or semi-chronic) in terms of health effects (Wang et al. 2013).

The presence of pesticides in agriculture is common in food for animal or human consumption; it is of course a topic of interest which concerns the quality of produce and the health of consumers and is of particular importance for regional agricultural products. The aim of this study was to determine the concentration of organochlorine pesticides on the EPA priority list in the peel of Xoconostle.

Methods
Sampling
Xoconostle fruits were sampled in five 50 m^2 plots from the state of Hidalgo, Mexico. We collected fruits in January from the upper, medium and lower levels of the plants in a zigzag sampling pattern to obtain a composite sample of 3 kg from each production unit according at NOM-007-RECNAT-1997; ripe fruits were picked fresh from the plant. The samples were transported to the laboratory at 4 °C and stored at this temperature until their analysis.

Of the pests that affect both the prickly pear plant and the fruit, the most important are: weevil borer (*Cactophagus spinolae*) with high densities in May to August;

weevil thorns (*Cylindrocopturus birraddiatus*), with high populations in April and May; white worm, zebra, cochineal (*Dactylopius indicus*), which occurs all year round; bedbug gray (*Chelinidae tabulatus*), which is present all year with older populations from September to December; red bedbug (*Hesperolabops gelatops*), increased populations from August to November; prickly pear thrips (*Sericothrips opuntiae*) with high populations from February to April. Needless to say, the presence, incidence and severity of damage vary from region to region, and is also dependent on cultural practices and pest management. The application of pesticides to crops began using chlorinated compounds in the 1960s and these have largely been replaced by organophosphate compounds in recent decades.

Extraction of organochlorine pesticides
The peel was selected for analysis because of its lipid composition and its common use for animal consumption and in the fermentation process for alcoholic beverages. Prior to extraction the peel was removed with a knife cleaned with HPLC grade solvent to avoid cross contamination. The peel was dried at 40 °C in a drying oven for four days and reported results are presented as mass per mass of dry matter. The organochlorine compounds in peel of Xoconostle were extracted by Soxhlet extraction (hexane/dichloromethane 1:1 v/v) according 3540C EPA (1996), and purified with chromatographic columns (desactived Florisil, method 3620C EPA 2014). The eluent from the columns was concentrated by rotatory evaporation to 3 mL and then transferred to a small vial for gas chromatography analysis according to USEPA methodology (Method 8081B) (EPA 2007).

Chromatographic analysis
The samples were analyzed by gas chromatography with electron capture detection (GC-ECD, HP 6890 Series, Wilmington; DE, USA). (Column HP-5, 30 m, 0.25 mm ID, 0.25 μm film thickness, crosslinked 5% PH ME Siloxone, Waltham, MA, USA). The oven temperature was initially set at 90 °C and held for 2 min, then ramped at 30 °C min^{-1} to 180 °C (0 min), then ramped at 1 °C min^{-1} to 200 °C (0 min), and ramped at 10 °C min^{-1} to a final temperature of 300 °C (6 min), with a total analysis time of 41 min. Detector and injector temperatures were 320 °C. The carrier gas was high purity helium (99.99%) with flow of 6 mL/min and make up of nitrogen gas with a flow of 60 mL/min. A sample of 1 mL was injected in splitless mode.

The organochlorine pesticides analyzed were 16 compounds identified according to an EPA priority list. They were: 1) alpha, 2) beta, 3) gamma (Lindane) and 4) delta HCH; 5) heptachlor; 6) aldrin; 7) heptachlor epoxide; 8) endosulfan I; 9) p,p'-DDE; 10) dieldrin; 11) endrin; 12) endosulfan II; 13) p,p'-DDD; 14) endrin aldehyde; 15) endosulfan sulfate and 16) p,p'-DDE (PPO-8JM, Chem Service Inc, West Chester, PA, USA). An external method for quantification of organochlorine compounds based on retention times was used.

Quality control

The analysis included solvent blanks, spiked recoveries, replicate analysis and comparison of standards. Solvent blanks were run with each sample batch. All solvents and others materials contacting samples were not contaminated with the analytes, as confirmed by analysis of solvent blank samples. Spiked recovery tests were performed periodically during analyses. All of these showed acceptable recoveries between 80 and 95% and limits of detection were between 0.10 and 0.30 ng/g (Table 1).

Comparison of chromatograms of fortified and unfortified blanks provided further confidence of the quality of the method. The specificity of the method was checked daily through comparison of chromatograms at the retention times of the analyses. The middle standard of the calibration curve was injected daily to evaluate the performance of the GC. Where necessary, blank subtraction was used for areas of the chromatogram where interference occurred in samples with values were close to limits of detection. The areas of the samples were corrected by subtraction of the appropriate contaminant area, adjusted for recovery values.

Table 1 Values of limit of detection for Xoconostle analysis

Pesticides	Retention times (min)	Limit of detection (ng/g)	Recovery (%)	Standard deviation (+/−)
Alpha-HCH	9.83	0.10	90	7
Beta-HCH	10.64	0.10	88	5
Gamma-HCH	10.87	0.16	85	8
Delta-HCH	11.74	0.21	80	6
Heptachlor	13.92	0.30	90	9
Aldrin	15.81	0.30	88	4
Heptachlor epoxide	18.42	0.22	89	10
Endosulfan I	21.14	0.20	85	2
DDE	23.39	0.18	95	8
Dieldrin	23.65	0.15	92	7
Endrin	25.31	0.10	90	9
Endosulfan II	26.12	0.12	85	5
DDD	27.11	0.18	90	6
Endrin aldehyde	27.42	0.13	82	5
Endosulfate sulfate	28.73	0.20	95	2
DDT	29.05	0.30	85	9

Note: HCH = Hexachlorociclohexane, DDE = 1,1-dichloro-2,2-bis (4 chlorophenyl) ethylene, DDT = 1,1,1-trichloro-2,2-bis (4 chlorophenyl) etano, DDD = dichlorodiphenyldichloroetane

The method detection limits (DLs) were defined by the injection of standard solutions whose signal-to-noise (S/N) ratios were at least three; values ranged from 0.1 to 0.3 ng/g. Those samples with concentrations detected less than DLs were treated as not detected (n.d.).

Results and discussion

Table 2 shows the concentrations of the organochlorine compounds in the plots. Plots 1, 4 and 5 generally showed a greater abundance of pesticides. Heptachlor, endosulfan, p,p'-DDE, endrin, beta and gamma HCH were dominant in plot 1; p,p'-DDE, heptachlor, endrin aldehyde, gamma HCH (lindane), endrin, heptachlor epoxide and aldrin in plot 4, and aldrin, heptachlor, gamma, delta and beta HCH in plot 5. The order of total concentration of the organochlorine pesticides was Plot4 > Plot1 > Plot5 > Plot3 > Plot2. There was no significance difference among plots ($p = 0.05$). The plots were located in one municipality of Hidalgo.

In general, heptachlor was present at the greatest concentrations in most of the plots followed by aldrin, epoxide and p,p'-DDE. The isomers of HCH were present in all of the plots. Heptachlor is used to control pests such as *Sericotrips opuntiae* Hood, *Melanotus sp* and *Phyllophaga spp* that are common in Xococnostle plantations. Endosulfan use is permitted in Mexico to control *Cactophagus spinolae* Gyll, and lindane has restricted use status to help control *Cylindrocopturus birradiatus* Champ and *Chelinidea tabulata* Burn (INE, 2003). The presence of organochlorine compounds in the plantations is indicative of use to control pests in vegetables and livestock. The use of aldrin in Mexico is forbidden because of its high risk to human health, although it is thought that some illegal use occurs due to its low cost.

The presence of isomers of DDT as p,p'-DDE and p,p'-DDD are indicative of use in the past to control pests. DDT is now prohibited for this use and has been replaced by technical lindane that contains contaminants. This could be seen by the presence of isomers of HCH in the plots. Xoconostle peel acts as natural barrier due to its lipidic composition in the form of waxes; pesticides deposited on the fruit peel remain there due to their lipophilic characteristics (Gutiérrez et al. 2010). It was assumed that all contaminants remained in the peel.

Gutiérrez et al. (2010) determined the organochlorine pesticides in peel and fruit of pineapple. They found heptachlor (7.13 ng/g), aldrin (9.9 ng/g), DDE (4.6 ng/g) and heptaclor epoxide (12.4 ng/g), mostly in the peel. Low levels of these compounds were found in peel of Xoconostle in this study, in terms as average concentration of the same compound for the plots studied in the study area. For example, concentrations were heptachlor (0.34 ng/g), aldrin (0.25 ng/g), DDE (0.18 ng/g) and heptachlor epoxide (0.12 ng/g).

Table 2 Concentrations of organochlorine pesticides in plots from Hidalgo, Mexico

Compounds	Plots (ng/g)				
	1	2	3	4	5
Alpha HCH	0.10 ± 0.02*	LD	LD	LD	0.10 ± 0.06
Beta HCH	0.18 ± 0.08	LD	LD	LD	0.20 ± 0.09
Gamma HCH	0.18 ± 0.07	LD	0.18 ± 0.09	LD	0.23 ± 0.14
Delta HCH	LD	LD	LD	0.27 ± 0.12	0.21 ± 0.09
Heptachlor	0.60 ± 0.09	LD	LD	0.45 ± 0.15	0.50 ± 0.28
Aldrin	0.43 ± 0.10	LD	LD	LD	0.55 ± 0.19
Epoxide of heptachlor	0.37 ± 0.11	LD	LD	0.22 ± 0.06	LD
Endosulfan I	0.20 ± 0.05	LD	LD	LD	LD
p,p'-DDE	0.19 ± 0.05	LD	LD	0.72 ± 0.35	LD
Dieldrin	LD	LD	LD	0.18 ± 0.08	LD
Endrin	0.16 ± 0.09	LD	LD	0.25 ± 0.11	LD
Endosulfan II	LD	LD	LD	0.13 ± 0.07	LD
p,p'-DDD	LD	LD	LD	0.18 ± 0.08	LD
Endrin aldehyde	LD	LD	LD	0.43 ± 0.25	LD
Endosulfan sulfate	LD	LD	LD	LD	LD
p,p'-DDT	LD	LD	LD	0.30 ± 0.06	LD
Total	2.41 ± 0.66	LD	0.18 ± 0.09	3.13 ± 1.33	1.79 ± 0.85

Note: LD below of limit of detection, *Standard deviation σ

Fenik et al. (2011) analyzed the presence of organochlorine pesticides and their risks in vegetables; their results showed not authorized or restricted pesticides according to the National Reference Center for Pesticides and Contaminants in 2005 and 2007 in several vegetables including the stem cactus. For example, the presence of endosulfan in cilantro and guava was indicative of the continued use of organochlorine compounds by farmers. They also found organophosphate pesticides that were not authorized for use on the crops examined. Considering Mexico, Pérez et al. (2013) showed that even though several harmful pesticides have been banned, while some are authorized for use despite being prohibited in many countries. There are reports of a number of obsolete products in use or being stored inappropriately that pose risks to the environment further of inadequate regulation and management, trade, weak enforcement of bans on pesticides importation, illegal use/application and lack of logistics to monitor these pesticides (Okoffo et al., 2016). The presence of pesticide residues in vegetables and fruits is frequent, being reported in over 50% of the samples tested in national monitoring and in products exported to the United States in specific studies.

The Codex Alimentarius does not state any residual permissible limit for organochlorine pesticides in Xoconostle, however there are values for a few vegetables (Table 3) that can be used for comparison. It is important that national and international institutions responsible for control of contaminants in food consider the presence of organochlorine pesticides in Xoconostle, and the use in alternative diets. In addition, other chlorine compounds such as lindane do not have a permissible limit in vegetables.

Pesticide residues in fruits and vegetables are a major concern to consumers due to their negative impacts on health. They have been found in both raw and processed fresh produce. However, food processing techniques have been found to significantly reduce the pesticide residues in fruits and vegetables (Chen et al. 2011). The main processes that reduced pesticide residue levels were blanching, cooking, frying, peeling and washing. Washing has been found to reduce pesticides that are loosely attached to the surface while peeling removes even those that have penetrated the cuticles of the fruits or vegetables (Latif et al., 2011). Food safety is an important consideration because of domestic/international trade and the potential health implications for the consumer.

Table 3 Residual permissible limit for some plantations

Pesticide	Food	Residual Permissible Limit (mg/kg)
Heptachlor	Pineapple	0.01
Aldrin	Leaf vegetables	0.05
DDT	Carrot	0.2
Endrin	Cucurbits and vegetables	0.05

Codex Alimentarius, 2015

In addition, many investigations refer to the study of fruits and vegetables with high consumption such as lettuce, radishes, carrots, squash, spinach, apple, orange, pineapple, which are among those that are most traded worldwide (Fenik et al., 2011; Narayan et al. 2012). The recent commercialization and promotion of consumption of other fruits and vegetables (named traditional or regional) by international organizations such as the FAO, thought of as giving benefits to health either through fresh consumption or after processing, has generated the need for regulations to national and international standards to ensure safety (FAO, 2015).

Figure 1 shows the total distribution of organochlorine pesticides in the Xoconostle peel, where heptachlor was present at a relatively high concentration in the plantation, followed mainly by, aldrin, epoxide of heptachlor, p,p'-DDE, dieldrin and endrin. These compounds are typical of control of pests in zones related to agricultural and livestock activities.

The distribution of pesticides in the current study indicate the use of HCH for control of pests but the product is probably a technical grade due to the presence of contaminants such as alpha and beta isomers (Fenik et al., 2011, Pérez et al. 2013). This compound is employed extensively in rural zones dedicated to agriculture, employed in human health (gamma HCH/lindane to control lice on children), and in livestock activities. According to INE (2003), there several denominations for this pesticide employed for sale among farmers in an attempt to use old stock. Many studies concerning the toxicology of organochlorine compounds in Mexico have focused on cow's milk, breast milk, fat tissue, sediments and mollusks. Despite the intention to replace their use by organophosphorus compounds, they are still used frequently as a cheap alternative in developing countries by food producers. Mexico is a signatory of an international treaty that only allows the use of DDT in special cases for health reasons (control of malaria vectors). The concentration of DDE surpasses that of DDT in Xoconostle plantations, an indication that degradation of residues already present is occurring, not of continued use. The presence of other organochlorine compounds such as endrin, endosulfan, aldrin and dieldrin is also likely to be due to their residual presence as their use is forbidden or restricted in agriculture (Fenik et al., 2011; Narayan et al. 2012).

Organochlorine pesticides can be classified according to their chemical structure as alicyclic, cyclodienics and, aromatic. The family groups that were measured in the plots from the municipality were: Plot 1, alicyclics 0.46 ng/g, cyclodienics 1.76 ng/g and aromatics 0.19 ng/g; Plot 2 no detectable organochlorine compounds present; Plot 3 only alicyclics 0.18 ng/g; Plot 4 alicyclics 0.27 ng/g, cyclodienics 1.66 ng/g and aromatics 1.20 ng/g, and finally plot 5 alicyclics 0.74 ng/g and cyclodienics 1.05 ng/g. The cyclodienics were present at the greatest concentrations followed by alicyclics and aromatics.

The cyclodienics include heptachlor, aldrin, epoxide of heptachlor, endosulfan, dieldrin, endrin, endrin aldehyde and endosulfan sulfate. Heptachlor, within its group, had a relatively high concentration, probably because it is used in cactus plantations mainly for the control of pests. The alicyclics include isomers of HCH, used widely to control agricultural and livestock pests. The aromatics include DDT, DDE and DDD, which as previously mentioned, have been forbidden since 1980 in Mexico by international

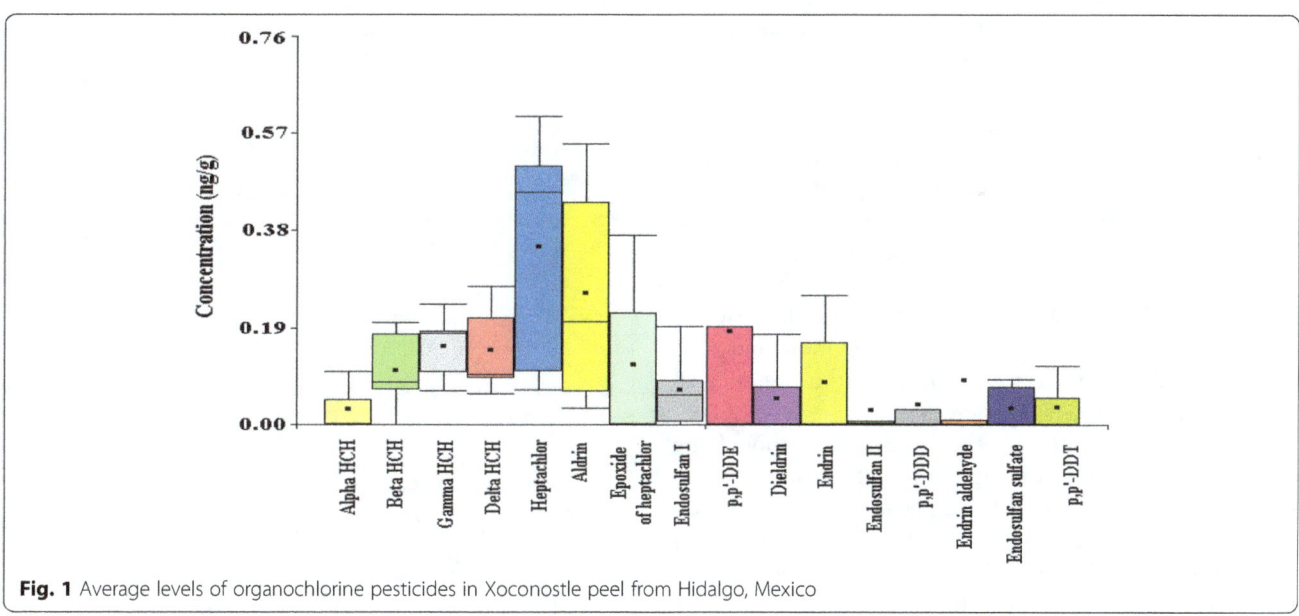

Fig. 1 Average levels of organochlorine pesticides in Xoconostle peel from Hidalgo, Mexico

convention and the presence of DDE and DDD is the result of degradation of DDT, which is persistent in the environment.

Organochlorine pesticides enter biota by the respiratory and digestive systems, and through the skin. The chemical structure determines the dominant mode of entry. For example, DDT is absorbed minimally through the skin, while aldrin and endrin are absorbed relatively quickly by this method. The speed of adsorption after digestion is increased for substances dissolved in animal fat or vegetable wax (Keikotlhaile et al., 2010).

The presence of organochlorine pesticides in agricultural crops and processed foods also indicates the need for improving residue control at production, tighter regulation of pesticide spraying, and also tighter regulation in the sale of pesticides, as well as for education of farmers and the implementation of integrate pest management methods. Nevertheless, monitoring programs are increasingly important and essential to ensure minimal pesticide residue levels in food (Narayan et al. 2012; Wang et al., 2013; Okoffo et al., 2016).

The presence of organochlorine pesticides in food as fruits, vegetable leaf and seed is a major topic due to the human health risks such as reproductive impairment and suppression of the immune system which can have long-term consequences for population viability.

Conclusion

The occurrence of organochlorine pesticides in Xoconostle is not considered amongst the maximum permissible residue levels considered by the Codex Alimentarius, so there is a need for regulation of this food source. The use of Xoconostle peel for fermentation and fodder is acceptable and poses little risk to the health of consumers according to comparison with other vegetables. The differences of occurrence of organochlorine pesticides in the study area were probably due to differences in environmental conditions such as topography, wind direction and vegetation that protect the plots.

Organochlorine pesticides are undoubtedly candidates for continuing monitoring studies due to their persistence in the environment. Their presence in Xoconostle peel and fruit should be monitored as typically consumption is of the raw fruit without lengthy periods of storage. There is great potential for the use of traditional vegetables for human consumption and it is highly desirable to understand and limit the presence of contaminants in these food sources.

In addition, farmers in the study area should be educated with respect, the right amounts to use and the frequency of application of these pesticides in Xoconostle plantation. This will help prevent the use of unapproved or banned pesticides as well as reduce the amounts of pesticide residues in the environment.

Acknowledgements
This research was supported by Autonomous Metropolitan University by financial support in the development of the research. The authors also wish to thank Dr. Richard Gibson for his comments for this manuscript.

Authors' contributions
JJP and RO wrote the manuscript, thay have research interests in organic pollutants in food. MLR and DR carried out much of practical work in the laboratory. JO and DM collected samples in the field. All authors read and approved the final manuscript.

Competing interests
All involved authors declare that they have no personal economic or private institutional interest related to this research. All data reported was generated by research in the laboratories of the researchers involved.

Author details
[1]Departamento de Producción Agrícola y Animal, Laboratorio de Análisis Instrumental, Universidad Autónoma Metropolitana, Unidad Xochimilco, Calzada del Hueso No 1100, Colonia Villa Quietud, Delegación Coyoacán C.P. 04960, D.F., México. [2]Departamento de Producción Agrícola y Animal, Laboratorio de Fitopatología, Universidad Autónoma Metropolitana, Unidad Xochimilco, Calzada del Hueso No 1100, Colonia Villa Quietud, Delegación Coyoacán C.P. 04960, D.F., México.

References
Chen CH, Qian Y, Chen Q, Tao CH, Li CH, Li Y. Evaluation of pesticide residues in fruits and vegetables from Xiamen, China. Food Control. 2011;22:1114–20.
Codex Alimentarius. Residuos de plaguicidas en Alimentos y Piensos. http://www.fao.org/fao-whocodexalimentarius/standards/pestres/pesticides/es/. Accessed 14 Nov 2015.
EPA (Environmental Protection Agency US). Method 3540c Soxhlet extraction. 1996. www.epa.gov/hw-sw846/sw-846-test-method-3540c-soxhlet-extraction. Accessed 25 Aug 2015.
EPA (Environmental Protection Agency US). Method 3620C Florisil cleanup. 2014. https://www.epa.gov/sites/production/files/2015-12/documents/3620c.pdf. Accessed 25 Aug 2015.
EPA (Environmental Protection Agency US). Method 8081B Organochlorine pesticides by Gas Chromatography. 2007. https://www.epa.gov/sites/production/files/2015-12/documents/8081b.pdf. Accessed 25 Aug 2015.
FAO (Food and Agriculture Organization of the United Nations). The State of Food Insecurity in the World. Meeting the 2015 international hunger targets: taking stock of uneven progress. Rome: FAO. 2015. p. 56.
Fenik J, Tankiewicz M, Biziuk M. Properties and determination of pesticides in fruits and vegetables. Trends Anal Chem. 2011;30(6):814–26.
Gutiérrez JA, Pinzón MI, Londoño A, Blanch D, Rojas AM. "Residuos de plaguicidas organoclorados, organofosforados y análisis fisicoquímico en piña (Ananas comosus L.)". Agro Sur. 2010;38(3):199–211.
Guzmán MSH, Morales MAL, Mondragón JC, Herrera HG, Guevara LF, Reynoso CR. "Physicochemical, Nutritional and Functional Characterization of Fruits Xoconostle (Opuntia matudae) Pears from Central-México Region". J Food Sci. 2010;75(6):C485–92.
Instituto Nacional de Ecología. "Nopal tunero Opuntia spp. Cultivo alternativo para las zonas áridas y semiáridas de México" Comisión Nacional de las Zonas Áridas. 2003. http://www.inecc.gob.mx/descargas/publicaciones/70.pdf. Accessed 25 Aug 2015.
Keikotlhaile BM, Spanoghe P, Steurbaut W. Effects of food processing on pesticide residues in fruits and vegetables: A meta-analysis approach. Food Chem Tox. 2010;48:1–6.
Latif Y, Sherazi STH, Bhanger MI. Assessment of pesticide residues in commonly used vegetables in Hyderabad, Pakistan. Ecotoxicol Environ Saf. 2011;74: 2299–2303.
Morales P, Ramírez-Moreno E, Cortes SMM, Carvalho AM, Ferreira ICFR. Nutritional and antioxidant properties of pulp and seeds of two xoconostle cultivars (Opuntia joconostle F.A.C. Weber ex Diguet and Opuntia matudae Scheinvar) of high consumption in Mexico. Food Res Int. 2012;46:279–285.

Narayan-Sinha S, Vardhana-Rao MV, Vasudev K. Distribution of pesticides in different commonly used vegetables from Hyderabad, India. Food Res Int. 2012;45:161–9.

NORMA Oficial Mexicana NOM-007-RECNAT-1997 Que establece los procedimientos, criterios y especificaciones para realizar el aprovechamiento, transporte y almacenamiento de ramas, hojas o pencas, flores, frutos y semillas. DOF 30 de mayo de 1997. www.ordenjuridico.gob.mx/Publicaciones/CDs2006/CDAmbiente/pdf/NOM3.pdf. Accessed 1 Oct 2015.

Okoffo ED, Fosu-Mensah BY, Gordon CH. Persistent organochlorine pesticide residues in cocoa beans from Ghana, a concern for public health. Int J Food Contamination. 2016;3:5. doi:10.1186/s40550-016-0028-4.

Osorio-Esquivel O, Ortiz-Moreno A, Álvarez V, Dorantes-Álvarez L, Giusti M. Phenolics, betacyanins and antioxidant activity in Opuntia joconostle fruits. Food Res Int. 2011;44:2160–68.

Pérez MA, Navarro H, Miranda E. "Residuos de plaguicidas en hortalizas: problemática y riesgos en México" Rev Int de Contaminación Ambiental 29 (Número especial sobre plaguicidas). 2013. p. 45–64.

Saenz C, Berger H, Corrales J, Galletti L, García V, Higuera I, Mondragón C, Rodríguez A, Sepúlveda E, Varnero MT. Utilización agroindustrial del nopal. Boletín de Servicios Agrícolas de la FAO. Roma, Italia. 2006. 84 p.

Wang S, Wang Z, Zhang Y, Wang J, Guo R. Pesticide residues in market foods in Shaanxi Province of China in 2010. Food Chem. 2013;138:2016–25.

8

Hygienic assessment of spontaneously fermented raw camel milk (*suusa*) along the informal value chain in Kenya

Linnet Wanjiru Mwangi[1*], Joseph Wafula Matofari[1], Patrick Simiyu Muliro[1] and Bockline Omedo Bebe[2]

Abstract

Background: *Suusa* is a spontaneously fermented milk product from raw camel milk used by the pastoral communities of Northern and Eastern Kenya. The product can be as a result of intentional fermentation at ambient temperature for 3 days where it is prepared by women specifically for home consumption. The product can also result from unintentional fermentation where raw camel milk intended for sale, undergoes coagulation at any node of the informal value chain. Since no heat treatment is involved in preparation, microbial safety and quality of *suusa* is completely dependent on the raw milk inherent flora and handling practices. Therefore, the aim of this study was to determine the microbiological quality and safety of *suusa* along the informal value chain in relation to the raw camel milk handling practices. The study was carried out in Isiolo County where production, bulking and cooling of raw camel milk is done and in Nairobi County where there is biggest market for that milk. A total of 59 milk samples were obtained from the production, bulking, cooling and marketing nodes and analysed for, Titratable Acidity (TA), Total Viable Count (TVC), Coliform Count (CC), Spore Count (SC) and Yeast and Moulds Count (YM). The microbial load of TVC, CC and YM, increased significantly ($P < 0.05$) by 1 log increase, while SC increased by 3 log increase from production to market. The lactic acid increased from 0.07 % to 0.23 % for the unintended suusa. The microbial load comprised of 67 % Gram Negative Rods (GNR), 62 % Gram Positive Cocci (GPC) and 28 % YM from production, processing and marketing. Hygienic practices in raw camel milk and *suusa* production potentially expose the product to microbial contamination associated with reduced shelf life and public health concern.

Results: A total of 59 milk samples were obtained from the production, bulking, cooling and marketing nodes and analysed for, Titratable Acidity (TA), Total Viable Count (TVC), Coliform Count (CC), Spore Count (SC) and Yeast and Moulds Count (YM). The microbial load of TVC, CC and YM, increased significantly (P < 0.05) by 1 log increase, while SC increased by 3 log increase from production to market. The lactic acid increased from 0.07 % to 0.23 % for the unintended suusa.The microbial load comprised of 67 % Gram Negative Rods (GNR), 62 % Gram Positive Cocci (GPC) and 28 % YM from production, processing and marketing.

Conclusion: Hygienic practices in raw camel milk and suusa production potentially expose the product tomicrobial contamination associated with reduced shelf life and public health concern.

Keywords: Camel milk, Microbial quality, *Suusa*, Contamination, Value chain, Isiolo

---

* Correspondence: lynmwng@gmail.com
[1]Department of Dairy and Food Science and Technology, Egerton University,
P.O. Box 536-20115, Egerton, Kenya
Full list of author information is available at the end of the article

Background

In Kenya, the camel population is approximated to be 3.1 million (Corman et al. 2014). They are all one-humped (*Camelus dromedarius*), found mainly in the low lands of Northern Kenya. The camel lives in areas not suitable for crop production and where other livestock species can hardly thrive (Noor et al., 2012). Due to its outstanding performance in arid and semi arid areas (ASAL) of northern Kenya, camels play a central role to the livelihoods and culture of nomadic pastoralists (Guliye et al. 2007). They provide milk, meat and means of transport. The camel milk production in Kenya is estimated at 937 thousand tonnes in 2013 (FAOSTAT, 2015) which translates to about US$ 107.1 million. This quantity of milk represents about 12 % of the total national Kenyan milk production (Musinga et al. 2008). During prolonged droughts, camel milk may contribute up to 50 % of total nutrient intake by pastoral groups (Wayua et al., 2012). The most popular camel milk product among the pastoral groups is *suusa*. *Suusa* is spontaneously fermented raw camel milk. The fermentation is carried out at room temperature ranging from 26–29 °C, for 1–2 days in a gourd (Lore et al., 2005). The product is a white, low-viscosity product with a distinct smoky flavour and astringent taste (Lore et al., 2005). However, due to demand, the gourds have become small to produce the amount needed and the women pastoralists have turned to recycled plastic oil containers. *Suusa* is prepared by the Borana and Somali communities of North and Eastern Kenya, by storing milk in plastic containers which is allowed to slowly coagulate over a period of 1–3 days.

Spontaneous fermentation of raw milk takes advantage of the action of naturally occurring mixed microflora inherent in the milk in the plastic containers as well as factors such as temperature and pH provide the necessary selective factors for evolution of lactic acid bacteria (LAB) that impart desirable attributes to the product (Lore et al., 2005).However, *suusa* production process faces several handicaps; these include unpredictable production environment, unknown microbiology in processing, lack of process control and, unknown toxicological status (Chinyere and Onyekwere, 1996). Under pastoral production of *suusa*, factors such as unknown udder health, plastic milking and storage containers, milking personnel practices like tying the quarters to prevent suckling by the calf and dusty milking environment, and lack of water may act as points of contamination (Mulwa et al., 2011). During transportation, either by walking, on donkeys or occasionally on open pickups, long distance from milking point to collection or storage point, poor roads and lack of cooling facilities affects the microbial load by providing conditions for rapid

multiplication in the milk(Mulwa et al., 2011). At the collection centres, milk from different suppliers is pooled without prior quality control tests and this acts as a source of contamination and affects the safety and quality of the milk (Momanyi and Jenet 2010; Noor et al., 2013). The only test carried out is organoleptic (taste, sight and smell) (Noor et al., 2013) which is insufficient to detect for other detrimental quality and safety issues related to milk. The variables involved the fermentation include lack of heat treatment of the milk, storage at ambient temperature, lack of known culture composition and hence lack of process control which results in a product of variable quality (Eyassu, 2007). The final product is sold in open air markets and this has an effect on the quality of the product. *Suusa* produced under pastoral environment faces these challenges along the value chain from production to consumption.

This practice of spontaneously fermenting raw camel milk into *suusa* is highly valued by the pastoral communities for nutritional and cultural reasons. Earlier studies on *suusa* product have majorly focused on the handling practices (Wayua et al., 2012) and Lactic Acid Bacteria (LABs) responsible for the fermentation (Lore et al., 2005). However, there exists no information on the microbial quality and safety of *suusa* along the value chain. The aim of this study was to determine the microbiological quality of *suusa* along the informal value chain with the aim of enhancing food and income security among the pastoral communities.

Methods
Study site
The study was carried out in Isiolo County which is camel milk producing area and in Eastleigh, Nairobi County which is a major urban consumption centre for camel milk.Isiolo County is located in Eastern Kenya, approximately 285 km North of Nairobi. The County is located at coordinates are 0° 21' 0" North and 37° 35' 0" East and an altitude ranging from 200 to 300 meters above sea level (ASL) although there are some areas in the county that go up to 1000 ASL. Its annual average temperature ranges between 12 and 28 °C and receives low rainfall ranging between 300 and 500 mm per year.

Milk handling practices survey
Semi structured questionnaires were administered to 90 respondents. At production, 30 respondents were purposively selected who comprised of herders, herd owners and transporters. At collection/bulking centres 30 respondents were selected who comprised of cooling hub attendants, traders and consumers. At Nairobi, Eastleigh market 30 respondents were selected who comprised of

consumers and traders. Focus Group Discussions (FGD) held with women group of 15 members who were randomly selected to represent camel milk producers, herders, milk traders, and consumers at Isiolo and Eastleigh, Nairobi. This was done to identify handling practices along the value chain.

Milk sampling

Samples of raw camel milk and *suusa* were collected in triplicates at each representative point of the value chain for microbiological analysis. At production, pooled milk samples from the herds. Milk samples were collected from each container at the herd level and later pooled to make a representative sample of 20. At bulking centres, a total of 12 random samples were collected from each cooling hub. A total of 7 unintended *suusa* samples were collected at the time of the study while a total of 10 intended *suusa* were collected. Intended *suusa* was collected from pastoral women who were requested to prepare it since the commodity is rare. A total of 10 *suusa* samples were collected from Eastleigh, Nairobi market from 10 traders. Pooled *suusa* sample was made by pooling milk from as many containers as each woman trader had to obtain a representative sample. At each sampling point, 50 ml of milk sample was taken and transferred into sterile screw-capped sampling bottles, securely capped, clearly labelled and immediately transported to the laboratory for analysis under ice (4 °C). A total of 59 samples were obtained for titratable acidity and microbial analysis.

Sample analysis

The raw camel milk and *suusa* samples analysis was done at Egerton University, Food microbiology laboratory. Serial dilution of up to 10^{-6} was done using peptone water and 1 ml of homogenate of sample was aseptically transferred into a sterile petri dish. Total Viable Counts (TVC) was enumerated on Plate Count Agar (PCA) (Oxoid, UK) using pour plating method and the plates incubated at 37 °C for 48 hours. The Coliforms counts (CC) were enumerated on Violet Red Bile Agar (VRBA) (Oxoid, UK) using pour plating technique and plates incubated at 37 °C for 24 hours(AOAC, 1995). The Spore Counts (SC) were enumerated by heat treating milk samples in a water bath at 80 °C for 10 - minutes and 1 ml of appropriate dilution pour plated on (PCA) (Oxoid, UK) and the plates incubated at 37 °C for 24 hours (AOAC, 1995). While the yeast and mould were determined on Potato Dextrose Agar (PDA) (Oxoid, UK) by spread plating technique and the plates incubated at 25 °C for 5 days (AOAC, 1995).

Discrete colonies grown on plates after incuation were selected randomly and purified by repeated plating on the same agar according to Lore et al. (2005). The colonies were then subjected to morphological (cell shape, motility, cell grouping and endospores), biochemical (catalase, oxidase, carbohydrate utilization, indole, and Methyl red-Vosges-Proskauer) and physiological tests and identified to genus level (AOAC, 1995).

Developed acidity in the samples was determined according to the method described by the International Dairy Federation (I.D.F.) (1990). 9 ml of the milk samples were measured into the conical flasks, and 1 ml 0.5 % alcoholic phenolphthalein indicator added then titrating with 0.1 N sodium hydroxide (NaOH) until a faint pink colour appears. The results were then expressed as % lactic acid where 1/10 ml NaOH is equal to 0.09 % w/v lactic acid.

Statistical analysis of data

The microbial counts for the total viable count (TVC), coliform count (CC), spore count (SC), yeast and mould count (YM) were transformed to base-10 logarithm of colony forming units (cfus) per millilitre (ml) of the milk samples ($\log_{10}$ cfu/ml). The transformed data was tested for normality using PROC NPAR1WAY procedure of Komolgorov–Smirnoff's test and also tested for homogeneity of variances using Levene's test before assumption of analysis of variance (ANOVA) was done using the General Linear Model (GLM) procedure of SAS version 9.1 (SAS Institute, Inc., Cary, NC). The independent variable was the milk quality and value chain points (production, bulking, processing and marketing). The significance of the means was determined using Tukey's Honestly Significance Difference (HSD) test at $P < 0.05$.

Results

Mapping the *suusa* value chain

The value chain for *suusa* mapped from surveys and FGD is represented in Fig. 1, revealing the handling practices along the value chain. Camels are milked at "*boma*" (similar to a kraal) by herders. Fresh milk is bought by women groups or individuals to make *suusa* or sell in open air market. Soured milk is downgraded and sold as unintended *suusa*. Fresh camel milk and *suusa* is consumed by both pastoralists and non-pastoralists.

The common camel milk handling practices that influence the contamination levels in milk is shown in Table 1. It was found that all herders neither wash their hands nor wash the camel udder before milking and all camel milk handling containers were plastic. After milking and bulking at the herd level, all milk was found to be transported by the either motorbikes when the herd was near Isiolo town or trucks when the herds were far from the town e.g. from Kulamawe which was about 100 KM form Isiolo town. Isiolo town is the main collection center for raw camel milk where there are cooling facilities for the milk. At Isiolo, milk

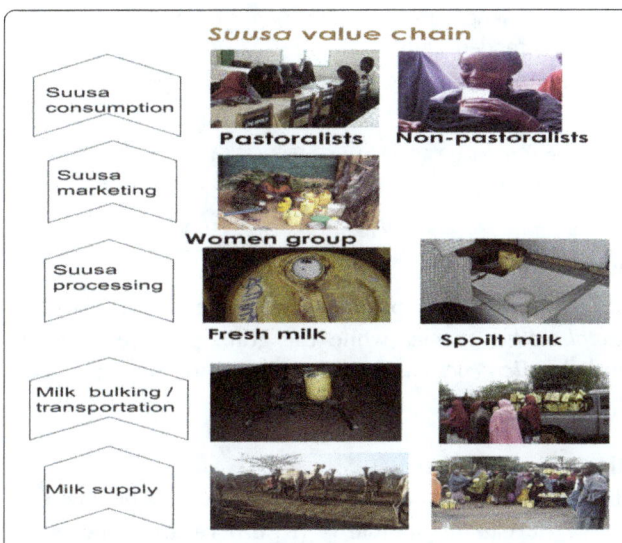

Fig. 1 Mapped *suusa* value chain from Isiolo camel producing points to Eastleigh market, Nairobi

was either transported to Nairobi using Buses or sold in the town through open air vending or milk bars. Of interest was that *suusa* was made through 2 identified processes. One process is that raw camel milk that gets sour/coagulates on transit to collection or market centres from the production area- *Boma* would be sold as *suusa* referred to as unintended *suusa* by the authors. The fermentation in this case would take from 4-24hours. The other process is intentional spontaneous fermentation of raw camel milk for a period of 3 days mostly done by pastoral women for home consumption referred as intended

suusa. However, preparation of *suusa* through the 2 processes involved no heat treatment or quality control tests before fermentation. Unintended *suusa* was found to be vended by the roadside (80 %), sold in the milk bars (10 %) and also sold in hotels (10 %).

Practices associated with handling of camel milk by the pastoralists

The challenges faced by the actors along the camel informal milk value chainsthat affect the quality of the milk as identified from the FDGs are shown in Table 2. Sanitation problems were identified as the biggest challenge due to the lack of potable water along the value chain. Lack of milk cooling facilities, interrupted electricity supply where these facilities are available, poor road network to the market, long distance to the market and pooling of milk from different sources were the main challenges identified for the spoilage of milk to unintended *suusa*.

Microbial load and acidity of *suusa* along the value chain

Figure 2 shows the microbial load for unintended *suusa* along the value chain. For unintended *suusa*, the Total Viable Count (TVC) increased from $\log_{10}$ 7.79 at production to $\log_{10}$ 8.51 at the market, Coliform Counts (CC) increased from log10 6.31 to log10 7.99, Spore Count (SC) increased from log10 4.53 to log10 7.56 and Yeast and Moulds (YM) counts increased from log10 4.85 to log10 5.70 cfu/ml. The microbial load of intended *suusa* at the production and market levels is shown in Fig. 2. The TVC increased from log10 7.79 at production to log10 8.41 at the market, CC increased

Table 1 The common handling practices for the camel milk

Practice	N	Frequency (%)
Hand washing	30	0
Udder washing	30	0
Plastic milk storage containers	30	100
Means of transport to cooling centres:		
Truck	30	20
Motorbike		70
Donkeys		10
Means of transport to the market-Nairobi:		
Truck	30	0
Buses		100
Non-performance of quality control tests before bulking	30	100
Boiling milk before making *suusa*	30	0
Sale area of *suusa* : Roadside	30	80
Dairy		10
Hotel		10

Table 2 The Challenges faced in the camel milk value chain

Value chain node	Challenges
Production	Lack of water
	Lack of cooling facilities
	Personal, equipment and environment hygiene
	Mixing of milk from diseased camels
	Lack of veterinary service due to high mobility
Cooling centres	Spoilage/unexpected fermentation
	Interrupted power supply to coolers
	Pooling milk
	Lack of clean water
	Lack of quality control tests
	Lack of knowledge on hygiene and quality checks
Transportation	Lack of refrigerated tankers for transporting the milk
	Poor state of roads
Marketing	Sale in open air-roadside
	Long distance to market
	Lack of cooling facilities
	Spoilage/unexpected fermentation

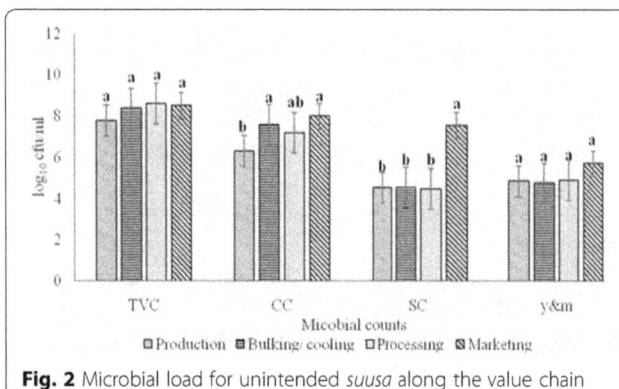

Fig. 2 Microbial load for unintended *suusa* along the value chain

Table 4 Incidence of main groups of microorganisms isolated from the intended *suusa*

Value chain node	N	G+ rods	G- rods	G+ cocci	Spores	YM
Production	10	2	5	6	2	2
Processing/marketing	10	3	7	4	3	3
Total	20	5	12	10	5	5
Incidence		25 %	60 %	50 %	25 %	25 %

Enterobacter. Gram positive rods included *Bacillus* and *Lactobacillus* species while the gram positive cocci included *Micrococcus*, *Streptococcus* and *Staphylococcus* species.

from log10 6.31 to log10 7.75, SC increased from log10 4.53 to log10 7.24 and YM increased from log10 4.85 to log10 5.41 cfu/ml. The SC was significantly higher at $P < 0.05$ for the *suusa* at the market level than at the production level. Also, the microbial load for the TVC, CC and YM were significantly high in *suusa* at the market than at the production.

Figure 4 shows the change in developed lactic acid in unintended *suusa*. Lactic acid increased significantly ($P < 0.05$) along the value chain from 0.07 to 0.23 %. The lactic acid (% LA) for the intended *suusa* increased from 0.07 % to 0.60 % from production to the market as shown in figure.

Main types of organisms in *suusa* along the value chain

Table 3 shows the main type of microorganisms that were isolated from unintended *suusa*. There was high incidence of Gram negative rods of 67 % from production to the market followed by Gram positive cocci with an incidence of 62 %. YM had the least incidence at 28 %.

Table 4 shows the main type of microorganisms that were isolated from intended *suusa*. The incidence of Gram negative rods and gram positive cocci were highest at 60 % and 50 %, respectively.

Table 5 shows the different types of microorganisms isolated from the two *suusa* value chains. Gram negative rods were identified to be *E. coli*, *Pseudomonas* and

Table 3 Incidence of Main groups of microorganisms isolated from the unintended *suusa*

Chain node	N	G+ rods[a]	G- rods[a]	G+ cocci[a]	Spores[a]	Y&M[a]
Production	10	2	6	7	2	2
Cooling\bulking	12	6	8	7	3	2
Processed product	7	4	4	4	3	4
Market	10	8	8	6	5	3
Total	39	20	26	24	13	11
Incidence (%)		51	67	62	33	28

Key: N is the number of samples; G+: Gram positive, G-: Gram negative, Y&M: yeast and moulds: [a]is the number of positive observed for a specific group of organisms.

Discussion

Microbial load and acidity along the *suusa* value chain

Kenya Bureau of Standards (KEBS) regards raw whole camel milk as good when the total viable counts (TVC) are between $0\text{-}5 \times 10^5$ cfu/ml for grade I and II (KEBS, 2007). The raw camel milk at production was above the recommended range and therefore, the milk can be regarded as of poor quality. High TVC at production can be attributed to handling practices like not washing hands before milking, no washing of the camel's udder before milking, use of plastic containers for milking and storage of milk in plastic containers which are not easy to clean. Use of recycled plastic containers which are not easy to clean harbours spoilage microorganisms, unrefrigerated transportation, long distance, poor roads to cooling centres and pooling of milk from different suppliers at the cooling centres are risk factors to the growth and multiplication of the indigenous microflora, resulting in reduction of milk quality and safety (Wayua et al., 2012).There was significant increase($P < 0.0.5$) in TVC from production to marketing which is attributed to absence of heat treatment of milk prior to fermentation coupled with spontaneous fermentation. Results agree with those reported by Odongo et al. (2016) and Matofari et al. (2013). The GNR were the most prevalent types of microorganisms for both the intended and unintended *suusa* while the YM were the least prevalent for both.

Lactic acid increased significantly ($P < 0.05$) from production to the market with no effect on microbial load reduction as shown in Figs. 4 and 5.With fermentation, lactic acid bacteria break down lactose into lactic acid. Presence of GNR which are fermentative organisms had an influence on increased acid content. Intended *suusa* is fermented over a period of 3 days (72 hrs) and this further explains why the percentage lactic acid was higher than unintended *suusa* which takes less than 24 hours.

Coliforms increased significantly ($P < 0.05$) from production to the market by 1 log increase as shown in

Table 5 Types of microorganisms isolated from the *suusa* value chain

Chain node	N	*E. coli*	*Enterobacter*	*Pseudomonas*	*Micrococcus*	*Staphylococcus*	*Streptococcus*	*Bacillus*	Yeast and molds
Production	10	5	6	-	5	2	3	2	2
Bulking/cooling	12	4	4	8	3	2	4	3	2
Processing	7	4	4	-	2	2	3	3	4
Marketing	10	4	3	2	6	2	2	5	3
Total	39	17	17	10	16	8	12	13	11
Incidence (%)		44	44	27	41	21	31	33	28

Key: N is the number of samples analysed

Figs. 2 and 3. Coliforms are found in the soil, mud, dust, plant materials and can be dispersed into the atmosphere by dust into the product. With natural fermentation, the coliforms will multiply and cause problems in the final product because lactic acid bacteria will initially be very low Gadaga et al. (2004). Coliforms also have adaptation strategies that range from temperature evasions, acid tolerance and production of probiotics like colicins that inhibit growth of other microorganisms (Abee et al. 1995; Gadaga et al., 2004). Occurrence of coliforms more so *E.coli* in the final product despite high lactic acid (Figs. 4 and 5) is probably due to induced acid tolerance by the organism through production of acid shock proteins which enhance its survival through neutralization of the external environment, adjusting catabolism to the new environment, performing DNA repair and membrane biogenesis and contribute to microbial pathogenesis (Bearson et al., 1997). Karagözlü et al. (2007) found that stationary phase cells of *E.coli* strains were able to survive and multiply in *kefir* (Caucasian fermented camel milk). It has also been found to survive in fermented goat milk, *amasi* (Bearson et al., 1997). Isolation of coliform bacteria along the *suusa* value chain is an indication of presence of enteric pathogens in the *suusa* value chain as shown in Table 5. This shows hygienic conditions during handling and processing of camel milk into *suusa* are low. *Suusa* therefore has public health risk potential for spread of foodborne illnesses such as *Escherichia coli O157:H7* illnesses.

There was a significant ($p < 0.05$) increase in spore counts from production to market samples by 3 log increase as shown in Figs. 2 and 3. Spore forming bacteria are environmental microorganisms such as *Bacillus* and *Clostridium* species. At production, they may originate from water used to wash the milking equipment and dust from the milking area. Spore formers like *Bacillus cereus* display a mechanism of acid tolerance response (ATR) and can survive below pH 4.0 favourable for spore formation (Gadaga et al., 2004). This explains the existence of increased spore forming bacteria in *suusa* at the market level. Gram positive spore forming rods were identified as *Bacilli*. High incidence at the market could be attributed to the marketing environment characterised by sale in the open with heaps of waste material, dust and mud close to where the product is sold. Spores are carried by wind into the atmosphere and into the product. *Bacilli* are aerobic whose typical habitat is soil although they are widely distributed in nature and gain access to milk and *suusa* through air, water, fodder and feed. Spore-forming bacteria are known to cause food spoilage and food-poisoning by producing heat labile enterotoxins. Therefore, their presence in *suusa* poses a risk of food poisoning by the enterotoxins to the consumers of the milk product.

Yeast and moulds increased by 1 log increase from production to the marketas shown in Figs. 2 and 3. High contamination by yeasts and moulds may be due to poor

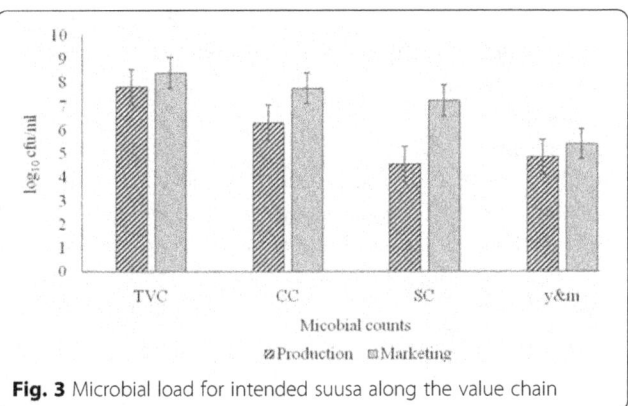

Fig. 3 Microbial load for intended suusa along the value chain

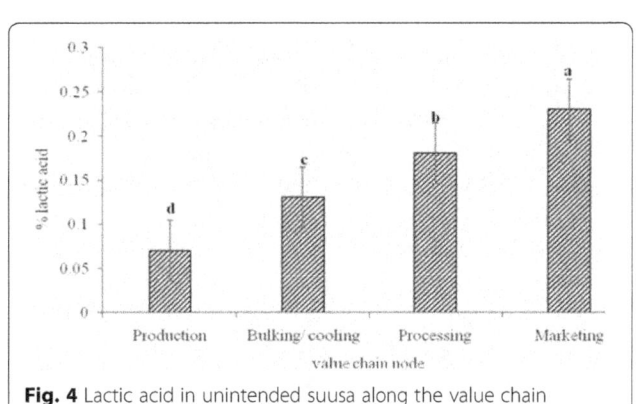

Fig. 4 Lactic acid in unintended suusa along the value chain

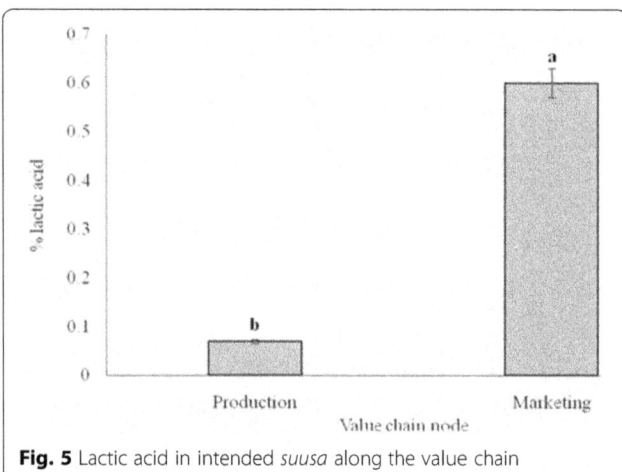

Fig. 5 Lactic acid in intended *suusa* along the value chain

processing and marketing conditions and uncontrolled fermentation which led to contamination. During spontaneous fermentation of *suusa*, organic acids such as lactic, acetic and propionic acids are produced which lower the pH. The lower pH is favourable for growth of yeast and mould species which causes these species to become competitive in the immediate medium (*suusa*) hence the significant increase ($P < 0.05$) in yeast and mould count in marketed *suusa* (Lefoka, 2009).

Results in Tables 3 and 4 display the main type of microorganisms that were isolated from intended and un-intended *suusa* along the value chain. Gram negative rods had the highest incidence of 88 % from production to the market followed by Gram positive cocci with an incidence of 84 %. *Micrococci* were isolated and could account for the high count of Gram positive cocci which is mostly found in water, soil and dust. Yeast and moulds had the least incidence at 28 %. Gram negative rods were identified to be *E. coli*, *Pseudomonas* and *Enterobacter*. Presence of *E.coli* is an indication of faecal contamination by handling from production to the market Rochelle-Newall et al. (2015). This indicates possibility of presence of other enteric pathogens. Gram positive spore forming rods were identified as *Bacillus* and the high incidence at the market could be attributed to the environment in which the product is produced and sold. The environment was characterised by heaps of waste material, dust and mud and this could be the source of the spores which are carried into the atmosphere and into the product.

Gram positive cocci isolated from camel milk and *suusa* included *Streptococcus* and *Micrococcus* species. *Streptococcus* species especially the *Streptococcus lactis* group originate from equipment that is contaminated due to insufficient sanitation. Organisms like *Micrococci*, coliforms and enteric pathogens originate from hand milking and milk handling that might contaminate the milk via the skin, nose and mouth(Cui et al., 2016). Risk

factors identified in the field of study that have contributed to the specific organisms associated with *suusa* were: not washing hands or camels' udder before milking, dusty milking environment, use of organoleptic tests for quality assessment of milk, bulking milk from different suppliers, use of not easy to clean plastic containers, delayed milk delivery, lack of refrigeration during transport and sale of *suusa* in open air markets (Noor et al., 2013). Other authors (Akweya et al. 2012; Wanjohi et al. 2010) have also detected *Staphylococcus aureus*, in camel milk.

Spores of *Bacillus* species as well as the organism were also detected. The genus *Bacillus* are typical habitats of the soil and are widely distributed in nature and may gain access to milk and dairy products through the air, water, fodder and feed thereby present on the skin and hair of cattle (Loralyn and Robert, 2009). *Bacillus* species produce heat stable protease and lipase which may eventually cause spoilage to camel milk and milk products (Samaržija et al. 2012). Other species such as *Bacillus cereus* produce toxins which lead to food intoxication from ingestion of contaminated food. Presence of the *Bacillus* species in camel milk makes the food a potential risk to consumers. Spores increased significantly ($P < 0.05$) along the value chain in Fig. 2. This is attributed to handling practices observed along the value chain that have led to contamination of both raw and fermented product. Risk factors identified include: dusty milking environment, use of plastic milking and storage containers and unrefrigerated transport of raw milk from production to cooling centres. These factors most probably led to the growth and multiplication of spore forming bacteria hence high counts.

Among the major isolates of the microorganisms was the *Pseudomonas* species. *Pseudomonas* species produce heat stable proteases and lipases keeping their activity even after pasteurization thereby producing off-flavours in milk as well sweet curdling of pasteurized milk (Perko, 2011). *Pseudomonas* species are the main psychrotrophic bacteria isolated from refrigerated raw milk, being among the major spoilage agents in the dairy industry (Paula Ana et al. 2011). Presence of the genus *Pseudomonas* indicates improper cooling and refrigeration of camel milk.

Conclusions

Handling practices along the *suusa* value chain influences microbial load and consequently, contributes to spoilage of camel milk and *suusa*. Extension, training and regular monitoring to improve handling practices along the *suusa* value chain is essential. Statutory (legal) institutions like Kenya Bureau of Standards need to address handling practices of camel milk and milk products along the chain to improve safety, quality and acceptability hence better market. As a consequence of lack of process control in *suusa* fermentation, the microbial diversity in the product is uncontrolled. *Suusa* has presence of high loads of microorganisms of safety concern. More in depth studies on *suusa* are needed to quantify and identify organisms of safety concern to the species level using molecular techniques.

Abbreviations
ANOVA: Analysis of Variance; CC: Coliform Count; FGD: Focus Group Discussions; GLM: General Linear Model; GNR: Gram Negative Rods; GPC: Gram Positive cocci; HSD: Honest Significant Difference; KEBS: Kenya Bureau of Standards; LAB: Lactic Acid Bacteria; LSD: Least Significant Difference; NC: North Carolina; PCA: Plate Count Agar; PDA: Potato Detrose Agar; RUFORUM: Regional Universities Forum for capacity building in Agriculture; SAS: Statistical analysis software; SC: Spore count; TA: Titratable acidity; TVC: Total viable count; UK: United Kingdom; VRBA: Violet red bile agar; YM: Yeast and Molds

Acknowledgements
The authors acknowledge the women group of Isiolo County for availing samples and participating in the study. Acknowledgement also goes to Steven Wakoli and Pauline Madete for their assistance in sample collection. We also acknowledge Dr. Olivier Kashongwe for assistance in data analysis as well Mr. Nobert Wafula for the invaluable assistance in the editing of the manuscript.

Funding
This research work was financially supported by Regional Universities Forum for Capacity Building in Agriculture (RUFORUM) under research grant number RU2011GRG11 titled: Consumer and regulator concerns about pastoral indigenous knowledge food processing: participatory analysis for producers' and consumers' benefits in Kenya.

Authors' contributions
This research work was part of LM Thesis research for the award of MSc. Food Science degree of Egerton University and supervised by JM, PM and BB. LM was in-charge of FGDs and questionnaire administration, samples collection and analysis under the guidance and supervision of the three supervisors. All the authors were involved in the designing of the experiment, data analysis, interpretation of the results and manuscript drafting. All authors read and approved the final manuscript.

Authors' information
Ms. Linnet Mwangi was a graduate Student in the Department of Dairy and Food Science and Technology in Egerton University, Kenya, undertaking MSc. Food Science. Ms. Mwangi has a BSc. Food Science and Technology and MSc in Food Science from the same university. She has several years of working experience in the dairy and food manufacturing industries in Kenya. Prof. Joseph Matofari is an associate professor and Dr. Patrick Muliro is a senior lecturer in the Department of Dairy and Food Science and Technology in Egerton University while Prof. Bockline Bebe is a professor at the of Department of Animal Science, Egerton University, Kenya.

Competing interests
Authors of this research article declare that they don't have any financial or non-financial competing interests.

Author details
[1]Department of Dairy and Food Science and Technology, Egerton University, P.O. Box 536-20115, Egerton, Kenya. [2]Department of Animal Sciences, Egerton University, P.O. Box 536-20115, Egerton, Kenya.

References
Abee TL, Krokel L, Hill C. Bacteriocins: Modes of action and potentials in food preservation and control of food poisoning. Int J Food Microbiol. 1995;28:169–85.

Akweya BA, Gitao CG, Okoth MW. The acceptability of camel milk and milk products from north eastern province in some urban areas of Kenya. Afr J Food Sci. 2012;6(19):465–73.

Bearson S, Bearson B, Foster JW. Acid stress responses in enterobacteria. FEMS Microbiol Lett. 1997;147:173–80.

Chinyere II, Onyekwere SE. Nigerian indigenous fermented foods: their traditional process operation, inherent problems, improvements and current status. Food Res Int. 1996;29:527–40.

Corman VM, Jores J, Myer B, Younan M, Liljander A, Said MY. Antibodies against MERS Coronavirus in dromedary camels, Kenya, 1992–2013. Emerg Infect Dis. 2014;20:1319–22.

Cui H, Zhou H, Lin L. The specific antibacterial effect of the Salvia oil nanoliposomes against *Staphylococcus aureus* biofilms on milk container. Food Control. 2016;61:92–8.

Eyassu S. 2007. Handling, preservation and utilization of camel milk and camel milk products in Shinile and Jijiga Zones, eastern Ethiopia. Livest. Res. Rural Dev. 2016;19(6). from http://www.lrrd.org/lrrd19/6/seif19086.htm.

FAOSTAT 2015. World camel population. At http://faostat3.fao.org/browse/Q/QA/E. Accessed 12 Feb 2016.

Gadaga TH, Nyanga LK, Mutukumira AN. The occurrence, growth and control of pathogens in african fermented foods. Afr J Food Agric Nutr Dev. 2004;4:1.

Guliye AY, Noor IM, Bebe BO, Koskey IS. Role of camels (*Camelusdromedarius*) in the traditional lifestyle of Somali pastoralists in northern Kenya. Outlook Agric. 2007;36(1):29–34.

International Dairy Federation (I.D.F.). 1990. Handbook on milk collection in warm developing countries. IDF Bulletin special issue 9002

Karagözlü N, Karagözlü C, Ergönül B. Survival characteristics of *E. coli* O157:H7, *S. typhimurium* and *S. aureus*during *kefir* fermentation. Czech J Food Sci. 2007; 25:202–7.

KEBS. 2007. Raw whole camel milk — Specification, KS 2061:2007 (confirmed 2013). Kenya

Lefoka Mamajoro. 2009. Survival of microbial pathogens in dairy products. Msc dissertation, University of the Free State. scholar.ufs.ac.za:8080/xmlui/bitstream/handle/.../LefokaM.pdf?. Accessed 14 Oct 2016.

Loralyn H. Ledenbach and Robert T. Marshall. 2009. Microbiological spoilage of dairy products. Compendium of the microbiological spoilage of foods and beverages, p,27.

Lore TA, Samuel KM, John W. Enumeration and identification of microflora in *Suusac*, a Kenyan traditional fermented camel milk produc. Lebensm-Wissu-Technol. 2005;38:125–30. marketing practices in an emerging peri-urban production system in Isiolo County, Kenya. Pastoralism: Research, Policy and Practice, 3:28.

Matofari JW, Shalo PL, Younan M, Nanua NJ, Adongo A, Qabale A, Misiko BN. Analysis of microbial quality and safety of camel (Camelus dromedarius) milk chain and implications in Kenya. J Agric Ext Rural Dev. 2013;5(3):50–4.

Momanyi S. and Jenet A. 2010. Study on hygiene practices and market chain of milk and milk products in Somalia. http://fex.ennonline.net/39/study.aspx. Accessed 14 Oct 2016.

Mulwa WD, Schelling E, Wangoh J, Imungi KJ, Farah Z, Meile L. Microbiological quality of raw camel milk across the Kenyan market chain. Global Sci Book. 2011;5(1):79–83.

Musinga M., Kimenye D. and Kivolonzi P. 2008. The Camel Milk Industry in Kenya http://www.ebpdn.org/download/download.php?table. Accessed 14 Oct 2016.

Noor IM, Bebe OB, Abdi YG. Analysis of an emerging peri-urban camel production in Isiolo County, Northern Kenya. J Camelid Sci. 2012;5:41–61.

Noor I.M., Gulye A.Y., Tariq M. And Bebe B.O. 2013. Assessment of camel and camel milk

Odongo N. O, Lamuka P. O, Abong G. O, Matofari J. W, Abey K. A. 2016. Physiochemical and Microbiological Post-Harvest Losses of Camel Milk Due Deterioration Along the Camel Milk Value Chain in Isiolo, Kenya. Current Research in Nutrition and Food Science; 4(2)

Patricia Curiff. 1995. Official Methods of Analysis of AOAC International 16th Edition. Virginia, USA: AOAC International; p 108–110.

Paula Ana FC, Daniel JD, Renata W, Adriano B. Hydrolytic potential of a psychrotrophic *Pseudomonas* isolated from refrigerated raw milk. Braz J Microbiol. 2011;42:1479–84.

Perko B. Effect of prolonged storage on microbiological quality of raw milk. Mljekarstvo. 2011;61(2):114–24.

Rochelle-Newall E, Nguyen TMH, Le TPQ, Sengtaheuanghoung O, Ribolzi O. A short review of fecal indicator bacteria in tropical aquatic ecosystems: knowledge gaps and future directions. Front Microbiol. 2015;6:308.

Samaržija D, Šimun Z, Tomislav P. Psychrotrophic bacteria and milk and dairy products quality. Mljekarstvo. 2012;62(2):77–95.

Wanjohi, GM, CG Gitao, and LC Bebora. 2010. The hygienic quality of camel milk marketed from North Eastern Province of Kenya and how it can be improved. Garissa: A paper presented at International Camel Symposium

Wayua FO, Okoth MW, Wangoh J. Survey of post-harvest handling, preservation and processing practices along camel milk chain In Isiolo District, Kenya. Afr J Food Agric NutrDev. 2012;12(7):6897–912.

Bacteriological milk quality: possible hygienic factors and the role of *Staphylococcus aureus* in raw bovine milk in and around Gondar, Ethiopia

Betelihem Tegegne[1] and Shimels Tesfaye[2*]

Abstract

Background: In Ethiopia, around 97% of the annual milk production is accounted by the traditional milk processing system using on-farm traditional milk processing materials that are generally poor in processing capacity, causing high product loss and risky for public consumption. A cross-sectional study was carried out in and around Gondar town, Amhara Regional State of Ethiopia from October 2014 to may 2015 with the objective to assess the bacteriological milk quality, possible hygienic factors and status of *S. aureus* as contamination of bovine raw milk. The study employed questionnaire survey and raw bacteriological load analysis and cow milk samples for isolation and detection of *S. aureus* from raw cow milk. Sixty (60) randomly selected dairy farms were interviewed for the survey-based study of farm hygienic practices and 72 raw milk samples [60 from directly from teats and 12 from collecting tanks (buckets) were aseptically collected and tested for bacteriological load analysis and isolation of *S. aureus*.

Results: The overall average total bacterial count (TBC) were $4.59 \pm 0.118\log10$ (38,904.51 cfu/ml) and $4.77 \pm 0.23\log10$ (58,884.37 cfu/ml) for milk samples collected directly from teat during milking and milking buckets at farm level respectively. Accordingly, the count increased by 0.18 ± 0.23 log10 or 19,979.86 cfu/ml (51.36%) increase from teat to milking buckets. Results showed very significant differences in plate counts ($P < 0.05$) between the two milk collection points. 73.30% of the milk samples collected directly from the teat were found (>100,000 bacteria per ml), evidence of poor milk hygiene when compared to international standards. In this study hygienic and management factors like udder cleaning, water and soap using for cleaning of udder, hand washing and water and soap using for milking vessels were significantly ($P < 0.05$) affects the bacteriological count of the milk.

Conclusions: The results of the current study indicated that the cow milk produced and distributed in the study area can generally be considered as substandard in quality for consumption unless pasteurized. Therefore, this risk assessment study with similar different studies reported from different regions in Ethiopia might provide a foundation for the establishment of national milk quality standards that currently do not exist in Ethiopia.

Keywords: Bacteriological quality, Milk hygiene, *S. aureus*, Total plate count

* Correspondence: shimelsrich@gmail.com
[2]Faculty of Veterinary Medicine, Department of Para-Clinical Studies (Veterinary Microbiology), University of Gondar, Gondar, Ethiopia
Full list of author information is available at the end of the article

Background

Livestock represents major national resources and form an integral part of agricultural production system in Ethiopia (Gebrewold et al. 2000); cows contribute about 95% of the total annual milk produced by dairy cows, goats and camels at national level (CSA 2010).

In Ethiopia, milk production systems can be categorized into urban, peri-urban and rural, based on location (Reda 1998). Dairying constitutes an important sector of the agricultural production system. For smallholder farmers, dairying provides the opportunity to efficient use land, labour and feed resources and generates regular income in Ethiopia (Yitaye et al. 2009). In sub Saharan countries the traditional dairy sector, which is characterized by small herd size dominated by indigenous zebu breeds. These breeds, normally known by their low milk production with very little or no-specialized inputs, accounts 70–80% of Africa's cattle population (Ibrahim and Olaluku 2000).

In Ethiopia around 97% of the annual milk production is accounted by the traditional milk processing system using on-farm traditional milk processing materials (Felleke 2003), which is likewise dominated by indigenous breeds. In almost all areas in Ethiopia, the milk produced are traditionally processed to naturally fermented sour whole milk (ergo), traditional butter (Kibe), butter milk (Arera), cottage cheese (ayib), whey (aguat) and ghee (nitir kibe) dairy products. The traditional milk processing materials used are also similar among different areas which generally poor in quality of processing, includes; plastic container, Bottle gourd (Lagenaria siceraria) and clay pot (Duguma & Janssens 2014; Wafula et al. 2016). Most of the very few enterprises currently operating in and around the capital entirely depend on the traditional sector for their milk intake, while others depend on it for the majority of their intake. These underscore the importance of understanding the traditional sector in order to make improvement interventions. Economically, in Ethiopia Milk and milk products are also very important farm commodities and dairy farming is an investment option for smallholder farmers (Tsehay 2001).

Microbial load is a major factor in determining milk quality (Fatine et al. 2012). It indicates the hygienic level exercised during milking, cleanliness of the milk utensils, condition of storage, manner of transport as well as the cleanliness of the udder of the individual animals. Milk from a healthy udder contains few bacteria but it picks up many bacteria from the time it leaves the teat of the cow until it is used for further processing. These microorganisms are indicators of both manner of handling milk from milking till consumption and the quality of the milk (Lunder and Brehne 1996).

In Ethiopia, the fresh milk is sold unpasteurized to the public either directly from small producers, via informal markets or through dairy farmers cooperatives. This informal marketing system has been a challenge for milk quality control in urban and peri-urban areas at all levels (Godefay and Molla 2000). Awareness and resources aiding for hygienic milk production, storage, and transportation are very limited, especially smallholder production system is under developed when compared with the institutional and urban producers in the in and around Gondar.

The consumption of raw milk, naturally sour/fermented milk (Erego)) and its derivatives is common in Ethiopia (Yilma 2012), which causes for harbouring of milk-transmitted zoonoses, including bovine tuberculosis (bTB), brucellosis and Staphylococcal Food Poisoning (SFP). Raw or processed milk is a well know food medium that supports the growth of several microbes with resultant spoilage of the product or infections (intoxications) in consumers (Oliver et al. 2009).

Staphylococcal Food Poisoning (SFP) is among the most prevalent causes of gastroenteritis worldwide (Wang et al. 2007; European Food Safety Authority 2010; CDC 2016). S. aureus has many potential virulence factors and staphylococcus enterotoxin (SE) is one of among several responsible for food poisoning. Ingestion of less than 1.0 µg enterotoxin causes SFP (Seo and Bohach 2007; Enquebaher et al. 2015). To date, more than 21 different SEs and SE-like super-antigens have been identified and designated classical as enterotoxins SEA/SEB/SEC/SED/ SEE (SEA-SEE), new (SEG-SEI) and new (SEIJ-SEIV) (Bennett and Hait 2011; Hennekinne et al. 2011).

In developing countries like Ethiopia, where high prevalence of clinical and subclinical mastitis mainly caused by S. aureus (Abera et al. 2010; Sori 2011) and high consumption of raw milk is common (Makita et al. 2012), Staphylococcal Food Poisoning (SFP) the most important target for study as risk of milk borne contaminations. Therefore, in order to protect consumers from unhygienic milk consumption and consequently expose to microbial contamination, it was found for us very important to study bacterial load and level of pathogenic microbes such as S. aureus in the milk production and collection. Such surveillance data may provide a basis for risk assessment study as well as give a foundation for the establishment of national milk quality standards that currently do not exist in Ethiopia. Based on these, the present study has been designed to give base line data for bacteriological quality of raw cow milk and status of S. aureus as raw milk contaminant in Northern Gondar, especially in and around the Gondar town.

Methods

Study area, study population and design

The study was conducted in and around Gondar town, which is located 740Km away North of Addis Ababa, the capital of Ethiopia. The town of Gondar is found at latitude of 12 °4'North, longitude of 27°2'east with an altitude of 1800–2500 m above sea level. The annual mean temperature of the area was 20.5 °C (17.2–23.9 °C) and annual rainfall of about 1000 mm (600–1400 mm). The region receives a bimodal rain fall, the average annual precipitation rate being 1000 that comes from the long and short rainy seasons. The short rainy season occurs during March, April, and May, while the long one extends from June through September (CSA 2010).

A cross sectional study was done from October 2014 to may 2015, in and around Gondar town by taking raw milk samples from lactating cows of selected dairy farms directly from teats during milking and milking buckets at farm level. The milk samples were collected from representative cows from each farm for milk samples collected directly from teat. The cow representing one farm was selected by simple random sampling method so that each farm supplying the milk to the local communities have representative.

The study subjects were milk samples collected from teat during milking and milking buckets at farm level and also questionnaire survey on milking personnel and farm attendants. In this study a total of 72 milk samples and 60 personnel for interview were included. The milk samples were collected from 60 farms selected from the sample frame by simple random sampling method and University of Gondar dairy farm included purposively. Two milk collection points (teat during milking and milking buckets at farm level) were considered for bacterial milk load and Staphylococcus aureus load too.

The sampling frame for farms selection was taken from agriculture office for Gondar Towns and its surroundings, mainly the members of Lame Bora milk producers association and institutional big farm (University Gondar dairy farm).

Questionnaire survey

Semi-structured questionnaires were used to assess the hygienic practices of dairy farms. Around 60 milking personnel and farm attendants related to the selected farms were interviewed. Consequently, hygienic practices employed in the study farms such as house cleaning, udder cleaning, hand washing practices and milking utensils and collecting vessels (buckets) hygiene and other conditions thought to affect the hygienic quality of raw milk were assessed.

Collection of samples and handling procedures

During sampling of raw milk directly from teats, the udder and teats were cleaned and dried before sampling; each teat end was scrubbed gently with cotton swabs moistened with 70% ethyl alcohol. The first 3–4 streams of milk were discarded, and approximately 10 ml of milk was collected into sterile sampling bottles. Each specimen

was labelled and placed in ice box and transported to Veterinary microbiology laboratory, University of Gondar, Faculty of Veterinary Medicine. After arrival at the laboratory, samples were preserved in refrigerator at +4 °C temporarily for 24 h for processing.

Analysis for bacterial load and detection of Staphylococcus aureus

The raw milk samples were assessed for bacteriological quality using the standard plate count. Total bacteria count was carried out by inoculation of serially diluted milk samples on standard plate count agar (Oxoid, England) and mannitol salt agar (Oxoid, England). All the samples positive for presumptive S. aureus contaminations on mannitol salt agar (Oxoid, England) were confirmed using Gram's staining, cultural and biochemical examinations.

Standard plate count

1 ml from each sample of raw milk was transferred to 9 ml sterile distilled water (10%) and thoroughly mixed to give 1:10 dilution. Serial dilutions were made by transferring 1 ml of the previous dilution in 9 ml of sterile distilled water up to 1:10,000 dilutions. Then only 0.1 ml sample from each dilution level was cultured by a glass spread method to the standard plate count agar (Oxoid, England). Total Bacterial Count was made by incubating cultured dilutions of milk samples on Plate Count Agar (Oxoid, England) plates. Colonies were counted after the culture media was incubated at 37 °C for 24 h. Total number of colonies on plates 25 to 250 per plates was selected and colonies were counted (Weldaragay et al. 2012).

Detection of Staphylococcus aureus

Serial dilutions method for total count on plate count agar also followed on Mannitol salt agar (Oxoid, England) for presumptive S. aureus load count. The presumptively identified S. aureus from mannitol salt Agar were subcultured to nutrient agar plate and after 24 h culture colonies of S. aureus was picked by bacteriological loop and placed on clean slide with a small drop of distilled water and emulsified. The test suspension was treated with a drop of rabbit plasma and mixed well with a needle for 5–10 s. Those forming Clumping of cocci were taken as positive (Quinn et al. 2002). Finally, slide coagulase positive samples were cultured on Purple agar base (PAB) (Difco, France) with the addition of 1% maltose and incubated at 37 °C for 24–48 h. The identification was based on the fact that S. aureus rapidly ferment maltose with in 24 h and the acid metabolic products cause the pH indicator (bromocresol purple) to change the medium and colonies to yellow. The rapid fermentation (24 h) was considered as S. aureus isolates (Quinn et al. 2002).

Data management and analysis

The data were entered into excel spread sheet and analysed using a statistical software (SPSS version 20.0). The $Log10$ transformation of bacterial count was done, before the analysis. Percentages were also used to assess the general farm activities and hygienic practices and to express the proportion of bacterial isolation and milk quality grade based on Indian standards, because we don't have an Ethiopian standard and difficult to follow European and USA standards. Analyzing the effects of hygienic practices on bacteriological count of the milk was also performed by linear Regression analysis. The differences in bacterial load between the raw milk directly from teats and milking buckets were compared by Mean [±S.E] $log10$ cfu/ml values between the two collecting points. The results were reported as significant for $p < 0.05$.

Results

Questionnaire survey

Back grounds of the farms

Around 60% of the farms were managed under intensive production systems which have milking cows that ranged from 1 to 28 in number. 8.3% of respondents owned <5 head of cows; 51.70% had 5–10 cattle and 40% of respondents had >10 milking cows. All respondents (100%) milk their cows twice a day -early in the morning and at evening. In the study 48.3% of the farms have separate milking barns and 43.3% have continuous water supply for hygiene of floor and equipments. House cleaning intervals of the farms were variable; 88.3% of them clean more than twice per week (Table 1, Fig. 1).

Small holders management and hygienic practices

The study also showed that 75% of respondents' washes the udder before milking and 68.3% washes their hands before milking. Cleaning was performed in all farms using tap water only, while 28.3% of the respondents use towel to dry udder after washing. The cleaning agents used for milking utensils and collecting tanks (buckets) were only in 98.3% of the respondents was water and soap (Table 1).

Linear regression analysis for hygienic Vs milk quality assessment

The result shows the independent variables Udder cleaning, Water and soap using for cleaning of udder, hand washing and Water and soap using for milking vessels were significantly ($P < 0.05$) affects the bacteriological count of the milk.

Table 1 Small holder's dairy production system, management and milking practices

Hygienic practices frequency of cleaning		Frequency ($N = 60$)	Percentage (in %)
House cleaning interval	Twice a week	7	11.70
	More than twice a week	53	88.30
Udder cleaning	Yes	45	75.00
	No	15	25.00
Cleaning agent for udder	Water	60	100.00
	Water and soap	0	0.00
Towel for drying udder	Yes	17	28.30
	No	43	71.70
Hand washing	Yes	41	68.30
	No	19	31.70
Cleaning agent for hand wash	Water	35	58.30
	Water and soap	25	41.70
Cleaning agent for milking utensils and collecting tankes (Buckets)	Water	1	1.70
	Water and soap	59	98.30

N total number of samples, % percentage
Buckets: used as milking utensils in larger farms where tanks used for collection; used as collecting vessels where small no of cows milked

Bacteriological analysis

Tables 2 and 3 Standard Plate Counts (SPC): Total bacterial count and *S. aureus* isolates count from Raw Milk collected directly from teat and milking buckets.

The mean ± standard error for standard plate counts [expressed in $log 10$ cfu/ml] of raw milk sampled directly from teat during milking and milking buckets at farm level are shown in Table 4. The overall average total bacterial count (TBC) were $4.59 \pm 0.118 log10$ (38,904.51 cfu/ml) and $4.77 \pm 0.23 log10$ (58,884.37 cfu/ml) for milk samples collected directly from teat during milking and milking buckets at farm level respectively. Accordingly, the count increased by 0.18 ± 0.23 $log10$ or 19,979.86 cfu/ml (51.36%) increase from teat to milking buckets. Results showed very significant differences in plate counts ($P < 0.05$) between each milk collection points.

There was significant ($P < 0.05$) milk contamination from direct teat collection to milking buckets (Table 4).

Discussion

Milk is virtually a sterile fluid when secreted into alveoli of udder. However, post-harvest handling like the milking personnel and milk handling containers might generally be source of microbial contamination for raw milk, the three main sources of bacterial contamination; within the udder, exterior to the udder from the surface of teats, milk handling and storage equipments (Abate et al. 2015: Reta et al. 2016).

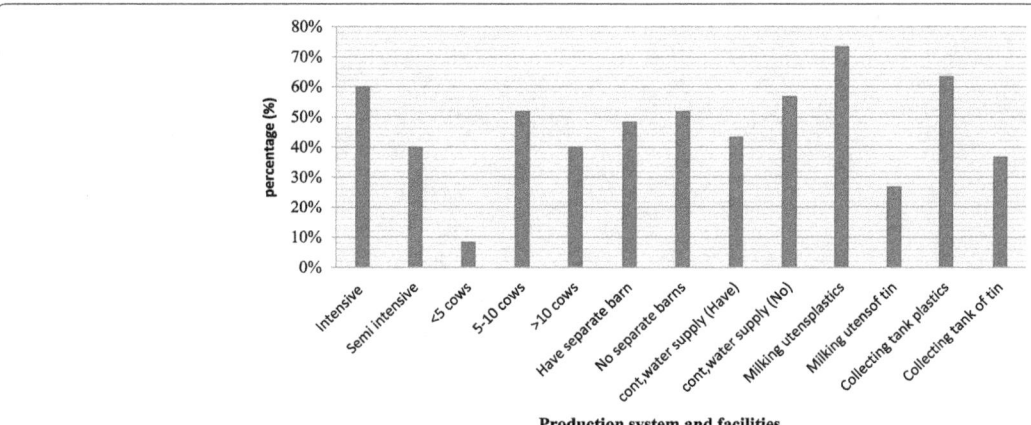

Fig. 1 Dairy production system and facilities of small holders in the study farms. [1]comparable variables: intensive and extensive farming systems; no of milking cows' owned by the farm owners; common and separate barns in the farm; availability of continuous water supply for hygiene of floor and equipments; milking utensils (buckets) and collecting tanks type (plastic or tin). [2]Cont. water supply = farms with continuous water supply; < 5 cows = farms having less than five cows

On average, aseptically drawn milk from healthy udders contains between 500 and 1000 bacteria per ml. But in our study only 5% of individual cows sampled directly from teat had the total bacterial count (TBC) of <1000 cfu/ml which indicates microbiological quality of the raw milk was very poor when compared with Theodore et al. (2016), which reported 95% of the cow milk with TBC <1000 cfu/ml from western Zambia. According to European milk bacteriology standards and USA legal limits for milk collected on the farm level (<100,000 cfu/ml) only 26.70% of the samples can fit this standard. But 73.30% of the milk samples were found high initial counts (>100,000 bacteria per ml), evidence of poor milk hygiene when compared to international standards.

In this study udder milk, had a better bacteriological quality because it was not subjected to further contamination after milking. The milk produced under hygienic conditions from healthy cows should not contain more than 4.7 log10 cfu/ml (O'Connor 1994). The current study revealed mean bacterial counts lower than this standard in which the mean ± standard error (SE) bacterial count was 4.59 ± 0.12 log10 cfu/ml from milk collected directly from teat and 4.77 + 0.23 log10 cfu/ml in

Table 2 Linear Regression analysis for effect of hygienic practices of farms on bacteriological count of the milk

Hygienic practices		Freq. (No = 60)	Percentage	p-value	95.0% CI	
					Lower	Upper
House cleaning interval	Twice a week	7	11.70	0.253	−1.153	0.309
	> twice a week	53	88.30			
Udder cleaning	Yes	45	75.00	0.050	−1.059	0.002
	No	15	25.00			
Cleaning agent for udder	Water	60	100.00	0.050	−1.059	0.002
	Water and soap	0	0.00			
Towel for drying udder	Yes	17	28.30	0.308	−0.791	0.254
	No	43	71.70			
Hand washing	Yes	41	68.30	0.039	−1.010	−0.026
	No	19	31.70			
Cleaning agent for hand wash	Water	35	58.30	0.202	−3.009	0.649
	Water and soap	25	41.70			
Cleaning agent for milking utensils and collecting tanks (Buckets)	Water	1	1.70	0.001	−0.963	−0.250
	Water and soap	59	98.30			

[a]N total number of samples, % percentage
[b]CI confidence interval

Table 3 Standard Plate Counts (SPC): Total bacterial count and *S. aureus* isolates count from Raw Milk collected directly from teat and milking buckets

Sample No.	Bacterial load count (CFU/ml)	S. aureus (CFU/ml)	Sample No.	Bacterial load count (CFU/ml)	S. aureus (CFU/ml)
Raw Milk collected directly from teat			Raw Milk collected directly from teat		
1	3.36×10^4	0	38	1.60×10^5	7.56×10^2
2	2.74×10^4	0	39	2.03×10^5	0
3	2.10×10^4	3.4×10^2	40	4.10×10^4	0
4	2.25×10^2	0	41	1.20×10^4	0
5	6.20×10^3	0	42	8.00×10^3	0
6	6.40×10^4	2.41×10^2	43	5.80×10^3	0
7	2.48×10^5	0	44	2.60×10^4	0
8	7.20×10^4	0	45	9.40×10^5	0
9	2.55×10^6	5.00×10^2	46	2.09×10^6	0
10	2.25×10^5	0	47	4.17×10^5	2.20×10
11	7.04×10^5	2.50×10	48	4.86×10^4	0
12	6.49×10^5	0	49	1.37×10^4	1.75×10^2
13	1.84×10^4	0	50	3.00×10^3	0
14	1.05×10^6	3.45×10^2	51	3.40×10^4	0
15	5.40×10^4	8.6×10	52	7.05×10^4	0
16	4.00×10^3	0	53	4.30×10^4	0
17	2.30×10^4	0	54	4.20×10^3	0
18	6.50×10^3	0	55	8.27×10^4	0
19	2.38×10^6	0	56	4.00×10^5	2.50×10^2
20	4.05×10^4	0	57	2.05×10^4	0
21	9.70×10^4	0	58	3.52×10^3	0
22	5.94×10^4	0	59	2.45×10^3	0
23	2.62×10^5	0	60	5.63×10^4	0
24	3.95×10^5	0	Milking buckets		
25	4.92×10^5	0	61	7.10×10^3	
26	5.38×10^4	0	62	1.17×10^4	
27	2.09×10^4	0	63	2.39×10^5	
28	1.28×10^4	0	64	2.67×10^4	5.25×10^2
29	9.60×10^3	0	65	5.10×10^3	
30	7.20×10^2	0	66	1.46×10^5	
31	2.60×10^3	2.76×10^2	67	5.80×10^4	3.12×10^2
32	1.36×10^4	0	68	4.6×10^4	
33	1.20×10^2	0	69	2.7×10^6	2.3×10^2
34	3.00×10^3	0	70	3.1×10^5	
35	6.22×10^3	0	71	3.08×10^5	
36	8.20×10^3	0	72	6.4×10^4	
37	6.80×10^3	0			

milk collected from milking buckets slightly higher than the standard given. This result is also much lower than the findings of Worku et al. (2012) and Yilma (2012) about 7.59 log10 cfu/ml and 8.87log10 cfu/ml respectively. Even if the means of bacterial counts seem to be lower than the standards given, more than 42% from all samples were found to have greater log10 bacteria counts than the standard stated.

In this study, an increase in the bacterial counts between the two milk collection points which indicates

Table 4 Descriptive statistics for the standard plate count between two points

Milk collection points	N	MIN	Max	Mean[±S.E] log10 cfu/ml	Log10 increment	Df	95% CI for mean		P-value
							lower	upper	
Directly from teat	60	2.079	6.407	4.5907 [0.117]	0.18 [0.23]	1	4.352	4.832	**0.011**
Milking buckets	12	3.707	6.436	4.7706 [0.232]			4.369	5.206	

Total numbers of samples (N); maximum count (MAX) vs. minimum (MIN) log10 cfu/ml plate counts
Mean [±S.E] log10 cfu/ml = log 10 of colony-forming unit (CFU) above or below standard error (±S.E) in one millilitre of milk sample
Df degree of freedom, *I* confidence interval

decreasing of the hygienic conditions between milk collection points. Based on the linear regression analysis which was performed to investigate whether certain identified factors i.e., farmers' hygienic practices contributed to the bacteriological quality of the milk or total bacterial counts from raw milk directly from teats and collecting buckets. Udder cleaning, Water and soap using for cleaning of udder, hand washing and Water and soap using for milking vessels were found to be significantly ($P < 0.05$) affecting the standard plate counts. This is in agreement with the study up on the hygiene measures on raw milk by Abdalla and Elhagaz (2011) in Khartoum state, Sudan who showed that there was a significant effect on application of hygiene practices prior to milking in total bacterial count. Generally, this implies that the sanitary conditions in which milk has been produced and handled are substandard subjecting the product to microbial contamination and multiplication due to lack of and improper cooling systems at milk vending area. It is indicated that total bacterial count is a good indicator for monitoring the sanitary conditions practiced during production, collection, and handling of raw milk (Fatine et al. 2012).

In developing countries like Ethiopia, where high prevalence of clinical and subclinical mastitis mainly caused by *S. aureus* (Sori 2011) and high consumption of raw milk is common, Staphylococcal Food Poisoning (SFP) the most important target for study as risk of milk borne contaminations. Now days, it is not uncommon to here an extensive outbreak of staphylococcal food poisoning reports from both developed and developing nations from raw milk, powdered skim milk and reconstituted milks (Asao et al. 2003; Ikeda et al. 2005 and Johler et al. 2015). But in Ethiopia SFP outbreak investigations, identification of the causative strain are challenging and scarce data/or information available to estimate its magnitude may be due to limited commercial kits available for diagnosis of causative strains and of enterotoxins (SEs) and week disease outbreak investigation capacity.

Out of 60 samples of raw milk collected directly from teat and 12 collected from milking buckets, 18.33 and 25% were contaminated by *Staphylococcus aureus* respectively, with averages varying between 2.20×10 to 7.56×10^2 cfu/mL, as shown in Table 3. The result is higher than the figure studied by Worku et al. (2012), which was only 7. 29%.

Other lower results were also reported by Shunda et al. (2013) in which about 13.3% of samples were positive for *S. aureus*. According to Wallace (2009), even if the presence *S. aureus* in milk is known to cause spoilage of raw milk, it is not thought to be a frequent contributor to total collecting buckets counts and also he found that this organism is mainly associated with contagious mastitis.

Equipment used for milking, collecting and storage determine the quality of milk and milk products. The use of plastic, tins and traditional containers (clay pots and Bottle gourd) are the dominants in most part of Ethiopia which can be a potential source for the contamination of milk by bacteria, because these allow the multiplication of bacteria on milk contact surfaces during the milking process and their difficult nature for cleaning is also very crucial for contamination of milk. The result in this study also confirmed that 73.30% farmers use plastic containers for milking and collecting milk which can be compared with findings of Abate et al. (2015) which showed over 60% of farms used plastic containers and 40% used pots for milking and collecting milk. Higher figures were also reported by Yilma (2012) in which 81% use plastics the remaining 3.4 and 6.6% used tins and pots respectively.

Maintaining the sanitary condition of milking area is important for the production of good quality milk. The current study showed that about 88.3% of the farms clean the house more than twice per week usually on daily bases but 11.7% of the farms clean the barn twice per week due to shortage of water. Other study Yilma (2012) in Addis Ababa, reported that about 87% of the respondents cleaned their barn on daily basis, while few (9%) of them cleaned only once or twice a week. Contrary to this study Abebe et al. (2012) showed low proportion (47%) of the respondents cleaned the barn three times a week, while 39% cleaned two times and only 11.7% of them reported to clean daily Abate et al. (2015) also report more than 90% farms cleaned their houses once daily.

The study also shows 75% of respondents did not use udder washing before milking and only about 25% of respondents had washed the udder before milking reports. Contrary to the current findings of Weldaragay et al. (2012) in Hawasa reported that >80% households practicing pre milking udder washing (FSA [Food Standards Agency] 2006) reported cleaning of the udder before milking is important to remove both visible dirt and bacteria from the outer surface of the udder and to minimize contamination and produce good quality milk. Cleaning agent used for cleaning the udder in this study was only water with no any detergents. This result has an agreement with the study in Shashemene by Gemechu et al.(2014) in which most of the farms didn't use detergents for cleaning udder but only 2% reported by (Abate et al. 2015).

In this study about 71.1% of farms participated in the survey didn't use separate towels almost similar to the figures (71.79 and 71.0%) found by Gemechu et al. (2014). Hand washing practice before milking of cows in the current study is assessed to be about 68.3% which is not satisfactory with respect to keep the quality of milk. This result is lower than the reports in Jimma (>94%) by Yilma (2012). Most of (98.3%) of the dairy cow owners used water and detergent for cleaning milk handling equipment which is in agreement with the reports of Weldaragay et al. (2012).

Conclusions

The hygienic conditions of the farms studied in Gondar town can be judged as poor, in which most of the farm hygienic practices and parameters like hygienic condition of the milking environment, sanitation of the milk containers, udder and teats cleaning, use of separate towel for each cow and the personal hygiene of the milkers were not fully performed by most of the farm owners. Even if the mean bacterial counts seem to be fair according to the standard, the high proportion of sample having total bacterial counts higher than the USA maximum legal limit ($>1.00\times10^5$cfu/ml) indicate that the quality of milk produced in the study area had unacceptable levels of contamination with microorganisms that profoundly increase across the milk collection points. This risk assessment study with similar different studies in different regions in Ethiopia might provide a foundation for the establishment of national milk quality standards that currently do not exist in Ethiopia. For more, milk safety increment small scale pasteurization with continuous hygienic education for the farmers should be focused.

Acknowledgements

The authors gratefully acknowledge University of Gondar, faculty of veterinary medicine for their assistance in completing this study and to the farmers, for their invaluable input without which this work would not have been completed.

Authors' contributions

ST carried out the conception of the research concept and designed the methodology, data analysis and interpretation and preparation of the manuscript for publication. BT carried out the laboratory work, sample collection and revision of the manuscript. Both authors read and approved the final manuscript.

Competing interests

The authors declare that there is no financial or non-financial competing interest from anybody or institute. We also want to assure that we did not receive any technical assistant in developing the research concept or preparation of the manuscript.

Author details

[1]Wollo University, School of Veterinary Medicine, Dessie, Ethiopia. [2]Faculty of Veterinary Medicine, Department of Para-Clinical Studies (Veterinary Microbiology), University of Gondar, Gondar, Ethiopia.

References

Abate M, Wolde T, Nigussie A. Bacteriological quality and safety of raw cow's milk in and around Jigjiga City of Somali region, Eastern Ethiopia. Int J Res Stud Biosci. 2015;3:48–55.

Abdalla MOM, Elhagaz FMM. The impact of applying some hygiene practices on raw milk quality in Khartoum state, Sudan. Res J Agric Biol Sci. 2011;7:169–73.

Abebe B, Yilma Z, Ajebu N. Hygienic and microbial quality of raw whole cow's milk produced in Ezha district of the Gurage zone, Southern Ethiopia. Wudpecker J Agric Res. 2012;1:459–65.

Abera M, Demie B, Aragaw K, Ragassa F, Ragassa A. Isolation and identification of Staphylococcus aureus from bovine mastitic milk and their drug resistance pattern in Adama town, Ethiopia. J Vet Med Animal Health. 2010;2:29–34.

Asao T, Kumeda Y, Kawai T, Shibata T, Oda H, Haruki K, Nakazawa H, Kozaki S. An extensive outbreak of staphylococcal food poisoning due to low-fat milk in Japan: estimation of enterotoxin A in incriminated milk and powdered skim milk. Epidemiol Infect. 2003;130(1):33–40.

Bennett RW, Hait JM. BAM. Staphylococcal Enterotoxins. 2011. USA Food and Drug Adminstaration, FDA, 10903 New Hampshire Avenue, Silver Spring, MD 20993, 1-888-INFO-FDA (1-888-463-6332) . Avalaible: http://www.fda.gov/Food/FoodScienceResearch/LaboratoryMethods/ucm073674.htm .

CDC. Surveillance for food-borne diseases outbreaks Uinted States. 2016. p. 609–15.

CSA. Federal Democratic Republic of Ethiopia Central Statistical Authority. Livestock and livestock characterization. Ethiopia: Ethiopian central Statistical Authority; 2010. p. 107.

Duguma B and Janssens G.P.J. Smallholder Milk Processing and Marketing Characteristics at Urban Dairy Farms in Jimma Town of Oromia Regional State, Ethiopia. Global Veterinaria, 2014; 13 (3): 285–292.

Enquebaher T, Skeie S, Rudi K, Skjerdal T & Narvhus JA (2015) Staphylococcus aureus and Other Staphylococcus Species In Milk and Milk Products From Tigray Region, Northern Ethiopia., pp. 567–576.

European Food Safety Authority. The community summary report on trend and sources of zoonoses and zoonotic agents and food-borne outbreaks in the European Union in 2008. 2010. European Food Safety Authority (EFSA), Parma, Italy Journal. 2010;8(1):1496. [410 pp.]. doi: 10.2903/j.efsa.2010.1496 . Available from: https://www.efsa.europa.eu/en/press/news/zoonoses100128 .

Fatine H, Abdelmoula E, Doha B, Hinde H. Bacterial quality of informally marketed raw milk in Kenitra City, Morocco. Pak J Nutr. 2012;11(8):662–9.

Felleke G. Milk and dairy products, post-harvest losses and food safety in Sub Saharan Africa and the near east. In: FAO prevention of food losses programme. Rome: FAO; 2003.

FSA [Food Standards Agency]. Milk hygiene on the dairy farm.practical guides for milk producers. England: Food Standards Agency; 2006. Avaliable at: http://adlib.everysite.co.uk/adlib/defra/content.aspx?id=1QQUSGMWSS.0KG7VC4PSI9MAI.

Gebrewold A, Alemayehu M, Demeke S, Dediye S, Tadesse A. Status of dairy research in Ethiopia. In: Small holder dairy development. Addis Ababa: Ministry of Agriculture (MOA); 2000.

Gemechu T, Beyene F, Eshetu M. Handling practices and microbial quality of Raw Cow's milk produced and marketed in Shashemene Town, Southern Ethiopia. Int J Agric Soil Sci. 2014;2:153–62.

Godefay B, Molla B. Bacteriological quality of raw cow's milk from four dairy farms and a milk collection centre in and around Addis Ababa. Berl Munch Tierarztl Wschr. 2000;113:1–3.

Hennekinne JA, Byser MLD, Dragacci S. Staphylococcus aureus and its food poisoning toxins: characterization and outbreak investigation. FEMS Microbiol Rev. 2012;36: 815–836. doi: 10.1111/j.1574-6976.2011.00311.x.

Ibrahim H, Olaluku E. Improving cattle for milk, meat and traction. Nairobi: International Livestock Research Institute (ILRI); 2000. p. 135.

Johler, Weder, Bridy, Huguenin, Robert, Hummerjohann, Stephan. Outbreak of staphylococcal food poisoning among children and staff at a Swiss boarding school due to soft cheese made from raw milk. J Dairy Sci. 2015;98(5):2944–8. doi:10.3168/jds.2014-9123.

Lunder T, Brehne E. Factors in the farm production affecting bacterial contents in raw milk. In: Symposium on bacteriological quality raw milk passing Austria. 1996. p. 13–5.

Makita K, Desissa F, Teklu A, Zewde G, Grace D. Risk assessment of staphylococcal poisoning due to consumption of informally-marketed milk and home-made yoghurt in Debre Zeit, Ethiopia. Intl J Food Microbiol. 2012;153:135–45.

O'Connor CB. Rural dairy technology, ILRI training manual 1. Addis Ababa: ILRI (International Livestock Research Institute); 1994. p. 133.

Oliver SP, Boor KJ, Murphy SC, Murinda SE. Food safety hazards associated with consumption of raw milk. Foodborne Pathog Dis. 2009;6:793–806.

Quinn PJ, Markey BK, Carter ME, Donnelly WJC, Leonard FC, Maghir D. Veterinary Microbiology and Microbial Disease. Blackwell Science Ltd., London, 2002; Pp 191–208.

Reda T. Milk processing and marketing options for rural small scale producers. In: National conference of the Ethiopian society of animal production. Addis Ababa: ESAP; 1998. p. 61–7.

Reta MA, Bereda TW, Alemu AN. Bacterial contaminations of raw cow's milk consumed at Jigjiga City of Somali Regional State, Eastern Ethiopia. Int J Food Contam. 2016;3(1):1–9. doi:10.1186/s40550-016-0027-5.

Seo KS, Bohach GA. Staphylococcus aureus. In: Food microbiology fundamentals and frontiers. Washington, DC: ASM Press; 2007. p. 493–518.

Shunda D, Habtamu T, Endale B. Assessment of bacteriological quality of raw cow milk at different critical points in Mekelle, Ethiopia. Int J Livest Res. 2013;3:42–8.

Sori. Prevalence and Susceptibility assay of Staphylococcus aureus isolated from Bovine Mastitis in Dairy Farms of Jimma Town. South West Ethiopia. Journal of Animal and Veterinary Advances. 2011;10(6):745–749. doi: 10.3923/javaa. 2011.745.749.

Tetsuya I, Tamate N, Yamaguchi K, Makino S. Mass outbreak of food poisoning disease caused by small amounts of staphylococcal enterotoxins a and H. Appl Environ Microbiol. 2005;71(5):2793–5.

Theodore JD, Knight-Jones, et al. Microbial contamination and hygiene of fresh cow's milk produced by Smallholders in Western Zambia. 2016. p. 737.

Tsehay R. Small-scale milk marketing and processing in Ethiopia. In: Workshop on smallholder dairy production and marketing constraints and opportunities. 2001.

Wafula WN, Matofari WJ, Nduko MJ, et al. Effectiveness of the sanitation regimes used by dairy actors to control microbial contamination of plastic jerry cans' surfaces. Int J Food Contam. 2016;3:9. doi:10.1186/s40550-016-0032-8.

Wallace RL. Bacteria count in raw milk. 2009. http://www.livestocktrail.illinois.edu/ uploads/dairynet/papers/ Bacteria%20Counts%20in%20Raw%20Milk%20DD%202008.pdf.

Wang S, Duan H, Zhang W, Li JW. Analysis of bacterial foodborne disease outbreaks in China between 1994 and 2005. FEMS Immunol Med Microbiol. 2007;51:8–13.

Weldaragay H, Yilma Z, Tekle-Giorgis Y. Hygienic practices and microbiological quality of raw milk produced under different farm size in Hawassa, southern Ethiopia. Wudpecker J Agric Res Rev. 2012;4:132–42.

Worku T, Negera E, Nurfeta A, Welearegay H. Microbiological quality and safety of raw milk collected from Borana pastoral community, Oromia Regional State. Afr J Food Sci Technol. 2012;3:213–22.

Yilma Z. Microbial Properties of Ethiopian Marketed Milk and Milk Products and Associated Critical Points of Contamination: An Epidemiological Perspective. In: Epidemiology Insights. Edited by D. L. Dr.Maria, D. Ribeiro, Sauza D.C. (Ed.). 2012. ISBN:978-953-51-0565-InTech. Availablefrom: http://www.intechopen. com/books/epidemiology-insights/microbial-properties-of-marketed-milk-and-ethiopian-fermented-milk-products-and-associated-critica.

Yitaye AA, Azage T, Zollitsch W. Performance and limitation of two dairy production System in the North western Ethiopian highlands. Trop Anim Health Prod. 2009;41:1143–50.

10

Effects of per-household processes on the levels of chlorpyrifos residues in lettuce (*Lactuca sativa*)

Osei Akoto[1*], Fredrick Addai-Mensah[2] and Eric K. K. Abavare[3]

Abstract

Background: The study was organized to evaluate residue levels of chlorpyrifos on *Lactuca sativa* that are likely to have accumulated on crop during cultivation and also examine the effect some pre-household treatment procedure on residue levels on the crop grown in a tropical and humid environment. Chlorpyrifos residue was extracted using acetonitrile. Subsequent detection and a quantification were done using GC with PFPD. Concentrations of Chlorpyrifos applied at different stages of growth of lettuce were examined at different time intervals of 1 h, 24 h and 7 days after pesticide application.

Results: The results showed that residue levels detected at 1 h and 24 h after application were all above the MRL of 0.05 $mgkg^{-1}$ and can pose health risk to consumers while that recorded 7 days after application were far below the MRL. Accumulation of chlorpyrifos on crop during cultivation was not observed since no significant differences were observed 7 days after application at all the different stages. There is therefore the need for farmers to allowed 7 days' re-entry intervals before harvesting. The results also showed that, all the pre-household treatment procedures caused significant reduction in residues levels of chlorpyrifos on the crop.

Conclusions: Mild detergent treatment was however more effective compared with the other treatments. Hence to reduce the risk associated with intake of chlorpyrifos through lettuce, mild detergent washing procedures should be followed before consumption.

Keywords: Chlorpyrifos, Consumers, Health risk, Removal, Treatments

Background

Consumption of diets high in fresh vegetables by Ghanaians is growing. This is because vegetables are universally endorsed as healthy. Thus, they supply high amount of vitamins, antioxidants and dietary fiber which is linked to lower incidence of cardiovascular disease and obesity (Slavin and Lloyd, 2012). Lettuce (*Lactuca sativa*) is an important vegetables cultivated and consumed by both urban and rural dwellers in Ghana. Because of this increase in consumption, lettuce is now cultivated all year round in large quantities. However, pests and diseases militate against the successful cultivation of the crop. As a result, pesticides are use in the production of the crop (Akoto et al., 2013; Gerken et al., 2013).

Due to the inability of farmers to apply appropriately the prescribed dosage of the pesticides at every stage of the crop production, and unaware of their possible health effect on humans and the environment, large quantities of pesticides are used (Obeng-Ofori et al., 2002). Again the high demand for vegetables, causes farmers to harvest their produce without taking into account the pesticide withholding period (Amoah et al., 2006; Bhanti and Taneja, 2007). The increased and inappropriate use of pesticides in vegetable production in Ghana has led to an increase of residues on the crop above the MRL (Akoto et al. 2015; Obeng-Ofori et al., 2002). In Ghana, pesticide residues including that of chlorpyrifos have been found at concentrations above the acceptable limits in lettuce (Amoah et al., 2006), shallots (Kotey et al., 2008) and tomatoes (Essumang et al., 2008) that have not been subjected to any household treatment procedures.

Chlorpyrifos (O,O-diethyl O-3,5,6-trichloro-2-pyridyl phosphorothioate) a contact organophosphate insecticide

* Correspondence: wofakmann@yahoo.com
[1]Department of Chemistry, Kwame Nkrumah University of Science and Technology, Kumasi, Ghana
Full list of author information is available at the end of the article

is intensively used by farmers as a protective measure against pests and diseases on lettuce (Clark et al., 1997). Chlorpyrifos is toxic to humans and can be absorbed through all routes of exposure. Signs of severe toxicity include increased heart rate, unconsciousness, convulsions, respiratory depression, and paralysis (Reigart and Roberts, 1999; Thompson and Richardson, 2004).

Studies have shown that certain types of post-harvest treatments or pre-household preparations such as washing, drying and boiling at different temperatures may help to reduce pesticide residues (Kim et al., 2015; Dhiman et al., 2006; Krol et al., 2000). The effects of these household treatments on residue levels are extremely important in evaluating the risk associated with ingestion of pesticides residues through vegetable consumption. But information on household treatment procedure on chlorpyrifos residue in vegetables such as lettuce is very scarce. The objectives of the study were to measure chlorpyrifos residue levels in lettuce and examine the effect of pre-household treatment procedures on the residues.

Methods
Sampling
Pesticides were applied at three different stages during growth. At stage 1, pesticides were applied week 1 after the seedlings transplant, at stage 2 pesticides application was done week 3 after transplant of seedlings and at stage 3 application was done week 5 after the seedlings have been transplanted. Sampling of lettuce for residue analysis was done at three different stages during growth (stage 1, stage 2 and stage 3). At each stage, fresh lecture samples were collected at different time intervals thus 1 h, 12 h and 7 days after pesticide application.

The samples were labeled, wrapped in aluminum foil and transported to the laboratory immediately after collection in ice box. Samples were then cut into small pieces, mixed thoroughly and 100 g sub-samples weighed for each pre-household treatment procedure. Samples were subjected to three (3) pre-household treatment procedures, thus: washing under running tap water for 2 min (Treatment 2), dipping in 500 mL 2 % salt solution at room temperature for 2 min (Treatment 3) and dipping in 500 mL 1 % detergent solution (v/v) at room temperature for 2 min (Treatment 4). The treated samples were left for 1 h in colander to drain at room temperature (Kim et al., 2015). The forth sample was analysed without any pre-household treatment and were considered as the control samples (Treatment 1).

Extraction and GC analysis of pesticide residue
Two (2) grams of homogenized sample was mixed with 5 mL of acetonitrile in a separating funnel. The mixture was shaken with for one hour on a rotatory shaker. After shaking, the separating funnel was allowed to stand to settle and the acetonitrile layer was decanted into a round bottom

flask. This procedure was repeated twice and the extracts combined. The combined acetonitrile phase was centrifuge for 2 min at 3000 rpm and the supernatant transferred in a graduated vial. The extracts were concentrated to about 2 mL using a rotary evaporator operating at temperature of 30 C at a reduced pressure. The extract was made up to 2 mL with acetonitrile before GC injection.

Levels of chlorpyrifos residue were analyzed using Agilent Tech. 6890 N GC equip with pulse flame photometric detector (PFPD), coupled with Chemito 5000 data processor with a HP-5 capillary column (30 m x 0.32 mm id.) of 0.25 μm film thickness. The GC analysis was performed under the following condition. Oven temperature was 210 °C, Injector temperature was 230 °C, column temperature was 160 °C and detector temperature was 300 °C. Carrier gas was nitrogen at a flow rate of 2 mL min^{-1}.

Quality control measures
Recovery analyses were carried out on samples fortified at 0.001 mg kg^{-1} by adding standard chlorpyrifos solution. The recovery values were calculated from calibration curves constructed from the concentration and peak area of the chromatograms obtained with standards of chlorpyrifos. The recovery of the pesticide was in the range between 80 and 110 %. Blank analyses were also performed in order to check interference from the sample. For quality control of the gas chromatographic conditions, a checkout procedure was performed before sample analysis.

Results and discussion
Physical removal of pesticide residue from a crop is a way of reducing the concentration of the pesticide below acceptable levels. Pesticide residues below acceptable levels implies that the crops are safer for consumption by humans and that there is no expected health implication associated with the consumption of the product. Pre-household treatment procedures are used commercially and in the home to reduce pesticide residues concentrations to levels below the MRL value. Pre-household treatment procedures used in this study were washing under running tap water (Treatment 2), dipping in salt water for 2 min (Treatment 3) and dipping in 1 % detergent solution for 2 min (Treatment 4).

Mean concentrations of chlorpyrifos residues detected in the lettuce samples after subjecting them to the various household treatment procedures and that of the control at all the three (3) different stages of growth are presented in Table 1. Residue levels detected in the control samples (Treatment 1), 1 h after treatment with chlorpyrifos were 0.059 ± 0.008, 0.055 ± 0.0014 and 0.055 ± 0.009 mgkg^{-1} for stage 1, stage 2 and stage 3 respectively. The detected mean residues in the lettuce samples at all

Table 1 Means and standard deviations of chlorpyrifos residue levels (mg kg^{-1}) in the treated and control lettuce samples at different stages of growth, $n = 6$

Treatments	Stage 1			Stage 2			Stage 3		
	Sample 1A	Sample 2A	Sample 3A	Sample 1B	Sample 2B	Sample 3B	Sample 1C	Sample 2C	Sample 3C
Control/Treatment 1	0.059 ± 0.008	0.052 ± 0.001	0.006 ± 0.002	0.055 ± 0.0014	0.053 ± 0.012	0.007 ± 0.002	0.055 ± 0.009	0.034 ± 0.003	0.002 ± 0.001
Treatment 2	0.049 ± 0.009	0.048 ± 0.004	0.006 ± 0.001	0.052 ± 0.004	0.046 ± 0.007	0.005 ± 0.0028	0.053 ± 0.0023	0.03 ± 0.006	0.002 ± 0.0014
Treatment 3	0.039 ± 0.006	0.023 ± 0.004	0.004 ± 0.001	0.041 ± 0.008	0.035 ± 0.004	0.004 ± 0.0014	0.043 ± 0.0014	0.025 ± 0.004	0.002 ± 0.0004
Treatment 4	0.018 ± 0.006	0.01 ± 0.004	0.003 ± 0.001	0.028 ± 0.009	0.027 ± 0.004	0.004 ± 0.002	0.027 ± 0.003	0.019 ± 0.008	0.001 ± 0.001

the 3 stages of growth were above the Maximum Residue Level (MRL) of 0.05 mgkg^{-1} (FAO/WHO, 2004). The mean concentrations of chlorpyrifos in the control samples, 24 h after application at all the stages of growth were also very high and were above the MRL as presented in Table 1. This is in agreement with the observation made by Amoah et al., 2006, where residues of chlorpyrifos in lettuce were observed to be above the MRL value. The levels of chlorpyrifos at all the different stages of growth were drastically reduced to mean concentrations of 0.006 ± 0.002, 0.007 ± 0.002 and 0.002 ± 0.00 mgkg^{-1} for stage 1, stage 2 and stage 3 respectively. These reductions in chlorpyrifos levels were observed 7 days after application. These mean levels were far below the MRL of chlorpyrifos in lettuce.

Levels of chlorpyrifos in the lettuce samples at all the different stages of growth after tap water treatment (T_1) are presented in Table 1. The mean concentrations of chlorpyrifos in samples from this treatment, 1 h after application were 0.049 ± 0.009, 0.052 ± 0.004 and 0.053 ± 0.0028 mgkg^{-1} for stage 1, stage 2, and stage 3 respectively. All the values were lower than the MRL value. The differences in the concentrations of chlorpyrifos in the control (Treatment 1) and Treatment 2 samples, 1 h after application were not significant ($P > 0.05$) at all the different stages of growth. Chlorpyrifos residues detected in Treatment 2 samples, 24 h after application were 0.048 ± 0.004 mgkg^{-1} at growth stage 1, 0.04 ± 0.007 mgkg^{-1} at growth stage 2 and 0.03 ± 0.006 mgkg^{-1} at growth stage 3 (Table 1). The differences between the control samples and the Treatment 2 samples 24 h after pesticides application were not significant at all the 3 stages of growth. Again no statistical differences were observed among the Treatment 2 samples 24 h after application at the different growth stages. Residues of chlorpyrifos, 7 days after application followed by Treatment 2 at the different stages of growth were 0.006 ± 0.001, 0.005 ± 0.0028 and 0.002 ± 0.0014 for stage 1, stage 2 and stage 3 respectively. Differences between Treatment 2 and control samples, 7 days after application were significant ($p < 0.05$) but no significant variations were observed between the treated (Treatment 2) samples at the different growth stages. It was noted that washing with tap water reduced chlorpyrifos levels to 4 - 16 % compared to those

in the untreated (control samples). This finding doesn't agree with that by Ling et al. (2011), who reported a decrease rate of 0.23 % of chlorpyrifos in lettuce. This difference may be due to the types of adjuvants added during formulation. Some formulation with binders held pesticides firmly onto the applied surface. Other studies have reported decreased rate of 3.65, 10.6, 36.3 and 46.6 % of chlorpyrifos residues in galic sprouts, cucumber, eggplant and tomatoes respectively after washing with tap water (Kim et al., 2015).

The effectiveness of washing as a household treatment procedure to remove pesticide residues on the surface of a vegetable varies but depends on the physicochemical properties of the compound, age of residue, surface area of leave and nature and thickness of the cuticle. (Lopez-Fern et al., 2013; Ling et al., 2011; Kumari, 2008). Chlorpyrifos is a non-systemic insecticide and therefore would be located on the surface of the leaves after its application. The loosely bound surface residues of chlorpyrifos were therefore removed by washing under tap water. However, 22.2 % of the samples that were collected 1 h after chlorpyrifos application and subjected to Treatment 2 had residue levels above the MRL of 0.05 mgkg^{-1} (FAO/WHO, 2004) for chlorpyrifos on lettuce. Treatment 2 was ineffective in removing chlorpyrifos from the surface of the lettuce. The variations in the levels of chlorpyrifos on the lettuce samples after Treatment 2 were not significant at 5 % significance level from that of the control treatment (Treatment 1). This may be due to the fact that chlorpyrifos is not soluble in water and again the surface of the lettuce leaves are lined with thing layer of cuticle and therefore making the binding of the compound to the surface stronger.

Mean and the standard deviation of chlorpyrifos residue levels detected in lettuce after salt water washing (Treatment 3) for all the different stages of growth are presented in Table 1. The mean residue levels observed for samples treated with salt water wash during the first stage of growth were 0.039 ± 0.006, 0.023 ± 0.004 and 0.004 ± 0.001 mgkg^{-1} for samples that were taken at 1 h, 24 h, and 7 days after pesticide application respectively. During the second stage of growth, the residue levels at the different sampling time after Treatment 3 were 0.049 ± 0.008, 0.03 ± 0.004 and 0.004 ± 0.0014 mgkg^{-1} for

1 h 24 h and 7 days after pesticide application respectively. Chlorpyrifos residues detected during the third stage of growth of lettuce were 0.043 ± 0.001, 0.025 ± 0.004 and 0.002 ± 0.0004 mgkg⁻¹ for samples that were collected at 1 h, 24, h and 7 days after pesticide application respectively. Levels of chlorpyrifos detected in all the samples after washing with salt solution for 2 min were below the MRL. Relatively lower levels of chlorpyrifos were detected in the samples that were taken at 1 h after application and treated with salt solution (Treatment 3) when compared with the control samples (Treatment 1) at all the different growth stages. The percentage reduction of chlorpyrofos was found to be in a range of 35-43 % in the samples that were collected 1 h after application and treated with salt solution (Treatment 3) at the different growth stages.

Chlorpyrifos residue detected in lettuce after detergent washing (Treatment 4) are presented in Table 1. The detected residue levels observed during the first stage of growth of lettuce after treatment T₃ was 0.015 ± 0.004, 0.01 ± 0.004 and 0.003 ± 0.0008 mgkg⁻¹ for the samples that were collected at 1 h, 24 h and 7 days after pesticide application respectively. During the second stage of growth of lettuce, the detected residue levels observed were 0.048 ± 0.0098, 0.027 ± 0.004 and 0.004 ± 0.0019 mgkg⁻¹ for samples that were taken at 1 h 24 h and 7 days after pesticide application respectively. The detected residue levels on lettuce that was observed during the third stage of the growth of lettuce were 0.048 ± 0.0056, 0.019 ± 0.0098 and 0.001 ± 0.00028 mgkg⁻¹ for samples that were collected at 1 h, 24 h and 7 days after pesticide application respectively.

Residue detected on foliage were high in the control samples that were collected 1 h and 24 h after pesticide application at all the different stages of growth, but the detected levels were very low and even below detection limit in some of the samples that were analysed 7 days after application.

This insecticide had direct contact with the leaves and remained on the leaves after application. This was evident from the results since the highest recorded residue level at all the different stages of growth were recorded 1 h after application (Table 1). A gradual and continuous deterioration of the chlorpyrifos residues on the lettuce leaves were observed as a function of time (days) after pesticide application. The decrease in the residue concentration of chlorpyrifos on foliage over time may be due to factors such as volatilization and/or photodegradation. Chlorpyrifos adsorbed on leaf surfaces are usually lost through volatilization especially in hot climates (Roberts and Hutson, 1999). Volatilization may have been a factor to the reduced levels of the chlorpyrifos on the lettuce since this study was carried on a hot climatic region where temperatures are high. Temperature influenced pesticide

dissipation on plant through volatilization and photodegradation. Obviously, a higher temperature tends to favour volatilization and photodegradation of pesticides from plants, because the vapour pressure of the pesticide compound is temperature-dependent and additionally the extent of adsorption to the leaf surface decreases with increasing temperature.

According to Smith (1968), chlorpyrifos when exposed to sunlight, undergoes hydrolysis in the presence of water to liberate 3, 5, 6-trichloro-2-pyridinol (TCP), which undergo further decomposition to diols and triols and ultimately cleavage of the ring to fragmentary products. Therefore, photodegradation may be a factor to cause reduction in residue concentration of chlorpyrifos since the crops were watered regularly after pesticides application.

Dissipation of chlorpyrifos took place causing a reduction in the detected residue levels that was observed 1 h after pesticide application. Thus after 1 h of application of chlorpyrifos, the mean detected residue levels were 0.059 ± 0.008, 0.055 ± 0.001 and 0.055 ± 0.009 mgkg⁻¹ for the first, second and third growth stages respectively in the control samples. After 24 h, the mean residue levels reduced to 0.052 ± 0.001, 0.053 ± 0.012 and 0.034 ± 0.003 mgkg⁻¹ for the first, second and third growth stages respectively. There was a mean percentage reduction of 17.9 % of initial residue levels detected 1 h after pesticide application. After 7 days of pesticide application, the detected residue levels further reduced to 0.006 ± 0.002, 0.007 ± 0.002 and 0.002 ± 0.00 mgkg⁻¹ for the first, second and third growth stages respectively. There was a mean percentage reduction of 91.1 % of initial residue levels detected 1 h after pesticide application. This shows that if farmers will allow 7 days' intervals after application of chlorpyrifos on lettuce before harvest the levels of residues will be reduced far below the MRL if the pesticide is properly applied at a hot climate region.

Stage of growth at which pesticide is applied and frequency of application during growth are important factor that affect residue levels and dissipation rate on a crop. In this work relatively higher concentration of chlorpyrofos residues were observer on stage 1 samples than stage 2 and stage 3 samples that were collected at 1 h after application. This may be due to high density of applied pesticide on the foliage which have small surface area. But at the second and third stages the leaves sizes were larger and have large surface area therefore the applied pesticides spread to cover the whole surface thereby reducing the concentration of the pesticide. The rete of pesticide degradation was fast with the second and the third stages of the growth than the first stage. This was because the foliage sizes at the second and the third stages were larger than the first stage this provided large surface areas for volatalisation. Mean percentage reduction of 38.1 % and

96.3 % were observed 24 h and 7 days after application respectively on the control samples.

Results from this study show that, detergent is more effective in reducing chlorpyrifos levels on lettuce than washing with tap water and washing with salt solution. Dipping lettuce in 1 % detergent (Treatment 4) reduced the residues by a mean percentage of 54.4 %, whiles salt water washing caused a mean percentage reduction of 30.8 %. It was 3.7 times more effective than tap water wash and 1.6 times more effective than salt water washing. The levels of chlorpyrifos in all the samples were far below MRL set for chlorpyrifos on lettuce (FAO/WHO, 2004) after Treatment 4.

According to Hui et al., 2010, the rate of hydrolysis of chlorpyrifos increases with pH and also the stability of chlorpyrifos decreases as the pH increases. Smith, 1968 also stated that, hydrolysis of chlorpyrifos occurs readily at pH > 7. Detergent which was used in Treatment 4 caused an increase in the pH of aqueous solution. This resulted in destabilizing chlorpyrifos residue levels on the lettuce resulting in higher removal of chlorpyrifos. For instance, the residue level detected 1 h after pesticide application for the control samples was 0.059 ± 0.008 mgkg^{-1} but after dipping in mild detergent followed by thorough washing under tap water there was a reduction in the levels to 0.018 ± 0.006 mg kg^{-1}.

The detergent used in this work was anionic detergent formulated for household dishwashing. This detergent is very soluble in water and therefore the treated lettuce can be wash under running water for 5 min to remove any traces of the detergent on the vegetable. And if there is any detergent residue, it may be too miniscule to cause any harm to the health of consumers (Swisher, 1987). This is because the acute toxicity of anionic surfactant which is used in the formulation of household dishwasher in animals is low after skin contact or oral intake (Madsen et al., 2001).

Conclusion

This study was carried out to determine the extent of removal of chlorpyrifos residues from lettuce through household treatment process such as washing with tap water, washing with salt solution and washing with 1 % detergent solution. And also measure extent of chrloripyrifos accumulation during the cultivation of lettuce. Residue levels of chlorpyrifos were above the Maximum Residue Levels on vegetables that were analysed 1 h after pesticide application. These levels were reduced by 17.9 % and 91. 1 % at 24 h and 7 days after application respectively. Despite the loss of some of the chlorpyrifos applied onto the lettuce 24 h after application, residues found on the foliage were high above the MRL in most of the samples. This could pose risk if lettuce is consumed as fresh vegetables and not properly wash. Therefore, it is

important to allow at least 1-week (7 days) withholding period. After such a period, the residue concentrations were found to be below the MRL in samples where the pesticide was detected.

A comparison of the effects of different pre-household treatment procedures on the levels of chlorpyrifos in the lettuce samples, indicated that levels of chlorpyrifos residues were reduced significantly by washing with salt solution and 1 % detergent solution. Tap water washing did not show any significant effect on the chlorpyrifos levels on the foliage of lettuce plant. Hence to reduce the risk associated with intake of chlorpyrifos through lettuce, mild detergent washing procedures should be followed before consumption. Nevertheless, lettuce should be rinsed thoroughly after the use of detergent, otherwise the detergent may be consumed together with the vegetables.

Competing interests
Authors declare that they have no competing interest.

Authors' contribution
Authors contributed equally. All authors have read and approved the final manuscript.

Acknowledgement
The work was supported by the Ghana Government Research Allowance and the MSc. Fellowship by the Scholarship secretariat in Ghana. The authors are to the staff of the Pesticide Residues Laboratory of Ghana Standards Authority for providing the laboratory assistance.

Author details
[1]Department of Chemistry, Kwame Nkrumah University of Science and Technology, Kumasi, Ghana. [2]Department of Theoretical and Applied Biology, Kwame Nkrumah University of Science and Technology, Kumasi, Ghana. [3]Department of Physics, Kwame Nkrumah university of Science and Technology, Kumasi, Ghana.

References
Akoto O, Andoh H, Dark G, Eshun K, Osei-Fosu P. Health risk assessment of pesticides residue in maize and cowpea from Ejura, Ghana. Chemosphere. 2013;92:67–73.
Akoto O, Gavor S, Appah MK, Apau J. Estimation of human health risk associated with the consumption of pesticide-contaminated vegetables from Kumasi, Ghana. Environ Monit Assess. 2015;187:244. doi:10.1007/s10661-015-4471-0.
Amoah P, Dreschel P, Abaidoo RC, Ntow WJ. Pesticide and pathogen contamination of vegetables in Ghana's urban markets. Arch Environ Contam Toxicol. 2006;50(1):1–6.
Bhanti M, Taneja A. Contamination of vegetables of different seasons with organophosphorus pesticides and related health risk assessment in Northern India. Chemosphere. 2007;69:63–8.
Clarke EEK, Levy LS, Spurgeon A, Calvert IA. The Problems Associated with Pesticide Use by Irrigation Workers in Ghana. Occup Med. 1997;47(5):301–8.
Dhiman N, Jyot G, Bakhshi AK. J Food Sci Technol. 2006;43(1):92–5.
Essumang DK, Dodoo DK, Adokoh CK, Fumador EA. Analysis of some pesticide residues in tomatoes in Ghana. Hum Ecol Risk Assess. 2008;14(4):796–806.
FAO/WHO. Food Standards Programme. Codex Alimentarius Commission, Twenty-seventh Session, Geneva, Switzerland. 28 June – 03 July, 2004. 2004.
Gerken A, Suglo JV, Braun M (2013) Crop Protection Policy in Ghana. Pokuase – Accra: Integrated Crop Protection Project, PPRSD/GTZ;.
Hui TJ, Ariffin MM, Tahir NM. Hydrolysis of chlorpyrifos in aqueous solutions at different temperatures and pH. Malaysian J Anal Sci. 2010;14(2):50–5.
Kim SW, Abd El-Aty AM, Rahman M, Choi JH, Lee YJ, Ko AY, Choi OJ, Shim JH. The effect of household processing on the decline pattern of dimethomorph in pepper fruits and leaves. Food Control. 2015;50:118–24.

Kotey AD, Gbewonyo WSK, Afreh-Nuamah K. High chlorpyrifos levels on vegetables in Ghana. Pesticide News. 2008;80:1–6.

Krol WJ, Arsenault TL, Pylypiw HM, Mattina MJI. Connecticut Agricultural Experiment Station. J Agric Food Chem. 2000;48(10):4666.

Kumari B. Effects of household processing on reduction of pesticide residues in vegetables. ARPN J Agric Biol Sci. 2008;3(4):46–51.

Ling Y, Wang H, Yong W, Zhang F, Sun L, Yang ML, Wu YN, Chu XG. The effects of washing and cooking on chlorpyrifos and its toxic metabolites in vegetables. Food Control. 2011;22:54–8. doi.org/10.1016/j.

Lopez-Fernandez R, Rial-Otero O, Simal-Gandara J. Factors governingthe removal of mancozeb residues from lettuces with washing solutions. Food Control. 2013;34:530–8.

Madsen T, Boyd HB, Nylén D, Pedersen AR, Petersen GI, Simonsen F. Environmental and Health Assessment of Substances in Household Detergents and Cosmetic Detergent Products. 2001. Environmental Project No. 615 2001 Miljøprojekt.

Obeng-Ofori D, Owusu EO, Kaiwa ET. Variation in the level of carboxylesterase activity as an indicator of insecticide resistance in populations of the 102 diamondback moth Plutella xylostella (L.) attacking cabbage in Ghana. J Ghana Sci Assoc. 2002;4(2):52–62.

Reigart JR, Roberts JR. Organophosphate Insecticides. Recognition and management of Pesticide Poisonings. 5th ed. 1999. p. 34–48.

Roberts TR, Hutson DH. Metabolic Pathways of Agrochemicals - Part 2: Insecticides and Fungicides. Cambridge: The Royal Society of Chemistry; 1999. p. 235–42.

Slavin JL, Lloyd B. Health Benefits of Fruits and Vegetables. Adv Nutr Int J. 2012;3:506–16.

Smith GN. Ultraviolet light decomposition studies with Dursban and 3,5,6-trichloro-2-pyridinol. J Econ Entomol. 1968;61(3):793–9.

Swisher RD (1987). Surfactant Biodegredation (Second.). New York, New York: Marcel Dekker Inc. Retrieved from http://books.google.com/books. Accessed 10 June 2016.

Thompson CM, Richardson RJ. Anticholinesterase Insecticides. In: Marrs TC, Ballantyne B, editors. Pesticide Toxicology and International Regulation. West Sussex: Wiley; 2004. p. 89–127.

11

Persistent organochlorine pesticide residues in cocoa beans from Ghana, a concern for public health

Elvis D. Okoffo, Benedicta Y. Fosu-Mensah[*] and Christopher Gordon

Abstract

Background: Residual levels of fifteen (15) organochlorine pesticides were determined in 32 cocoa bean samples collected from sixteen (16) selected cocoa farms in the Dormaa West District of Ghana to assess the levels of pesticides contamination.

Results: The results show that all cocoa bean samples analysed from the study area had one or more organochlorine pesticide residues detected in them. The study revealed the presence of eight organochlorine pesticide residues in the cocoa bean samples analysed at varying concentrations. The organochlorine pesticide residues detected were aldrin (0.02–0.03 mg/kg), dieldrin (0.02–0.04 mg/kg), lindane (0.03–0.05 mg/kg), beta-HCH (0.02–0.03 mg/kg), p,p'-DDE (0.02–0.03 mg/kg), p,p'-DDD (0.02–0.04 mg/kg), p,p'-DDT (0.04–0.05 mg/kg) and methoxychlor (0.02–0.04 mg/kg). The most frequently found and abundant pesticide residue was the metabolite of DDT (p,p'-DDT) which occurred in 62.5 % of the samples, followed by lindane (56.3 %) and then beta-HCH and p,p'-DDD occurring in 50 % of the samples. None of the detected pesticide mean residues recorded from the various study sites exceeded their European Union (EU) Maximum Residue Limits (MRLs) for cocoa beans except beta-HCH at Krakrom (S3).

Conclusion: The levels of organochlorine pesticide residues in the fermented dried cocoa beans analysed compared to the European Union (EU) commission regulations on pesticide residues showed no health risks to consumers of cocoa beans from Ghana and no threat to cocoa export to Europe. The occurrence of organochlorine pesticide residues in the samples analysed could be due to their illegal use by farmers in the study area or due to their past use, since these chemicals are prohibited from agricultural use in Ghana. There should be regular monitoring of pesticide residues especially in cocoa beans to protect consumers from health related risks. There is a need to check and enforce regulations on the use of banned/restricted and unapproved pesticides in cocoa production in Ghana.

Keywords: Cocoa beans, Pesticide residues, Organochlorine, Health risks, Environment

Background

Pests and diseases are recognised as a major factor responsible for the decline in crop productivity, particularly cocoa yield in Ghana. This has resulted in increased use of pesticides in an effort to increase productivity. However, the regular application and indiscriminate use of these chemicals can have unintended environmental and human health consequences. Generally, pesticides are broadly divided into many classes of which the most important are the organochlorine pesticides (Adeyemi et al. 2011; Kuranchie-Mensah et al. 2012). Organochlorine pesticides (OCs) which are among the agrochemicals that have been used extensively for long period in Ghana (Clarke et al. 1997), particularly in cocoa production are broad spectrum synthetic organic pesticides which are made up of predominantly carbon, hydrogen, chlorine and sometimes oxygen (Afful et al. 2010; Frimpong et al. 2012a; Idowu et al. 2013; Asiedu 2013). They are composed of three broad groups or types namely; dichlorodiphenylethanes (DDT) and analogues, cyclodienes and the chlorinated benzenes or cyclohexane's (Asiedu 2013). Organochlorine pesticides that have been

* Correspondence: yayramensah@staff.ug.edu.gh
Institute for Environment and Sanitation Studies (IESS), University of Ghana,
P. O. Box 209, Legon, Accra, Ghana

widely used in cocoa production in Ghana include aldrin/ dieldrin, chlordane, DDT, endrin, hexachlorobenzene, endosulfan, methoxychlor, lindane, diazinon and heptachlor (Afful et al. 2010; Owusu-Ansah et al. 2010; Frimpong et al. 2012a; Frimpong et al. 2012b; Kuranchie-Mensah et al. 2012).

Organochlorine pesticides break down slowly and can persist in the environment long after application and in organisms long after exposure (Botwe 2007; Dikshith 2008; Darko et al. 2008). They have the tendency for long range transport and trans-boundary dispersions, and their capacity to bio-accumulate in the food chain poses a great threat to the environment, wildlife and humans (Ntow 2005; Darko et al. 2008; Frimpong et al. 2012a; Frimpong et al. 2012b). They are lipophilic and may bio-accumulate in the fatty parts of biological beings such as breast milk, blood and fatty tissues when small amounts are taken up in food (William et al. 2008). Exposure to organochlorines through the consumption of a crop that has organochlorine residues may cause acute and chronic toxicities. Organochlorine pesticides have become ubiquitous contaminants and have been implicated in a wide range of adverse health effects in laboratory animals and humans (Adeyemi et al. 2011). These toxic effects include reproduction and birth defects/failures, deformities, neurological damage, immune system dysfunction, endocrine disruptions, and cancer (Ahlborg et al. 1995; Amoah et al. 2006; Sosan et al. 2008; Leena et al. 2012).

Although the agricultural use of organochlorine pesticides have been banned or seriously restricted in many developed nations and some developing countries like Ghana (due to their adverse effects on human health and persistence in the environment) (Darko et al. 2008; Botwe et al. 2012), there are evidences of their continuous use on crops particularly cocoa trees, vegetables and fruits, among others (Botwe 2007; Darko and Acquaah 2007; Darko et al. 2008; Bempah and Donkor 2011; Asiedu 2013). The continuous use of these chemicals may be due to a combination of factors including inadequate regulation and management, trade, weak enforcement of bans on pesticides importation, illegal use/application and lack of logistics to monitor these pesticides in Ghana.

Furthermore, the continuous use of these chemicals may be due to their effectiveness in controlling pest and diseases, low cost, their versatility against various pests, their availability as well as ignorance of their harmful effects by farmers in Ghana (Bempah et al. 2011b). According to Asiedu (2013), already-made pesticide formulations in soft drink bottles and other unlabelled liquid containers sold to farmers in Ghana may contain some of these restricted or banned pesticides. Research data available have indicated the presence of organochlorine pesticide residues in surface water and sediments (Darko et al. 2008; Kuranchie-Mensah

et al. 2012), fruits and vegetables (Bempah and Donkor 2011; Bempah et al. 2011a; Bempah et al. 2011b; Asiedu 2013), meat (Darko and Acquaah 2007) and cocoa beans (Botchway 2000; Apau and Dodoo 2010; Agyen 2011; Daanu 2011; Boakye 2012; Frimpong et al. 2012a; Frimpong et al. 2012b) in Ghana, which are emanating from current and past use of these chemicals.

Recent changes in regulations in the European Union (EU), North America and Japan have called for a reflection on crop protection practices in cocoa and other commodity crops (International Cocoa Organization (ICCO) 2007). The quality of cocoa beans imported into the EU and elsewhere is assessed based on traces of pesticides and other substances that have been used in the supply chain (Afrane and Ntiamoah 2011). Cocoa beans that are found to contain traces of these substances above the Maximum Residue Levels (MRLs) are rejected, which reduces the value of cocoa beans produced form the producing nation. In 2010, over 20,000 MT of cocoa beans was rejected by Japan due to the presence of high pesticide residues (Kaminaga 2011). For instance, Sarfo (2013) reported that a 2,000 tonne shipment of cocoa bean to Japan from Ghana was rejected in 2006, due to the detection of illegal pesticide residues. Ghana therefore risks being blacklisted by its major foreign cocoa buyers, if immediate action is not taken to halt the trend which will impact negatively on Ghana's economy.

The Brong Ahafo region is one of the major cocoa producing regions in Ghana. However, over the years, pests and diseases have been a major factor responsible for the decline in cocoa yield in the region. The Dormaa West District is one of the major cocoa producing districts in the Brong Ahafo Region of Ghana. In order to increase cocoa productivity in the district, there has been an increased use of pesticides to control pests and diseases. This practice may however result in high levels of pesticide residues in cocoa beans and hence reduce its value. In addition, the regular application and indiscriminate use of these chemicals can have unintended environmental and human health consequences, particularly when residues of these chemicals are taken up by cocoa beans and ingested by humans. Unfortunately, there is little information on levels of organochlorine pesticide residues in cocoa beans produced from the district. This paper, therefore, seeks to assess the contamination levels of organochlorine pesticide residues in cocoa beans produced from the Dormaa West District of Ghana to ascertain the potential health risks of the general public.

Methods
Study area
The study was carried out in the Dormaa West District located at the western part of the Brong Ahafo Region of Ghana with slightly hilly terrain (240–300 m above sea

level) (Fig. 1). It shares boundaries in the north with the Dormaa Central Municipality, in the east with Asunafo North Municipality, in the west with Côte d'Ivoire and in the south west with Bia East District (Ghana Statistical Service 2014). The district is generally an agrarian economy which contributes immensely to food security in the country. Agriculture is the main source of employment (82 %) in the district. The major economic activities in the district include the cultivation of food and cash crops (including cocoa), poultry and livestock farming, oil palm extraction, cassava processing and sand winning (Ghana Statistical Service 2014).

Sampling design

Four cocoa growing communities namely Nkrankwanta (S1), Diabaa (S2), Krakrom (S3) and Kwakuanya (S4) as shown in Fig. 1 were randomly selected from the district. Within each selected community, four cocoa farms were identified and selected. A total of sixteen (16) cocoa farms were selected with the age of the farm (farm not less than 8 years and not more than 20 years with a history of at least five years of pesticides application) and the density of cocoa production used as determining factors. Sampling of cocoa beans was done between December 2014 and February 2015.

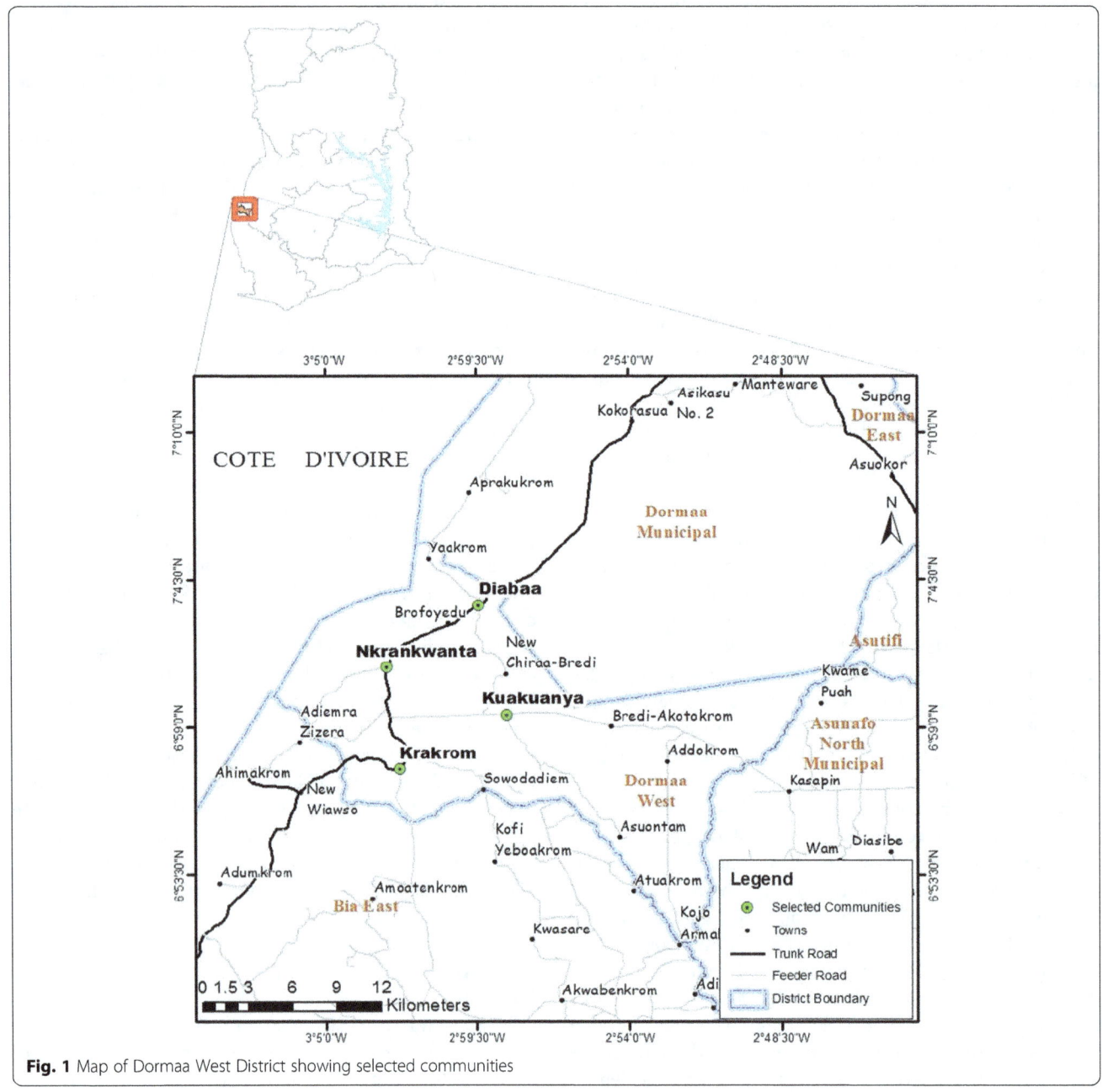

Fig. 1 Map of Dormaa West District showing selected communities

Sample collection

Each selected cocoa farm was divided into two quadrats (80 x 80). Within each quadrat, five (5) cocoa trees were randomly selected, where five (5) matured and ripped cocoa pods were randomly taken and kept in labelled bags. Two replicates of cocoa bean pod were collected from each cocoa farm giving a total of 32 cocoa bean pods (samples) from the study area. The pods were transported to the laboratory [Ecological Laboratory (ECOLAB), of University of Ghana, Legon] where they were broken and the cocoa beans fermented for seven days. The fermented beans were sun-dried for 21 days and later oven dried at 60 °C for 5 h until a constant weight was attained. The dried cocoa beans from each quadrat of a farm were then bulked together to form a composite sample. The composite cocoa bean samples were well mixed and 2 kg sub-samples were taken into clean labelled polythene bags and transported to the Ghana Standards Authority Pesticide Residue Laboratory in Accra, Ghana for analysis. In the laboratory, samples were frozen in the refrigerator and analysed within a week.

Sample preparation

Samples of fermented dried cocoa beans (2 kg) were thoroughly grounded into fine powder and homogenized. Approximately 10 g of the sample were used for pesticide analysis.

Chemicals and reagents used

The individual certified organochlorine pesticide standards used for the identification and quantification (β-HCH, γ-HCH, δ-HCH, heptachlor, aldrin, γ-chlordane, p,p'-DDE, p,p'- DDT, p,p'-DDD, dieldrin, endrin, α-endosulfan, β-endosulfan, endosulfan-sulfate and methoxychlor) were obtained from Dr. Ehrenstorfer GmbH (Augsburg, Germany) with certified purity of 98 % and stored in the freezer to minimize degradation. Pesticide residue grade acetonitrile (99.5 %) was purchased from BDH, England. ENVI-Carb/LC-NH$_2$ (500 mg / 500 mg / 6 mL) and Bond elute C-18 SPE cartridges (1000 mg / 6 mL) were purchased from Supelco Inc., USA and Phenomenex, USA, respectively. All other reagents and solvents used for analysis were of analytical grade purchased from BDH, England and included: acetone, ethyl acetate (99.8 %), anhydrous sodium sulphate (Na$_2$SO$_{4)}$ and sodium chloride (NaCl).

Preparation of standard solutions

Pesticides standard stock solutions (1000 μg / mL) of each of the certified reference organochlorine pesticide standards were prepared by pipetting the appropriate aliquot or weight of the certified reference pesticide into a 25 mL volumetric flasks, and then dissolving and diluting with ethyl acetate with the aid of a vortex mixer (Thermolyne Maxi Mix-Plus). Stock solutions were stored in refrigerator. Working solutions of the pesticide standards for use as fortification standards in the procedural recovery process, and as calibration standards in the instrument calibration were freshly prepared through the dilution of an appropriate aliquot of the stock solutions with ethyl acetate.

Analysis of organochlorine pesticide residues
Extraction and clean up

The Japanese multi-residue method for agricultural chemicals by GC/MS (agricultural products) released by the Department of Food Safety, Ministry of Health, Labour and Welfare (2006), (No. 0124001) as adapted in the Ghana Standard Authority (GSA) Pesticide Residues Laboratory Protocol were followed for extraction and clean-ups, with slight modifications. This method has been used by Frimpong et al. (2012a, b, c, d) to analysed pesticide residues in cocoa beans ready for export in Ghana.

Ten grams (10 g) of the homogenous cocoa bean sub-samples were weighed into 250 mL Nalgene jars. A 20 mL of distilled water was added to each sample and left to stand for 15 min. A 40 mL acetonitrile was added and homogenized using the Ultra Turrax T25 basic homogeniser (Staufen, Germany) for 2 min. They were then centrifuged (Jouan CR3i multifunction) at a speed of 3000 rpm for 3 min and decanted through filter papers into labelled 100 mL volumetric flasks. The extracts were placed back into the Nalgene jars and 20 mL acetonitrile was added, homogenized for 2 min and the dispersing element rinsed with 5 mL acetonitrile. The suspensions were then centrifuged (Jouan CR3i multifunction) at 3000 rpm for 3 min and filtered into each corresponding labelled 100 mL volumetric flasks. A 15 mL acetonitrile was used to rinse the jars and residues filtered. All filtrates were made up to 100 mL with acetonitrile in the 100 mL volumetric flasks. Aliquot of 20 mL of the filtrates were pipetted into labelled 250 mL separating flasks, and 10 g of NaCl and 20 mL of 0.5 mol/L phosphate buffer (pH 7.0) were added. The separating flasks were corked and shook for 20 min using the horizontal shaker (Ika-Werke HS 501 Digital) and left to stand for 10 min until the solutions were clearly separated into layers. The NaCl and lower aqueous layers in each separating flask were carefully removed and the organic layers (acetonitrile layers) transferred into labelled 50 mL beakers for clean-ups.

First clean up using bond elute C-18 cartridges

Bond elutes C-18 (1000 mg / 6 mL) cartridges were conditioned with 10 mL acetonitrile. Labelled 30 mL round bottomed flasks were placed under the cartridges to collect elutes. Sample extracts from the extraction stage were loaded onto each corresponding cartridge, and eluted with 2 mL acetonitrile. A 5 g portion of anhydrous Na_2SO_4 was placed on filter papers in funnels and the extracts dried over them. The receiving flasks were rinsed with acetonitrile and passed over the Na_2SO_4. Each filtrate was collected into a well labelled 100 mL round bottom flask and concentrated at 40 °C to dryness using the rotary evaporator (Bibby RE 200 and Buchi Ratovapor R-210). The residues were re-dissolved with 2 mL of acetonitrile in a ratio of 1:3 prior to the second clean-up step.

Second clean up using ENVI-Carb/LC-NH₂ cartridges

ENVI-Carb/LC-NH$_2$ (500 mg / 500 mg / 6 mL) cartridges were conditioned with 10 mL acetonitrile. A 50 mL round bottomed flasks were placed under the cartridges, and the extracts from the previous clean-up step loaded onto the corresponding cartridges. The extracts were allowed to filter and the eluates collected. The cartridges were then eluted with 20 mL acetonitrile in four portions with intermittent vacuum. All filtrates were then transferred into 100 mL round bottomed flasks and concentrated at 40 °C to approximately 1 mL using the rotary evaporator (Bibby RE 200 and Buchi Ratovapor R-210). A 10 mL of acetone was added to the concentrated solution and concentrated/evaporated to dryness. The extracts were re-dissolved in 2 mL ethyl acetate by pipetting, and transferred into labelled 15 mL screw capped tubes, closed and placed in a freezer for at least 30 min. The extracts were removed afterwards and immediately centrifuged (Jouan CR3i multifunction) at 3000 rpm for 5 min, and the top layers carefully transferred into labelled 2 mL GC standard opening vials prior to quantification by Gas Chromatography (GC) equipped with Ni electron capture detector (ECD). All extracts were kept frozen until quantifications were achieved.

Chromatographic conditions for organochlorine pesticides determination

The final extracts were analysed by a Gas Chromatography (GC) - Varian CP-3800 (Varian Association Inc. USA) equipped with ^{63}Ni Electron Capture Detector (GC-ECD) and Combi PAL auto sampler that allowed the detection of contaminants even at trace level concentrations from the matrix to which other detectors do not respond. The GC conditions and the detector response were adjusted so as to match the relative retention times and response as spelt out by Japanese analytical methods for agricultural chemicals. The GC conditions used for the analysis were capillary column (fused silica capillary) coated with VF-5 ms (30 m + 10 m EZ guard x 0.25 mm internal diameter, 0.25 μm film thickness). The injector and detector-ECD temperatures

were set at 270 °C and 300 °C, respectively. The oven temperature was programmed as follows: 70 °C held for 2 min, ramped at 25 °C min^{-1} to 180 °C, held for 1 min, and finally ramped at 5 °C min^{-1} to 300 °C. Nitrogen was used as carrier gas at a constant flow rate of 1.0 mL min^{-1} and detector make-up gas of 29 mL min^{-1}. The injection volume of the GC was 1.0 μL. The total run time for a sample was 31.4 min.

Quantification and limit of reporting

An external standard method using peak area was employed in the determination of the quantities of residues in the sample extracts. A standard mixture of known concentration of organochlorine pesticides was run and the response of the detector for each compound ascertained. The area of the corresponding peak in the sample was compared with that of the standard and the peak areas whose retention times coincided with the standards were extrapolated on their corresponding calibration curves to obtain the concentration. All analyses were carried out in triplicates and the mean concentrations computed accordingly.

The limit of reporting of the residues determined was based on the extract of the fortified samples that were serially diluted by factor of two to give different concentrations. One out of each concentration that gave a response three times the standard deviation of the least fortified sample was noted. This was used to estimate the statistical significance of differences between low level analyte responses and the combined uncertainties in both the analyte and the background measurement. The limit of reporting for the organochlorine pesticide residues was 0.005 mg/kg.

Quality control and quality assurance

Quality control and quality assurance were incorporated in the analytical scheme. To ensure the validity of results, the following measures were taken. All glass ware used for analysis (extraction and clean ups) were rigorously washed with detergent and tap water. They were then rinsed with distilled water, thoroughly rinsed with analytical grade acetone and dried overnight in an oven at 150 °C. The glass ware were then removed from the oven and allowed to cool down and stored in dust free cabinets.

The quality of organochlorine pesticide residues was assured through the analysis of solvent blanks, procedural matrix blanks and triplicate samples. All reagents used during the analysis followed the same extraction procedures. Solvent used were run to verify any interfering substances within the runtime. Prior to sample analysis, standards for all 15 organochlorine pesticides were analysed to verify adequate system performance. There was satisfactory agreement of data with analysed standards prior to sample analyses and between sample runs. In order to prevent background contamination, reagent blank in triplicate were included in each batch of sample analysis and three spiked samples in triplicate were also done to confirm satisfactory pesticide recovery. No organochlorine pesticide residues were detected in the blank for each extraction procedure. A fortification level of 0.005 mg/kg of standard mixtures was chosen based on the limit of reporting of the pesticide residues before analysis to evaluate the recovery of compounds in the cocoa bean samples analysed. Fortified samples were determined with good recoveries. Procedural recoveries were analysed concurrently with each batch of analytical extracts.

In addition, recalibration curves were run with each batch of sample to check that the correlation coefficient was kept above $r^2 > 0.995$. All analyses were carried out in triplicates and the mean concentrations were calculated based on the number of samples that tested positive to each residue. The efficiency of the analytical methods (the extraction and clean-up methods) was determined by recoveries of internal standards. The recoveries of the internal standards ranged between 70 and 98 % for all the pesticide residues detected which shows that the method used is reproducible.

Data analysis

Statistical Package for Social Sciences (SPSS) software version 20.0 was used to determine the means and standard deviation for the pesticide residues detected in the cocoa bean samples. One-way Analysis of variance (ANOVA) was used to test for the significant differences and similarities between the pesticide residues detected in cocoa beans from the various sampled sites. Differences were considered significant at $p < 0.05$.

Results and discussion

The monitored organochlorine pesticides were β-HCH, γ-HCH, δ-HCH, heptachlor, γ-chlordane, p,p'-DDE, p,p'-DDD, p,p'- DDT, aldrin, dieldrin, endrin, α-endosulfan, β-endosulfan, endosulfan-sulfate and methoxychlor. However, cocoa bean samples analysed from the study area showed the presence of eight (8) organochlorine pesticide residues (lindane (gamma-HCH), beta-HCH, dieldrin, aldrin, p,p'-DDE, p,p'-DDD, p,p'-DDT, and methoxychlor) as shown in Tables 1, 2, 3, 4 and 5. The results show that all cocoa bean samples analysed were contaminated with one or more pesticide residue. The most frequently encountered pesticide residue was p,p'-DDT; which was detected in 20 (62.5 %) out of the 32 samples of the analysed cocoa beans. Cocoa beans samples from S1, S2 and S3 recorded the highest number of different pesticide residues. Each recorded the presence of 7 different pesticide residues.

Lindane (gamma-HCH) occurred in 56.3 % of the cocoa bean samples analysed and ranged from 0.03 mg/kg at S1 and S4 to 0.05 mg/kg at S3 with a mean value of 0.04 ± 0.01 mg/kg. There were no significant differences ($p > 0.05$) in mean values of lindane among the sampled sites. The measured concentrations of lindane were far less than the European Union (EU) Maximum Residue Limit (MRL) of 1.00 mg/kg for cocoa beans.

Although gamma-HCH is the isomer of HCH that possesses high pesticidal activity, beta-HCH was also detected in 50 % of the cocoa bean samples with a mean value of 0.02 ± 0.00 mg/kg. Its concentration value ranged from 0.02 mg/kg at S1 and S2 to 0.03 mg/kg at S3. Similarly, there were no significant differences ($p > 0.05$) in mean values of beta-HCH among the sampled sites. The mean concentrations of beta-HCH recorded in this study were below the EU MRL of 0.02 mg/kg for cocoa beans with the exception of S3. In addition, the mean residue concentration of beta-HCH recorded in this study was comparable to the EU MRL of 0.02 mg/kg for cocoa beans.

The presence of gamma-HCH and beta-HCH suggests previous/historical and/or current illegal use of technical HCH pesticide in the study area, since its use has been banned in cocoa production in Ghana. Lindane was marketed in Ghana as Gammalin 20 and was widely used on cocoa plantations until 2007 when it was banned for agricultural purposes due to their persistence and toxicity to untargeted organisms (Afful et al. 2010; Agbeve et al. 2014). The mean lindane concentration observed in this study was higher than the mean concentrations of 0.01 mg/kg, 0.01 mg/kg, 0.01 mg/kg and 0.0001 mg/kg reported by Botchway (2000), Frimpong et al. (2012a), Frimpong et al. (2012b), and Olayinka (2013) in cocoa bean samples, respectively. However, the mean lindane concentration observed in this study was lower than the concentration of 0.411 mg/kg reported by Apau and Dodoo (2010) in cocoa beans from the Central Region of Ghana. The findings of lindane in this study was contrary to that of Owusu-Ansah et al. (2010) which recorded no lindane residue in cocoa beans from the Twifo Praso district of Ghana. This could be due to none use of pesticides containing lindane as active ingredient in cocoa production in the district. On the other hand, the mean value of beta-HCH recorded in this study was higher than the mean value of 0.01 mg/kg reported by Frimpong et al. (2012b) in cocoa beans ready for export in Ghana. Also the findings of beta-HCH in this study was in line with those reported by Boakye (2012) and Frimpong et al. (2012a) in cocoa beans analysed from Ghana.

Aldrin is readily converted to dieldrin (once it enters either the environment or the body), which is considered one of the most persistent of all pesticides (Miles et al.

Table 1 Concentrations of organochlorine residues (mg/kg) in cocoa beans from farms in Nkrankwanta (S1)

Sites	FARM 1		FRAM 2		FARM 3		FARM 4			
Pesticides[c]	Farm 1A	Farm 1B	Farm 2A	Farm 2B	Farm 3A	Farm 3B	Farm 4A	Farm 4B	Mean[a]	EU MRL[b]
Lindane (gamma-HCH)	0.01	0.03	ND	ND	0.03	0.03	ND	ND	0.03	1.00
Beta-HCH	ND	ND	0.02	0.01	0.02	ND	0.02	0.01	0.02	0.02
Dieldrin	0.01	0.02	ND	ND	ND	0.02	ND	ND	0.02	0.50
Aldrin	0.01	ND	ND	0.02	0.01	ND	0.02	ND	0.02	0.05
p,p'-DDE	ND	0.02	0.01	0.02	ND	ND	ND	ND	0.02	0.50
p,p'-DDD	ND	ND	ND	ND	ND	ND	ND	ND	ND	
p,p'-DDT	0.05	0.02	0.03	0.04	0.03	0.04	ND	0.04	0.04	
Methoxychlor	ND	ND	0.02	0.04	0.04	ND	0.04	0.04	0.04	0.10

ND Non-detected
[a]Mean of samples where residues were detected
[b]European Union maximum residue limits for pesticides in cocoa beans
[c]Number of pesticides found in any sample

2009; Hogarh et al. 2014). These pesticides were used extensively in agriculture until their use was restricted. Although the use of aldrin and dieldrin is banned in many countries including Ghana, these pesticides were detected in the cocoa bean samples analysed from the study area. Aldrin was detected in 43.8 % of the cocoa bean samples analysed and ranged from 0.02 mg/kg at S1 to 0.03 mg/kg at S3 and S4 with a mean value of 0.03 ± 0.00 mg/kg. There were no significant differences ($p > 0.05$) in mean values of aldrin among the sampled sites. The mean concentrations of aldrin recorded in this study were below the EU MRL of 0.05 mg/kg for cocoa beans. Similarly, its breakdown product, dieldrin, was recorded in 37.5 % of the cocoa bean samples analysed with a mean value of 0.03 ± 0.01 mg/kg, which ranged from 0.02 mg/kg at S1 to 0.04 mg/kg at S2 and S3. Similarly, there were no significant differences ($p > 0.05$) in mean values of dieldrin among the sampled sites. The mean concentrations of dieldrin recorded in this study were below the EU MRL of 0.50 mg/kg for cocoa beans.

The occurrence of aldrin and dieldrin in the cocoa bean samples analysed suggests the illegal use of the pesticide (pesticides with aldrin and dieldrin as their active ingredient) by cocoa farmers in the study area, and/or previous use of the chemicals, since their use for agricultural purposes has been banned in Ghana (Afful et al. 2010; Hogarh et al. 2014). In addition, the occurrence of dieldrin in the samples confirmed the possible degradation and/or metabolism of aldrin to dieldrin (which is stable, lipophilic and bioaccumulate in fats) in the environment. A similar observation was made by Boakye (2012). However, the relatively high percentage of cocoa beans (43.8 %) with detectable aldrin compared to samples (37.5 %) with detectable dieldrin, might also suggest a more or high input of aldrin at present than their degradation to dieldrin. The mean value of aldrin recorded in this study was higher than the mean values of 0.01 mg/kg and 0.01 mg/kg reported by Frimpong et al. (2012a) and Frimpong et al. (2012b), respectively, but lower than the 0.11 mg/kg reported by Daanu (2011) in

Table 2 Concentrations of organochlorine residues (mg/kg) in cocoa beans from farms in Diabaa (S2)

Sites	FARM 1		FRAM 2		FARM 3		FARM 4			
Pesticides[c]	Farm 1A	Farm 1B	Farm 2A	Farm 2B	Farm 3A	Farm 3B	Farm 4A	Farm 4B	Mean[a]	EU MRL[b]
Lindane (gamma-HCH)	0.04	ND	ND	ND	0.01	0.04	0.05	0.04	0.04	1.00
Beta-HCH	0.01	0.02	ND	ND	0.02	0.02	0.01	ND	0.02	0.02
Dieldrin	ND	ND	ND	0.02	0.05	0.05	ND	ND	0.04	0.50
Aldrin	0.03	0.03	0.03	ND	ND	ND	0.01	0.03	0.03	0.05
p,p'-DDE	ND	ND	ND	ND	ND	ND	ND	ND	ND	0.50
p,p'-DDD	ND	ND	0.01	0.04	0.04	0.04	0.04	0.04	0.04	
p,p'-DDT	0.02	0.06	0.06	ND	0.06	0.05	0.04	0.06	0.05	
Methoxychlor	0.01	0.02	ND	0.02	0.01	0.02	ND	ND	0.02	0.10

ND Non-detected
[a]Mean of samples where residues were detected
[b]European Union maximum residue limits for pesticides in cocoa beans
[c]Number of pesticides found in any sample

Table 3 Concentrations of organochlorine residues (mg/kg) in cocoa beans from farms in Krakrom (S3)

Sites	FARM 1		FRAM 2		FARM 3		FARM 4			
Pesticides[c]	Farm 1A	Farm 1B	Farm2A	Farm 2B	Farm3A	Farm 3B	Farm 4A	Farm 4B	Mean[a]	EU MRL[b]
Lindane (gamma-HCH)	0.01	0.05	0.06	0.05	0.04	0.06	ND	ND	0.05	1.00
Beta-HCH	0.03	0.02	0.02	ND	0.02	ND	0.03	0.03	0.03	0.02
Dieldrin	0.05	0.05	ND	0.01	ND	ND	ND	ND	0.04	0.50
Aldrin	ND	ND	ND	ND	ND	ND	ND	ND	ND	0.05
p,p'-DDE	0.01	0.03	ND	ND	0.03	0.03	ND	ND	0.03	0.50
p,p'-DDD	0.03	0.03	ND	0.03	ND	0.03	ND	0.01	0.03	
p,p'-DDT	ND	0.04	0.04	0.04	0.01	ND	0.04	0.04	0.04	
Methoxychlor	0.03	0.01	0.03	0.03	ND	ND	0.03	ND	0.03	0.10

ND Non-detected
[a]Mean of samples where residues were detected
[b]European Union maximum residue limits for pesticides in cocoa beans
[c]Number of pesticides found in any sample

cocoa beans from Ghana. Similarly, the mean value of dieldrin recorded in this study was higher than the mean values of 0.01 mg/kg and 0.02 mg/kg reported by Frimpong et al. (2012a) and Frimpong et al. (2012b), respectively in cocoa beans from Ghana. This could be due to the differences in the sampling methods used (in that, cocoa bean samples for this study was taken from cocoa farms while samples used in the other studies were taken from cocoa bean samples ready for shipment in Ghana) or differences in the area of which the cocoa beans were coming from.

p,p'-DDE was detected in 34.4 % of the samples analysed with a mean concentration of 0.02 ± 0.00 mg/kg. The measured concentrations ranged from 0.02 mg/kg at S1 and S4 to 0.03 mg/kg at S3. However, the differences in means were not statistically significant ($p > 0.05$). p,p'- DDE, one of the metabolites of DDT is formed by the loss of hydrogen chloride (dehydrohalogenation) in DDT. They are fat soluble and have the capacity to build up in the fat of animals and humans, and are rarely excreted from the body

(Aikpokpodion et al. 2012). Studies have showed that exposure to p,p'- DDE can cause endocrine disruptions, oxidative stress, Alzheimer's and Parkinson's disease, contributes to breast cancer and damage the brain's dopaminergic system (Aikpokpodion et al. 2012). The mean concentration of p,p'- DDE recorded in this study was lower than the mean value of 0.04 mg/kg reported by Aikpokpodion et al. (2012) in cocoa bean samples from three cocoa ecological zones in Nigeria, but higher than the 0.001 mg/kg reported by Frimpong et al. (2012a) in cocoa beans ready for export in Ghana. Also, the mean value of p,p'- DDE observed in this study was similar to the 0.02 mg/kg reported by Daanu (2011).

p,p'-DDD residue was detected in 50 % of the cocoa bean samples analysed with a mean value of 0.03 ± 0.01 mg/kg, which ranged from 0.02 mg/kg at S4 to 0.04 mg/kg at S2. There were no significant ($p > 0.05$) differences in mean values of p,p'-DDD among the sampled sites. The mean p,p'-DDD residue observed in this study was lower than the mean value of 0.15 mg/kg

Table 4 Concentrations of organochlorine residues (mg/kg) in cocoa beans from farms in Kwakwanya (S4)

Sites	FARM 1		FRAM 2		FARM 3		FARM 4			
Pesticides[c]	Farm 1A	Farm 1B	Farm2A	Farm 2B	Farm3A	Farm 3B	Farm 4A	Farm 4B	Mean[a]	EU MRL[b]
Lindane (gamma-HCH)	0.03	0.04	ND	ND	0.01	ND	ND	ND	0.03	1.00
Beta-HCH	ND	ND	ND	ND	ND	ND	ND	ND	ND	0.02
Dieldrin	ND	0.03	ND	ND	0.02	0.03	ND	ND	0.03	0.50
Aldrin	0.02	0.03	0.03	ND	ND	ND	0.02	0.03	0.03	0.05
p,p'-DDE	ND	ND	0.01	0.02	0.02	0.02	ND	ND	0.02	0.50
p,p'-DDD	0.03	0.01	0.01	0.02	0.03	ND	ND	ND	0.02	
p,p'-DDT	ND	ND	ND	ND	ND	ND	ND	ND	ND	
Methoxychlor	ND	ND	ND	ND	ND	ND	ND	ND	ND	0.10

ND Non-detected
[a]Mean of samples where residues were detected
[b]European Union maximum residue limits for pesticides in cocoa beans
[c]Number of pesticides found in any sample

Table 5 Summary of organochlorine residues (mg/kg) in cocoa beans analysed from the study area

Pesticides	Minimum	Maximum	Mean[a]	SD[b]	EU MRL[c]	% Detected[d]
Lindane (gamma-HCH)	0.03	0.05	0.04	0.01	1.00	18 (56.3)
Beta-HCH	0.02	0.03	0.02	0.00	0.02	16 (50)
Dieldrin	0.02	0.04	0.03	0.01	0.50	12 (37.5)
Aldrin	0.02	0.03	0.03	0.00	0.05	14 (43.8)
p,p'-DDE	0.02	0.03	0.02	0.00	0.50	11 (34.4)
p,p'-DDD	0.02	0.04	0.03	0.01		16 (50)
p,p'-DDT	0.04	0.05	0.04	0.00		20 (62.5)
Methoxychlor	0.02	0.04	0.03	0.01	0.10	10 (31.3)

ND Non-detected

Limit of reporting = 0.005 mg/kg

[a]Mean of mean values recorded for a pesticide residue for each sample site (mean ofsamples where residues were detected)

[b]Standard deviation of samples where residues were detected

[c]European Union maximum residue limits for pesticides in cocoa beans

[d]Number (%) of samples with positive pesticide detections

reported by Aikpokpodion et al. (2012), but higher than the mean values of 0.01 mg/kg and 0.001 mg/kg reported by Daanu (2011) and Frimpong et al. (2012a), respectively.

p,p'-DDT was the most frequently detected residue in the samples analysed. It was observed in 62.5 % of the samples and ranged from 0.04 at S1 and S3 to 0.05 mg/kg at S2 with a mean value of 0.04 ± 0.00 mg/kg. There were no significant differences ($p > 0.05$) in mean values of p,p'-DDT among the sampled sites. The mean value of p,p'-DDT recorded in this study was higher than the mean values of 0.03 mg/kg, 0.003 mg/kg and 0.01 mg/kg reported by Daanu (2011), Frimpong et al. (2012a) and Frimpong et al. (2012b), respectively. However, Aikpokpodion et al. (2012) recorded a mean p,p'-DDT value of 0.06 mg/kg which was higher than the mean value recorded in this study.

The concentrations of p,p'- DDE, p,p'-DDD and p,p'-DDT expressed as total DDT in cocoa bean samples analysed ranged from 0.04 mg/kg at S4 to 0.10 mg/kg at S3 with a mean value of 0.07 ± 0.02 mg/kg. This average residue concentration of DDT (p,p'-DDD, p,p'-DDE and p,p'-DDT) was by far lower than the EU MRL of 0.50 mg/kg for cocoa beans. The mean value of DDT recorded in this study was however higher than the mean values of 0.06 mg/kg, 0.006 mg/kg, 0.01 mg/kg and 0.0003 mg/kg reported by Daanu (2011), Frimpong et al. (2012a), Frimpong et al. (2012b) and Olayinka (2013), respectively, but lower than the mean value of 0.56 mg/kg reported by Aikpokpodion et al. (2012).

DDT has been banned from agricultural use and restricted due to public health concern under the Stockholm convention in which Ghana is a signatory (Afful et al. 2010; Agbeve et al. 2014). Therefore, the occurrence of DDT and its metabolites in cocoa beans from the study area is an indication of the current illegal use of the pesticide by cocoa farmers in the study area. This is confirmed by the occurrence and high concentration of p,p'-DDT compared to its metabolites p,p'- DDE and p,p'-DDD, which is an indication that there might be recent input of DDT in the various cocoa plantations. The continues use of DDT products by the farmers could be due to their efficacy and lower price. It could also be that, new pesticides containing DDT as its active ingredient but un-familiar trade names were sold to the uninformed farmers. A similar observation was made by Aikpokpodion et al. (2012). Similarly, Adu-kumi et al. (2010) noted DDT contamination at some sampled sites in Ghana and attributed the presence to their recent use. However, the higher concentrations of DDT could also be attributed to the enormous past uses of the parent compound DDT and due to its longer half-life (i.e., slow degradation and its long term persistence in the environment).

Methoxychlor was detected in 31.3 % of the cocoa bean samples analysed with a mean value of 0.03 ± 0.01 mg/kg, which ranged from 0.02 mg/kg at S2 to 0.04 mg/kg at S1. There were no significant differences ($p > 0.05$) in mean values of methoxychlor among the sampled sites. However, the observed methoxychlor concentrations were below the EU MRL of 0.10 mg/kg for cocoa beans. The presence of methoxychlor in the cocoa bean samples analysed suggests that methoxychlor had either been used on cocoa in the past (presence in the environment had not degraded) and/or currently illegally used, since they are banned for agricultural purposes in Ghana. In addition, the detection of methoxychlor may be either as a result of historical use of DDT of which technically methoxychlor contains about 88 % of the p,p'-isomer together with more than 50 structurally related contaminants, which might have been added to the actual amount of methoxychlor present (WHO 1996; Bempah and Donkor 2011; Agbeve et al. 2014). The mean value of methoxychlor recorded in the study was higher than the mean value of 0.002 mg/kg reported by Frimpong et al. (2012a) in cocoa beans from Ghana. However, the finding of methoxychlor in this study was contrary to the findings of Frimpong et al. (2012b) which reported no methoxychlor residue in cocoa beans ready for export in Ghana.

The presence of organochlorine pesticides in the cocoa beans samples analysed is a major concern due to the human health risks such as reproductive impairment and suppression of the immune system which can have long-term consequences for population viability. The contamination of the cocoa bean samples could have occurred directly by treating the crop with pesticides before harvest, where pesticide residues were adsorbed by the cocoa beans through the cocoa pod. In addition, the contamination may have occurred through the translocation of these chemicals from contaminated soil through the root system. However, the aforementioned might only hold for water soluble pesticides, since non-water soluble pesticides remain in soil. These pesticides may run off to other areas and cause damage to untargeted animals and plants.

Conclusion

The results of this study revealed that cocoa bean samples from the study area were contaminated with organochlorine pesticide residues. Eight organochlorine pesticide residues namely, aldrin, dieldrin, lindane, beta-HCH, *p,p'*-DDE, *p,p'*-DDD, *p,p'*-DDT and methoxychlor were detected in the cocoa bean samples analysed and were among the banned pesticides of the Environmental Protection Agency (EPA) of Ghana (Afful et al. 2010). *p,p'*-DDT was the most frequently detected organochlorine pesticide residue. The occurrence of organochlorine pesticide residues in the samples analysed could be due to their illegal use by farmers or due to their previous use in cocoa production in the study area, since these chemicals are prohibited from agricultural use including cocoa production in Ghana. The observed mean values of all the pesticide residues detected in the cocoa beans analysed from the study sites were below their respective European Union Maximum Residue Levels for cocoa beans, with the exception of S3 which recorded mean concentration of beta-HCH to be above the EU MRL for cocoa bean. Considering levels of pesticide residues in the fermented dried cocoa beans against the European (EU) commission regulations on pesticide residues, cocoa beans analysed from the study area do not pose health risks and hence, threat to the cocoa industry in Ghana as far as export to Europe is concerned.

The enforcing agencies (Ministry of Agriculture, Ghana Cocoa Board (COCOBOD) and the Environmental Protection Agency of Ghana) should put in place more stringent measures to restrict the importation, sale and use of banned/restricted and unapproved pesticides in cocoa production in Ghana. Also, the National Cocoa Disease and Pest Control (CODAPEC) programme, popularly called "Mass spraying", (whose duty is to assist cocoa farmers in Ghana to combat pests and diseases on cocoa farms using the Ghana Cocoa board approved or recommended pesticides), must institute a monitoring mechanism to ensure that farmers who privately spray their farms do so with approved pesticides. In addition, cocoa farmers in the study area should be educated on the approved Ghana cocoa board pesticides to use on cocoa trees, the right amounts to use and the frequency of application of these pesticides in a farming season. This will help prevent the use of unapproved or banned pesticides as well as reduce the amounts of pesticide residues in the environment. If possible, the Quality Control Company Limited (QCC) of Ghana COCOBOD and the Government of Ghana should institute award systems and pay premium prices on cocoa beans from communities with low or non-detectable levels of unapproved pesticide residues to encourage compliance to good agricultural practices.

Abbreviations
ANOVA, Analysis of variance; CODAPEC, National Cocoa Disease and Pest Control Programme; CP, Gas Chromatograph; DDE, Dichlorodiphenyldichloroethylene; DDT, dichlorodiphenyltrichloroethane; ECD, electron capture detector; EU, European Union; GSA, Ghana Standard Authority; HCH, Hexachlorocyclohexane; MRLs, maximum residue levels; ND, below limit of reporting; OCs, Organochlorine pesticides; S1, Nkrankwanta; S2, Diabaa; S3, Krakrom; S4, Kwakuanya; SPSS, Statistical Package for Social Sciences; WHO, World Health Organization.

Competing interests
The authors declare that they have no competing interest.

Authors' contributions
EDO, FMBY and CG designed the study and wrote the protocol, EDO collected data and conducted data analysis, FMBY and EDO, drafted the manuscript, and FMBY reviewed and edited the manuscript as well as serving as the corresponding author. All authors read and approved the final manuscript.

Acknowledgment
The authors express their profound gratitude to the Pesticide Residue Laboratory of Ghana Standards Authority in Ghana for logistical assistance in the analytical work.

References
Adeyemi D, Anyakora C, Ukpo G, Adedayo A, Darko G. Evaluation of the levels of organochlorine pesticide residues in water samples of Lagos Lagoon using solid phase extraction method. J Environ Chem Ecotoxicol. 2011;3(6):160–166. Retrieved from http://www.academicjournals.org/journal/JECE/article-abstract/13D01A41850.

Adu-kumi S, Kawano M, Shiki Y, Yeboah PO, Carboo D, Pwamang J, Morita M. Organochlorine pesticides (OCPs), dioxin-like polychlorinated biphenyls (dl-PCBs), polychlorinated dibenzo- p -dioxins and polychlorinated dibenzo furans (PCDD/Fs) in edible fish from Lake Volta, Lake Bosumtwi and Weija Lake in Ghana. Chemosphere. 2010;81(6):675–84. doi:10.1016/j.chemosphere.2010.08.018.

Afful S, Anim AK, Serfor-Armah Y. Spectrum of Organochlorine Pesticide Residues in Fish Samples from the Densu Basin. Res J Environ Earth Sci. 2010;2(3):133–8.

Afrane G, Ntiamoah A. Use of pesticides in the cocoa industry and their impact on the environment and the food chain. In: Stoytcheva M, editor. Pesticides in the modern world-risks and benefits. InTech; 2011. pp. 51–68. doi:10.5772/17921. Retrieved from http://www.intechopen.com/books/pesticides-in-the-modern-world-risks-and-benefits/use-of-pesticides-in-thecocoa-industry-and-their-impact-on-the-environment-and-the-food-chain.

Agbeve SK, Osei-Fosu P, Carboo D. Levels of organochlorine pesticide residues in Mondia whitei, a medicinal plant used in traditional medicine for erectile dysfunction in Ghana. Int J Adv Agric Res. 2014;1:9–16.

Agyen KE. Pesticide residues and levels of some metals in soils and cocoa beans in selected farms in the Kade area of the Eastern Region of Ghana. Ghana. (A published Master's thesis), Kwame Nkrumah University of Science and Technology, KNUST, Kumasi, Ghana. 2011. Retrieved from http://hdl.handle.net/123456789/4081.

Ahlborg UG, Lipworth L, Titus-Ernstoff L, Hsieh CC, Hanberg A, Baron J, Trichopoulos D, Adami HO. Organochlorine compounds in relation to breast cancer, edometrial cancer, and endometriosis: an assessment for the biological and epidemiological evidence. Crit Rev Toxicol. 1995;25(6):463–531. doi:10.3109/10408449509017924.

Aikpokpodion P, Lajide L, Aiyesanmi AF, Lacorte S. Residues of Dichlorodiphenyltrichloroethane (DDT) and its Metabolites in Cocoa Beans from Three Cocoa Ecological Zones in Nigeria. Eur J Appl Sci. 2012;4(2):52–7.

Amoah P, Drechsel P, Abaidoo RC, Ntow WJ. Pesticide and pathogen contamination of vegetables in Ghana's urban markets. Arch Environ Contam Toxicol. 2006;50:1–6. doi:10.1007/s00244-004-0054-8.

Apau J, Dodoo D. Lindane and propoxur residues in cocoa from Central region of Ghana. J Sci Technol. 2010;30(3):3–8. doi:10.4314/just.v30i3.64624.

Asiedu E. Pesticide contamination of fruits and vegetables - a market-basket survey from selected regions in Ghana. (A published master's thesis). University of Ghana, Accra, Ghana. 2013. Retrieved from http://hdl.handle.net/123456789/5569.

Bempah CK, Donkor AK. Pesticide residues in fruits at the market level in Accra Metropolis, Ghana. A preliminary study. Environ Monit Assess. 2011;175(1–4):551–61. doi:10.1007/s10661-010-1550-0.

Bempah CK, Buah-Kwofie A, Denutsui D, Asomaning J, Osei-Tutu A. Monitoring of pesticide residues in fruits and vegetables and related health risk assessment in Kumasi Metropolis, Ghana. Res J Environ Earth Sci. 2011a;3(6):761–71.

Bempah CK, Donkor A, Yeboah PO, Dubey B, Osei-Fosu P. A preliminary assessment of consumer's exposure to organochlorine pesticides in fruits and vegetables and the potential health risk in Accra Metropolis, Ghana. Food Chem. 2011b;128(4):1058–65. doi:10.1016/j.foodchem.2011.04.013.

Boakye S. Levels of selected pesticide residues in cocoa beans from Ashanti and Brong Ahafo regions of Ghana. (A published master's thesis), Kwame Nkrumah University of Science and Technology, KNUST, Kumasi, Ghana. 2012. Retrieved from http://hdl.handle.net/123456789/5770.

Botchway F. Analysis of pesticide residues in Ghana's exportable cocoa. A higher certificate project submitted to Institute of Science and Tchnology London, UK. 2000. p. 44–5.

Botwe BO. Organochlorine pesticide residues contamination of Ghanaian market vegetables. 2007. p. 1–6.

Botwe BO, Ntow WJ, Nyarko E, Kelderman P. Evaluation of Occupational and Vegetable Dietary Exposures to Current-Use Agricultural Pesticides in Ghana. In: Pesticides–Recent Trends in Pesticide Residue Assay. InTech; 2012. pp. 46–62. doi:10.5772/80105.

Clarke EEK, Levy LS, Spurgeon A, Calvert IA. The problems associated pesticide use by irrigation workers in Ghana. Occup Med (Chic Ill). 1997;47(5):301–8. doi:10.1093/occmed/47.5.301.

Daanu PB. Concentration of pesticide residues in fermented dried cocoa beans in Asukese and its environs in the Tano north district of Brong Ahafo region, Ghana. (A published master's thesis), Kwame Nkrumah University of Science and Technology, KNUST, Kumasi, Ghana. 2011. Retrieved from http://hdl.handle.net/123456789/4750.

Darko G, Acquaah SO. Levels of organochlorine pesticides residues in meat. Int J Environ Sci Technol. 2007;4(4):521–4. doi:10.1007/BF03325989.

Darko G, Akoto O, Oppong C. Persistent organochlorine pesticide residues in fish, sediments and water from Lake Bosomtwi, Ghana. Chemosphere. 2008;72:21–4. doi:10.1016/j.chemosphere.2008.02.052.

Dikshith TSS. Safe use of chemicals. A practical guide. Boca Raton, Florida, USA: CRC Press (Taylor & Francis Group); 2008.

Frimpong KS, Yeboah P, Fletcher JJ, Adomako D, Osei-fosu P, Acheampong K. Organochlorine pesticides levels in fermented dried cocoa beans produced in Ghana. Elixir Agric. 2012a;44:7280–4.

Frimpong KS, Yeboah P, Fletcher JJ, Adomako D, Pwamang J. Assessment of organochlorine pesticides residues in cocoa beans from Ghana. Elixir Food Sci. 2012b;50:10257–61.

Frimpong KS, Yeboah PO, Fletcher JJ, Pwamang J, Adomako D. Assessment of synthetic pyrethroids pesticides residues in cocoa beans from Ghana. Elixir Food Sci. 2012c;49:9871–5.

Frimpong KS, Yeboah PO, Fletcher JJ, Pwamang J, Adomako D. Multi-residue levels of Organophosphorous pesticides in cocoa beans produced from Ghana. Elixir Food Sci. 2012d;47:8721–5.

Ghana Statistical Service. 2010 Population and Housing Census: District Analytical Report of Dormaa West. 2014.

Hogarh JN, Seike N, Kobara Y, Ofosu-Budu GK, Carboo D, Masunaga S. Atmospheric burden of organochlorine pesticides in Ghana. Chemosphere. 2014;102:1–5. doi:10.1016/j.chemosphere.2013.10.019.

Idowu GA, Aiyesanmi AF, Owolabi BJ. Organochlorine pesticide residue levels in river water and sediment from cocoa-producing areas of Ondo State central senatorial district, Nigeria. J Environ Chem Ecotoxicol. 2013;5(9):242–9. doi:10.5897/JECE2013.0293.

International Cocoa Organization (ICCO). Progress Report Action Programme on Pesticides, ICCO Executive Committee Meeting, EBRD Offices, London. Italy: 2007.

Kaminaga K. The Positive List System in Japan and Our Approach to the Issues of Pesticide Residues in Cocoa. Chocolate and Cocoa Association of Japan (CCAJ). International Workshop on the Safe Use of Pesticides in Cocoa, Kuala Lumpur, Malaysia. 2011.

Kuranchie-Mensah H, Atiemo MS, Maud L, Palm LMND, Blankson-Arthur S, Tutu AO, Fosu P. Determination of organochlorine pesticide residue in sediment and water from the Densu river basin, Ghana. Chemosphere. 2012;86(3):286–92. doi:10.1016/j.chemosphere.2011.10.031.

Leena S, Choudhary SK, Singh PK. Pesticide concentration in water and sediment of River Ganga at selected sites in middle Ganga plain. Int J Environ Sci. 2012;3(1):260–74. doi:10.6088/ijes.2012030131026.

Miles A, Mark K, Ricca A, Robert A, Anthony G, James A. Organochlorine contaminants if fishes from coastal waters West of Amukta pass, Aleutian islands, Alaska, USA. Environ Toxicol Chem. 2009;28(8):1643–54.

Ntow WJ. Pesticide residues in Volta Lake, Ghana. Lakes Reserv Res Manag. 2005;10(4):243–8. doi:10.1111/j.1440-1770.2005.00278.x.

Olayinka AI. Levels of Organochlorine Pesticides (OCPS) Residue in Selected Cocoa Farms in Ilawe – Ekiti, Ekiti State, Nigeria. Open J Anal Chem Res. 2013;1(3):52–8. doi:10.12966/ojacr.11.02.2013.

Owusu-Ansah E, Koranteng-Addo JE, Boamponsem LK, Menlah E, Abole E. Assessment of Lindane pesticide residue in Cocoa beans in the Twifo Praso district of Ghana. J Chem Pharm Res. 2010;2(4):580–7.

Sarfo JE. Behavioural responses of cocoa mirids, sahlbergella singularis hagl and distantiella theobroma dist. (heteroptera:miridae), to sex pheromones. (A published Doctoral thesis), University of Greenwich, London, England. 2013. Retrieved from http://gala.gre.ac.uk/id/eprint/10335.

Sosan MB, Akingbohungbe AE, Ojo IAO, Durosinmi MA. Insecticide residues in the blood serum and domestic water source of cacao farmers in Southwestern Nigeria. Chemosphere. 2008;72(5):781–4. doi:10.1016/j.chemosphere.2008.03.015.

WHO. Guidelines for Drinking-water Quality: Vol. 2 Health and other Supporting Criteria. 2nd ed. Geneva: World Health Organization; 1996.

William JN, Tagoe LM, Drechsel P, Kelderman P, Gijzen HJ, Nyarko E. Cumulation of persistent organochlorine contaminants in milk and serum of farmers from Ghana. Environ Reserv. 2008;106(1):17–26. doi:10.1016/j.envres.2007.05.020.

12

Destabilization and off-flavors generated by *Pseudomonas* proteases during or after UHT-processing of milk

Sophie Marchand[1,2], Barbara Duquenne[1*], Marc Heyndrickx[1,3], Katleen Coudijzer[1] and Jan De Block[1]

Abstract

Background: *Pseudomonads* play a major role in the spoilage of UHT processed dairy products, due to their growth-related protease production in raw milk.

Results: To assess the off-flavor generating capacity of these AprX proteases in milk after UHT-processing, six major milk spoiling *Pseudomonas* groups were investigated. Sensory evaluation of the different processed milk samples showed large differences in the degree of proteolysis related to onset of off-flavors. Nevertheless, it was illustrated that *P. fragi* has the greatest spoilage potential within the tested *Pseudomonas* groups, when it comes to generating off-flavors.

Conclusions: No clear correlation could be obtained between protein hydrolysis and the presence of off-flavors in UHT milk.

Keywords: *Pseudomonas*, Protease, Sensory analysis, Spoilage, Milk

Background

Refrigerated storage of raw milk is universally accepted for prolonging shelf life and preventing spoilage by mesophilic bacteria. Due to evolutions in the dairy market in which dairies have become more and more centralized, milk is now stored longer at refrigerated temperatures (Gaafar and Ali 1995). To ensure good dairy products, the Belgian legislation foresees, that milk on farm should be collected within 72 h post-production (Anonymous 2007). Indeed, quality problems may arise if milk is stored too long at these refrigerated temperatures. This is mainly due to an outgrowth of pyschrotrophic microorganisms in the raw milk. Psychrotrophic *Pseudomonas* (especially *P. fragi*, *P. lundensis* and members of the *P. fluorescens*-like group) are the dominant microbiota of raw milk and are known to compromise heat-treated milk (e.g., UHT) due to the production of heat-stable enzymes during their growth in raw milk (Marchand et al. 2009a). While pseudomonads are readily eliminated by UHT heating conditions (minimal 135 °C

for 1 s), their heat-stable proteases may remain active in the heat-treated products (Chen et al. 2003; Griffiths et al. 1981). The presence of heat-stable *Pseudomonas* protease, encoded by the AprX gene, may result in spoilage and destabilization of UHT milk during extended storage (Dufour et al. 2008). Although, this protease gene is widespread over numerous *Pseudomonas* spp. (Chessa et al. 2000), the production process of this protease is still not completely understood and appears to be very complex. Quorum sensing (Juhas et al. 2005), temperature (Nicodème et al. 2005), iron content (Woods et al. 2001) and phase variation (van den Broeck et al. 2005) regulate and influence the production process of proteases at different levels. Typically, within *Pseudomonas* spp. only one protease, AprX, an alkaline zinc metalloprotease with a pH optimum of 6.5–8, is produced (Woods et al. 2001). This AprX protease is solely responsible for casein hydrolysis as evidenced by casein zymography (Marchand et al. 2009b). The family of serralysin proteases, to which the AprX *Pseudomonas* protease belong, appears to be highly conserved in some domains. Typical similarities in amino acid sequence are observed: a zinc-binding motif (xxxQTLTHEIGHxxGLxxGLxHPx), a calcium binding domain characterized by the presence of four glycine rich

---

* Correspondence: Barbara.Duquenne@ilvo.vlaanderen.be
[1]Institute for Agricultural and Fisheries Research (ILVO), Technology and Food Science Unit, Brusselsesteenweg 370, 9090 Melle, Belgium
Full list of author information is available at the end of the article

repeats (GGxGxD), a high content of hydrophobic amino acids and no cysteine residues (Rawlings and Klostermeyer 1995; Kumeta et al. 1999). However, despite of this general interspecies conservation, genetic differences in the AprX sequence might be responsible for observed inter individual differences in proteolytic capacity and/or the specific activity of these sequence divergent AprX proteases.

Proteolysis of UHT milk causes the development of bitter off-flavors, through the generation of hydrophobic peptides by hydrolysis of casein (Chen et al. 2003; Datta and Deeth 2003). Next to *Pseudomonas* proteases, proteolysis in UHT milk may also be attributed to the native milk enzyme plasmin. Milk plasmin is associated with the casein micelle and the milk fat globular membrane and is also quite heat-resistant (Saint-Denis et al. 2001; Fox and Kelly 2006). It may even partially survive mild UHT-processing conditions. Plasmin exists in milk in both its active form as well as its inactive precursor plasminogen. Its activity in milk is controlled by a complex network of enzyme activators and inhibitors (Fox and Kelly 2006). In addition, *Pseudomonas* proteases may contribute to overall plasmin activity by acting as plasminogen activators to convert plasminogen to plasmin (Fajardo-Lira et al. 2000).

Former research of the authors of this paper identified six major *Pseudomonas* protease groups with great milk spoilage behavior (Marchand et al. 2009a, b). However, no data was gathered yet on the off-flavor generation in milk with relation to that proteolytic capacity. Therefore, this paper addresses the differences in off-flavor generation by six representatives of the major milk *Pseudomonas* protease groups and assesses the correlation between protein hydrolysis and of off-flavor perception in processed UHT milk.

Methods
Selection of milk *Pseudomonas* strains
The selection of the different *Pseudomonas* strains was based on the findings of Marchand et al. (2009a, b). They defined six major *Pseudomonas* AprX protease groups. From each of these groups (A-B-C1-C2-D and the *P. lundensis* group) a representative was chosen to use in this study: respectively, *Pseudomonas* sp. Z34a, *Pseudomonas* sp. W12b, *Pseudomonas* sp. Z34b, *Pseudomonas* sp. W2a, *P. fragi* W41b and *P. lundensis* W52b.

Growth media
Cryopreserved *Pseudomonas* strains were first recovered in Brain Heart Infusion Broth (BHI) (Oxoid, Basingstoke, Hampshire, England) before inoculation in UHT milk. The isolates were incubated in BHI at room temperature until growth was visually present. Next, 100 µL of incubated BHI broth was inoculated in 10 mL of UHT milk and incubated overnight at ambient

temperature. The strains were checked for purity and bacterial counts showed that 24 h incubation at room temperature in UHT milk resulted in approximately 10^8 cfu mL^{-1}. The six cultures were diluted in ringer solution (Oxoid) until 10^7 cfu mL^{-1}.

Raw milk collection, pasteurization, inoculation with *Pseudomonas* strains
575 L of raw milk was collected from a farm in East-Flanders, Belgium. This full fat milk was pasteurized (72 °C,15 s) and aseptically divided in seven batches of 60 L. One batch was used as the control milk for further follow up of the experiment. The other six pasteurized milk batches were each inoculated with 6 mL of the ringer solution containing approximately 7 log mL^{-1} pseudomonads, in order to reach a final concentration of approximately 3 log pseudomonads in the 60 L batch.

Total colony counts and *Pseudomonas* counts
Total colony counts of the raw and pasteurized milk were determined by pour plating serial dilutions on Nutrient agar (Oxoid) with incubation at 30 °C for 3 days. Immediately after inoculation, *Pseudomonas* counts of the six 60 L milk batches were determined on a selective medium for *Pseudomonas* that contains cetrimide (10 mgL^{-1}), fucidin (10 mgL^{-1}) and cephalosporin (50 mgL^{-1}) (CFC agar) (Oxoid) with incubation at 22 °C for 3 days.

Cold milk storage
The inoculated milk batches were further stored for 3 (t3), 4 (t4) and 5 (t5) days at 6,5 °C until skimming and further UHT-processing.

Skimming and UHT-processing
After cold storage, the seven different milk batches were further processed. Before UHT-processing the milk was skimmed using a Elecrem decreamer (Type 315 L/H,7800 tpm; Tomega, Marche-en-Famenne, Belgium). Indirect UHT-processing was performed on a Junior N326L apparatus, Process Pilot Plant, 200 Lh-1 (APV, Aartselaar, Belgium) under the following conditions: 2 steps homogenization: 200 bar, 65 °C; indirect UHT-processing: 5 s,140 °C; cooling to 20 °C. Milk was aseptically filled in high density poly ethylene (HDPE) bottles of 0.5 L and stored at 37 °C to accelerate possible proteolysis events. To ensure safe sensory evaluation, all produced milk samples were tested for sterility. Therefore 2 (HDPE) bottles of each milk batch were chosen randomly and incubated at 30 °C for 3 days. Total plate counts of the milk samples were determined by undiluted pour plating and incubation at 30 °C for 3 days.

Sensory evaluation

Preliminary sensory evaluation and proteolysis measurements started after 3 days storage at 37 °C. The six *Pseudomonas* protease milk samples were compared with the reference control milk (by a taste panel of 5 persons) on a daily basis. If off-flavors were experienced, a larger sensory analysis took place. The panel consisted of 35 people (8 men and 27 women) who were staff of the Institute of Agriculture, Fisheries and Food Research. Their mean age was 37 years (range 24–55 years). All panelists had earlier experience in sensory evaluation of milk. Evaluations were conducted in a sensorial cabinet that was equipped with individually partitioned booths. Milk samples (30 mL) were served at 14 °C. The set-up was as follows: Milk with off-flavor was diluted in the following way: A) *Pseudomonas* protease milk undiluted, B) 2/3 *Pseudomonas* milk + 1/3 control milk, C) 1/3 *Pseudomonas* milk + 2/3 control milk, D) Control milk undiluted. Next, the taste panel was asked to rank the milk samples (A-B-C-D) according to preference. Statistical evaluation of the results was based on the Rank Test to Kramer (Kramer 1960) for α = 0.05 but also compared with a newly developed sensory evaluation test. In short: The four milk samples (A-B-C-D) were presented at random to the tasting panel. The tasting panel was asked to arrange the milk samples on a line scale of 10 cm length according to preference. 10 cm was considered as an excellent taste, 5 cm satisfactory and 0 cm was considered as having a very bad taste. Two

correction parameters were added to this test; (an example is given in Fig. 1). First, a correction within the tasting panel: the sensory analyst was enabled to choose between five statements to indicate the difference degree between the most extreme marks on the line scale. Each of these statements was correlated with a corrected difference in cm going from 0 cm (no noticeable difference) till 4 cm (very obvious difference) (Fig. 1). And second, a correction for the sensory analyst: the reference milk (the control undiluted milk, which is unknown to the tasting panel) and the other milk samples under evaluation can be placed by the sensorial analyst anywhere on the line scale. However, for the evaluation of the test, the control undiluted milk (D), is considered as satisfactory and is thus arbitrary associated with 5 cm on the line scale. Therefore, regardless of the place where the sensorial analyst has placed the reference, this milk gets de facto scaling 5. To obtain the corrected position for the other milk samples under evaluation, the distances between them and the reference mark need to be measured accurately. If the taste of the other milk samples was considered worse than the reference, the analyst would have placed its mark left from the reference on the line scale. If the taste was better, on the other hand, it would have been placed on the right hand side from the reference. Dependent if the mark is left or right from the reference, the measured distance between the two of them will be subtracted or added, respectively, from the arbitrary 5 cm scaling, resulting in the

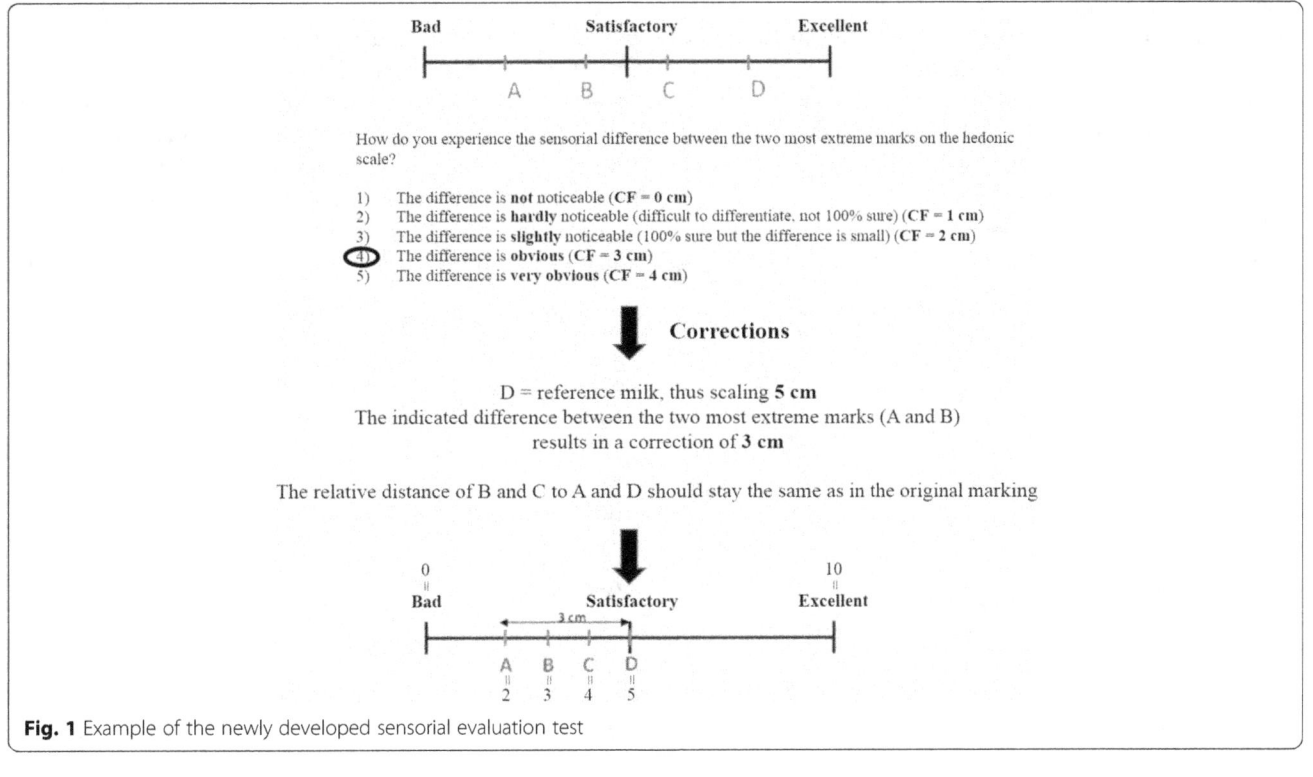

Fig. 1 Example of the newly developed sensorial evaluation test

corrected values. All values derived from the different sensorial analysts were grouped and mean values and standard deviations (sd) for each milk sample were determined. Milk samples with mean values (±1 x standard deviation) were considered significant according to the Rank test to Kramer (Kramer 1960). Simultaneously, proteolysis was determined in each milk dilution (A-B-C-D).

Measurement of proteolysis

Hydrolysis of proteins was measured by the determination of the release of a-amino groups directly in milk by the trinitrobenzenesulfonic acid (TNBS) method (Polychroniadou 1988), in which free amino groups react with the TNBS reagent (Sigma-Aldrich, Bornem, Belgium) at pH 9.2 in the absence of light. A yellow-orange color develops and its intensity is determined in duplicate by absorption measurements at 420 nm. The amount of proteolysis in the *Pseudomonas* milk samples and dilutions is calculated from the increase in absorption and expressed as µmol glycine equivalents mL^{-1} milk using glycine (Sigma-Aldrich) as a standard curve.

Results and discussion
Milk processing and bacterial counts

575 L raw (full fat) milk was pasteurized and aseptically divided in seven batches of 60 L. Total bacterial count of the raw milk was 33.000 cfu/mL. After pasteurization total bacterial counts were reduced to 1900 cfu/mL in the pasteurized milk. Six 60 L batches were subsequently inoculated with *Pseudomonas* strains. *Pseudomonas* counts and total bacterial counts were determined at the moment of inoculation (t0) and before UHT-processing (t3, t4) for every milk batch. Bacterial counts can be retrieved in Table 1. All produced milk samples were sterile and were used for further sensory evaluation. In addition, the experiments showed that milk, which had been stored for 5 days or longer cannot be processed anymore under UHT conditions, because of

destabilization of the milk, resulting in clogging of the heating exchanger.

Sensory evaluation of the processed milk samples

Sensory evaluation of the different processed milk samples showed large differences in the onset of off-flavors. The majority of the *Pseudomonas* inoculated milk samples (strains Z34b, Z34a, W2a and W52b) were stored 4 days at 6,5 °C before sufficient proteases were produced. In *Pseudomonas* sp. Z34b milk, *Pseudomonas* sp. Z34a milk and *Pseudomonas lundensis* W52b milk, off-flavors occurred after 13 days of storage at 37 °C post UHT-processing, while in *Pseudomonas* sp. W2a, off-flavors were already present after 10 days. However, it was illustrated that not all pseudomonads contain equal spoilage threats for the dairy industry: *Pseudomonas fragi* Z41b and *Pseudomonas* sp. W12b produced already sufficient proteases after 3 days of storage at 6,5 °C. Off-flavors were detected in those milk samples, after 15 days and 10 days storage at 37 °C, post UHT production, respectively (Table 2). From these results, it can be deduced that refrigerated storage of milk should be limited in order to prevent or reduce *Pseudomonas* protease production. In addition, the sensorial analysis evaluation by the method of Kramer (Kramer 1960) and the method described in this paper showed identical results (Table 2).

Correlation between off-flavors and protein hydrolysis in UHT milk by the six different *Pseudomonas* protease groups

In each sensory evaluated milk sample proteolysis was determined. TNBS-values of each milk sample can be retrieved in Table 2. Based on these grouped results the correlation between protein hydrolysis and off-flavors can be determined. First of all, control milk was checked for the occurrence of proteolysis events. During a period of 30 days, milk was monitored for possible proteolytic activity. The TNBS value remained constant over time and had a mean value of $1,01 \pm 0,04$ µmol glycine

Table 1 Bacterial counts in milk before the different processing steps (In bold: Pseudomonas counts sufficient to induce off-flavors in the processed milk samples)

MILK	After pasteurisation		Before UHT processing	Before UHT processing
	(t0)		3 days storage at 6,5 °C (t3)	4 days storage at 6,5 °C (t4)
	TBC log (cfu/ml)	Added *Pseudomonas* count log (cfu/ml)	*Pseudomonas* count log (cfu/ml)	*Pseudomonas* count log (cfu/ml)
CONTROL	3,28	0,00	0,00	0,00
Pseudomonas sp. Z34a	3,49	3,43	5,78	**6,96**
Pseudomonas sp. W12b	3,58	3,52	**6,43**	7,08
Pseudomonas sp. Z34b	3,57	3,45	6,23	**6,60**
Pseudomonas sp. W2a	3,61	3,41	6,63	**6,75**
P. fragi Z41b	3,18	3,04	**6,48**	6,93
P. lundensis W52b	3,38	3,20	6,48	**6,79**

Table 2 Oversight of the sensory evaluation and proteolysis results ([a]Milk samples with off-flavors have values within the range indicated by Kramer and are thus significantly different from the milk samples with values outside that range)

Pseudomonas culture	UHT production date	Time stored at 37 °C after UHT production / and time of sensory evaluation	Dilution with blanc milk	off-flavor	Sensorial evaluation method described in this paper value ± sd	Rank test to Kramer milk sample rank sum	(lowest - highest) insignificant rank sum[a]	# sensorial evaluators	Proteolysis (µmolglycine equivalents / ml)
Blanc	02/12/2010 (t3)	15 days	Undiluted	NO	5	33	(39–56)	19	1,01
Blanc	02/12/2010 (t3)	11 days	Undiluted	NO	5	18	(33–47)	16	0,96
Blanc	03/12/2010 (t4)	13 days	Undiluted	NO	5	24[Z34a] / 20[Z34b] / 26[W52b]	(35–50)[Z34a] / (33—47)[Z34b] / (35–50)[W52d]	17[Z34a] / 16[Z34b] / 17[W52d]	1,04
Blanc	03/12/2010 (t4)	10 days	Undiluted	NO	5	17	(33–47)	16	0,92
Z34a	03/12/2010 (t4)	13 days	1:3	NO	4,69 ± 1,72	25	(35–50)	17	1,18
			2:3	NO	4,62 ± 1,68	26	(35–50)	17	1,39
			Undiluted	YES	3,43 ± 1,53	35	(35–50)	17	1,73
W12b	02/12/2010 (t3)	11 days	1:3	NO	4,37 ± 1,29	22	(33–47)	16	1,35
			2:3	NO	4,55 ± 1,40	22	(33–47)	16	1,58
			Undiluted	YES	3,00 ± 1,88	35	(33–47)	16	1,84
Z34b	03/12/2010 (t4)	13 days	1:3	NO	4,39 ± 1,13	29	(33–47)	16	1,42
			2:3	YES	3,89 ± 1,13	37	(33–47)	16	1,78
			Undiluted	YES	2,16 ± 1,24	53	(33–47)	16	2,24
W2a	03/12/2010 (t4)	10 days	1:3	NO	4,1 ± 1,38	23	(33–47)	16	1,58
			2:3	YES	3,12 ± 1,79	38	(33–47)	16	2,04
			Undiluted	YES	1,64 ± 0,75	47	(33–47)	16	2,48
fragi Z41b	02/12/2010 (t3)	15 days	1:3	NO	4,82 ± 0,82	32	(39–56)	15	1,06
			2:3	YES	3,61 ± 1,35	41	(39–56)	15	1,16
			Undiluted	YES	3,38 ± 1,35	53	(39–56)	15	1,2
lundensis W52b	03/12/2010 (t4)	13 days	1:3	NO	4,53 ± 1,54	29	(35–50)	17	1,41
			2:3	YES	3,02 ± 1,44	43	(35–50)	17	1,81
			Undiluted	YES	2,78 ± 1,45	54	(35–50)	17	2,2

a

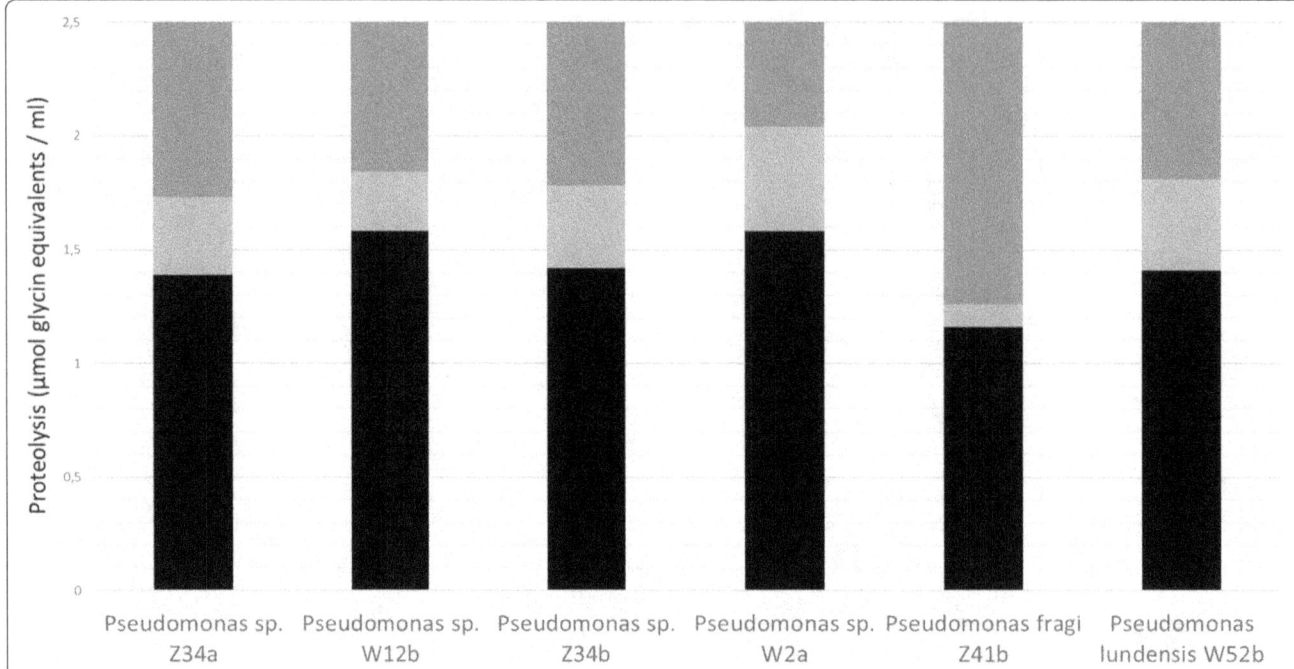

Fig. 2 Correlation between off-flavors and protein hydrolysis in UHT milk by 6 different *Pseudomonas* protease groups. * No significant proteolysis off-flavors indicated by black bars, the uncertainty range by light gray bars (the panel did not reject the milk samples; the lower limit of the bar is determined by the TNBS-value of the most diluted sample that was not rejected by sensory analysis) and the significant proteolysis off-flavors by dark gray bars (panel rejected the milk samples and tasted off-flavors; the lower limit of the bar is determined by the TNBS-value of the least diluted sample)

equivalents mL^{-1}. It can be concluded that no plasmin activity was present and milk (raw or processed) of good quality should thus have a TNBS-value in that range. Next, all data concerning sensorial and proteolysis analyses were compiled in Fig. 2. This graph shows that no clear correlation can be obtained between the onset of off-flavors and the rate of protein hydrolysis in milk by the different *Pseudomonas* protease groups. The TNBS-values of the milk samples in which off-flavors were significantly tasted were different for each *Pseudomonas* protease under evaluation. For example, with *Pseudomonas* sp. W2a proteases, the TNBS-value was allowed to rise with 1,03 µmol glycin equivalents mL^{-1} before any off-flavors were tasted. *P. fragi* proteases, on the other hand were capable in generating off-flavors after very limited proteolysis (a raise in TNBS-value of 0,15 glycin equivalents mL^{-1}). Therefore it can be speculated that not all *Pseudomonas* proteases have the same specificity for their casein substrates. The amino acid recognition sites within *Pseudomonas* proteases might thus be fundamentally different, resulting in peptide generation with a variable hydrophobic amino acid content. Further research, however is necessary to confirm this. Nevertheless, it is now clear that presence (of high numbers) of *P. fragi* strains prior to UHT-processing will severely compromise the shelf life of derived dairy products. To ensure good quality dairy products, milk should therefore be processed as quickly as possible or held

refrigerated (≤2 °C) (Griffiths 1989; Haryani et al. 2005) awaiting further processing.

Conclusions

High *Pseudomonas* counts and extended cold storage severely limits UHT-processing. Therefore, to ensure good quality dairy products, raw milk should be processed as quickly as possible or kept well refrigerated (≤2 °C) during the entire dairy chain (from farm to dairy). No clear correlation can be obtained between the degree of protein hydrolysis by the different *Pseudomonas* AprX proteases and the generation of off-flavors in UHT-milk. Nevertheless, *P. fragi* has the greatest spoilage potential within the tested *Pseudomonas* protease groups, when it comes to generating off-flavors.

Acknowledgements
The authors wish to thank Hans Bultinck for performing all pasteurization and UHT operations, Hans Steurbaut, Claudine Roels and Geert Goeteyn for the help with the flask filling operations. This research was supported by an OZM grant from the Agency for Innovation by Science and Technology (IWT).

Authors' contributions
SM carried out the samplings, the sample analyses and drafted the manuscript. BD and JDB developed the new sensory evaluation test. MH, KC, JDB and BD participated in the design of the study. All authors read and approved the final manuscript.

Competing interests
The authors declare that they have no competing interests.

Author details

[1]Institute for Agricultural and Fisheries Research (ILVO), Technology and Food Science Unit, Brusselsesteenweg 370, 9090 Melle, Belgium. [2]University Hospital Ghent, Metabolic and Cardiovascular Diseases, Ghent University, De Pintelaan 185, 9000 Gent, Belgium. [3]Department of Pathology, Bacteriology and Poultry Diseases, Ghent University, Salisburylaan 133, 9820 Merelbeke, Belgium.

References

Anonymous. Ministrieel besluit houdende goedkeuring van het document opgesteld door de erkende inteprofessionele organismen betreffende de modaliteiten van de controle van de kwaliteit van de rauwe koemelk. Belgisch Staatsblad. 2007;7679–83.

Chen L, Daniel RM, Coolbear T. Detection and impact of protease and lipase activities in milk and milkpowders. Int Dairy. 2003;7:255–75.

Chessa JP, Petrescu I, Bentahir M, van Beeumen J, Gerday C. Purification, physico-chemical characterization and sequence of a heat-labile alkaline metalloprotease isolated from a psychrophilic Pseudomonas species. Biochim Biophys Acta. 2000;1479:265–74.

Datta N, Deeth HC. Diagnosing the cause in proteolysis in UHT milk. LWT Food Sci Technol. 2003;36:173–82.

Dufour D, Nicodème M, Periin C, Driou A, Brusseaux E, Humbert G, et al. Molecular typing of industrial strains of Pseudomonas spp. isolated from milk and genetical and biochemical characterization of an extracellular protease produced by one of them. Int J Food Microbiol. 2008;125:188–96.

Fajardo-Lira C, Oria M, Hayes KD, Nielsen SS. Effect of psychrotrophic bacteria and of an isolated protease from Pseudomonas fluorescens M3/6 on the plasmin system of fresh milk. J Dairy Sci. 2000;83:2190–9.

Fox PF, Kelly AL. Indigenous enzymes in milk: overview and historical aspects - Part 1. Int Dairy J. 2006;16:500–16.

Gaafar AM, Ali AA. The role of psychrotrophic bacteria in raw milk on stability of milk proteins to UHT-treatment. Egypt J Food Sci. 1995;23:147–54.

Griffiths MW. Effect of temperature and milk fat on extracellular enzyme synthesis by psychrotrophic bacteria during growth in milk. Milchwissenschaft. 1989;44:537–43.

Griffiths MW, Phillips JD, Muir DD. Thermostability of proteases and lipases from a number of species of psychrotrophic bacteria of dairy origin. J Appl Bacteriol. 1981;50:289–303.

Haryani S, Datta N, Elliot AJ, Deeth HC. Production of proteinases by psychrotrophic bacteria in raw milk stored at low temperature. Aust J Dairy Technol. 2005;58:15–20.

Juhas M, Eberl L, Tümmler B. Quroum sensing: the power of cooperation in the world of Pseudomonas. Environ Microbiol. 2005;7:459–71.

Kramer A. A rapid method for determining significance of differences from rank sums. Food Technol. 1960;14:576–81.

Kumeta H, Hoshino T, Goda T, Okayama T, Shimada T, Ohgiya S, et al. Identification of the serralysin family isolated from the psychrotrophic bacterium, Pseudomonas fluorescens 114. Biosci Biotechnol Biochem. 1999;63:1165–70.

Marchand S, Heylen K, Messens W, Coudijzer K, De Vos P, Dewetinck K, et al. Seasonal influence on heat-resistant proteolytic capacity of P. lundensis and P. fragi, predominant milk spoilers isolated from Belgian raw milk samples. Environ Microbiol. 2009a;11:467–82.

Marchand S, Vandriesche G, Coorevits A, Coudijzer K, De Jonghe V, Dewettinck K, et al. Heterogeneity of heat-resistant proteases from milk Pseudomonas species. Int J Food Microbiol. 2009b;133:68–77.

Nicodème JP, Grill G, Gaillard JL. Extracellular protease activity of different Pseudomonas strains: dependence of proteolytic activity on culture conditions. J App Microbiol. 2005;99:641–8.

Polychroniadou A. A simple procedure using trinitrobenzenesulphonic acid for monitoring proteolysis in cheese. J Dairy Sci. 1988;55:585–96.

Rawlings ND, Klostermeyer H. Evolutionary families of metallopeptidases. Methods Enzymol. 1995;248:183–228.

Saint-Denis BT, Humbert G, Gaillard JL. Heat inactivation of native plasmin, plasminogen and plasminogen activators in bovine milk: a revisited study. Lait. 2001;81:715–29.

van den Broeck D, Bloemberg VG, Lugtenberg B. The role of phenotypic variation in rhizosphere Pseudomonas bacteria. Environ Microbiol. 2005;7:1686–97.

Woods RG, Burger M, Beven CA, Beacham IR. The aprX-lipA operon of Pseudomonas fluorescens B52: a molecular analysis of metalloprotease and lipase production. Microbiology. 2001;147:345–54.

13

Trace metal and aflatoxin concentrations in some processed cereal and root and tuber flour

Hayford Ofori*, Charles Tortoe, Paa Toah Akonor and Jonathan Ampah

Abstract

Background: Sweet potato, cocoyam, water yam, maize, millet, sorghum, and rice are major staple foods in Ghana. Flour from these roots and tubers and cereals are relatively cheaper, easy to produce and can be used in the manufacturing of wide range of products, including bread, cookies, meat pies, cake, chips and doughnut. However, due to the processing techniques used these flours may be contaminated with trace metals and aflatoxins. It was therefore necessary to determine the concentration of trace metal: As, Cu, Fe, Hg, Pb, Zn, and aflatoxin B_1, B_2, G_1, and G_2 in maize, sorghum, millet, rice, cocoyam, water yam and sweet potato flour samples used as composite flour for making bread and other pastries.

Results: The trace metal concentration of essential metals: Cu, Fe and Zn in all the seven flour samples analyzed ranged from 0.03 ± 0.01 to 6.63 ± 0.02, 10.97 ± 0.01 to 201.40 ± 0.14, and 6.04 ± 0.03 to 34.36 ± 0.06 mg/Kg, respectively. The maximum concentrations of toxic metals: As, Hg and Pb determined in all the seven flour samples were < 0.01, < 0.01, and 0.05 ± 0.01 mg/Kg, respectively. Aflatoxin B_1, B_2, G_1, and G_2 were not detected in any of the samples.

Conclusion: The concentrations of As, Cu, Hg, Pb and Zn determined in all seven flour samples were below the WHO set limits and therefore pose no health threat whiles Fe concentration in some flour samples was above the WHO set limit. Aflatoxin B_1, B_2, G_1, and G_2 were not detected in all flour samples and hence pose no health threat.

Keywords: Flour, Root and tubers, Cereals, Heavy metal, Aflatoxins, WHO

Background

Root and tubers as well as cereal are major staple foods in Ghana. They contribute significantly to the elimination of food insecurity in Ghana. Maize is a base for several food preparations in Ghana. Maize supply in Ghana has steadily been increasing over the past few years with the average supply at 1.5 million MT. In the northern Ghana, millet and sorghum are the main cereals produced and consumed, but maize becomes the substitute when millet and sorghum are not available (MoFA 2011). Sorghum together with millet and more recently maize is a fundamental crop for North-East Ghana farming systems. It is mainly cultivated by small farmers with average landholdings not more than 2 ha

(Kudadjie et al. 2004). Production of sorghum in Ghana between 2005 and 2010 ranged from 305,000 tons to 324,422 tons, respectively (FAO 2012). Beyond food security and provision of cash the value of sorghum is linked to the social context and religious ceremonies typical of Ghana rural areas. During these events the sorghum artisanal beer called *pito* is widely consumed. Sweet potato (*Ipomoea batatas*) is fast growing root crop that can be produced in all regions of Ghana. It matures within four months, and yields about 20-25 tons per hectare. Cocoyam (*Colocasia esculenta and Xanthosona spp*) is well adapted food crop across many agroecological zones of Sub-Saharan Africa. It is ranked third in importance, after cassava and yam among root and tuber crops cultivated and consumed in many West and Central Africa Countries (Onyeka 2014). Cocoyam is nutritionally superior to both cassava and yam in the possession of higher protein, mineral and vitamin contents

* Correspondence: oforihayford@yahoo.com
Council for Scientific and Industrial Research-Food Research Institute, P.O. Box M 20, Accra, Ghana

as well as easily digestible starch. Africa in the last three decades has consistently accounted for an increasing percentage of global cocoyam production, which currently stands at about 10 million tons per annum (FAO 2012). From 2008 to 2012, Africa accounted for 74 % of global cocoyam production with approximately 50 % of global output occurring in West and Central Africa. In Ghana cocoyam is predominantly grown in the wetter forest zones because of its high moisture content requirements for growth. Average yield of cocoyam in Ghana ranges between 4 and 7.5 tons per hectare (MoFA 2011). Yam (*Dioscorea sp*) is an annual or perennial climbing plant with edible underground tuber which is native to warmer regions of both southern and northern hemispheres (IITA 2004). A report by IITA indicated that, this tropical-vine tuber is popular in Africa, the West Indies, and parts of Asia, South and Central America (IITA 2004). Yam forms about 10 % of total roots and tubers produced in the World (FAO 2005). Seventy percent (70 %) of the total yam production in the World comes from Nigeria whilst the rest comes from Côte d'Ivoire, Ghana, Benin, Togo and Cameroon (FAO 2005). Over 600 yam varieties are grown throughout the World, but in West Africa the most economically viable species are the White yam (*Dioscorea rotundata*), Yellow yam (*Dioscorea cayenensis*), and Water yam (*Dioscorea alata*) (Vernier 1998).

Ghana is a net importer of rice and wheat. A report by Ministry of Food and Agriculture Ghana (MoFA 2011) put the consumption of wheat in Ghana at approximately 300,000 MT whiles the estimated per capita consumption stands at 12.5 kg. Almost 80 percent of wheat flour imported into Ghana goes into bread making whiles the remaining 20 percent is used for cakes and other pastries. To curtail the high cost of importation of wheat flour into Ghana, CSIR-Food Research Institute-Ghana under the West Africa Agriculture Productivity Program (WAAPP 2A) has trained bakers and pastry makers on how to use root and tuber and locally grown cereal flour as composite flour for the production of bread and other pastries. These flour are relatively cheaper, easy to produce and could be used in the manufacturing of wide range of products, including bread, cookies, meat pies, cake, chips and doughnut. Composite flour is a percentage blend of any root and tuber or cereal flour and wheat flour. The processes by which these flours are produced may lead to trace metal and aflatoxin contamination. Trace metal and aflatoxin contaminants in foods have become a matter of public health concern in recent times (Aradhna et al. 2009). Trace metals such as copper (Cu), iron (Fe), and zinc (Zn) are essential elements because of their role as cofactors in metabolic and biosynthetic processes. Arsenic (As), lead (Pb) and mercury (Hg) are non-essential for they have no known beneficiary roles and are toxic above

certain levels. The essential trace metals may produce toxic effect at high concentrations (Celik and Oehlenschlager 2007; Tuzen 2009). Trace metal accumulate in human organs such as liver, kidney, bone and causes severe health disorders. For instance Pb can cause renal masses, affect cognitive development and may lead to adult cardiovascular disease (Bandara et al. 2008). Mercury is considered by WHO as one of the top ten chemicals or group of chemicals of major public health concern. Mercury may have toxic effects on the nervous, digestive and immune systems, and on lungs, kidneys, skin and eyes (WHO 2003). For fetuses, infants and children, the primary health effect of mercury is impaired neurological development. Arsenic is also considered as toxic. Ingestion of As leads to gastrointestinal symptoms, and disturbances of cardiovascular and nervous systems functions. Long term exposure to As is casually related to increased risk of cancer (Xiong et al. 2013). Aflatoxin are known to be carcinogenic. Aflatoxin contamination in food is caused by the presence of *Asperigillus flavus*, *Asperigillus nomius* and *Asperigillus parasitus* (Essono et al. 2009). These organism usually contaminate the food product and synthesize the toxins as metabolites in the presence of high levels of carbohydrates and low levels of protein (Essono et al. 2009).

In the present study, the concentrations of trace metal including As, Cu, Fe, Hg, Pb, Zn and aflatoxin B_1, B_2, G_1 and G_2 were determined in sorghum flour, millet flour, maize flour, rice flour, water yam flour, sweet potato flour and cocoyam flour used as composite flour in bakery and pastry product.

Materials and methods

Materials

Raw sorghum, millet, maize, rice, water yam, sweet potato and cocoyam were bought from Agbogloshie market in Accra, Ghana and transported to the processing laboratory of CSIR-Food Research Institute, Accra for processing into flour.

Methods

Sample preparation

Cocoyam flour preparation: Matured and healthy cocoyam cormels were peeled with sharp stainless steel knife. Peeled cocoyam was washed three times to remove dirt and impurities. The washed peeled cocoyam was sliced into small pieces using the stainless steel knife for easy drying. It was blanched by steeping in hot water for 2-3 minutes and solar dried on thinly cleaned black plastic sheet till it became firm and brittle. Solar dried cocoyam slices were milled using hammer mill and milled flour sifted using sieve to remove lumps to ensure free-flow flour with even particles was obtained. The flour was packaged into sacs lined with moisture-proof polythene and sealed airtight using an electric impulse sealer till analysis was carried out (Tortoe et al. 2014). Sweet potato flour and water yam flour were prepared the same way as the cocoyam flour was prepared.

Maize flour preparation: All the three cereal flour samples were prepared using a method by Addo et al. 2015. Mold-free and weevil-free maize was selected. It was poured onto a tray for sorting. Foreign materials such as stones and insect infested maize were removed through winnowing. The maize was dehulled after sorting. The dehulled maize was solar dried on cleaned black plastic polythene sheet until moisture content was about 11 %. The dried dehulled maize was milled using hammer mill. The milled flour was sifted using a 250 microns sieve to remove lumps so as to obtain a free-flow flour. The maize flour was then put in sacs lined with moisture-proof polythene and sealed with air-tightened electric impulse sealer. The rice flour was prepared the same way as the sorghum flour.

Millet flour preparation: Mold-free and weevil-free millet was selected. Millet grains was poured into a plastic bowl containing enough water. Suspended water millet grain was decanted little at a time using calabash to ensure that millet grains are free of stones and impurities. This process was repeated several times till all stones and impurities are removed from the millet grains. Cleaned millet grains was solar dried by spreading it on a thin black plastic sheet. Millet grains was dehulled and solar dried the second time till moisture content of 11 %. The dehulled dried millet was milled into flour using hammer mill. The flour was sifted using a sieve of mesh size of 250 microns to remove all lumps.

Sorghum flour preparation: the variety of sorghum processed was the Beiko peleg which is suitable for the preparation of local food (*tuo zaafi*). Clean and mold-free sorghum grains was selected for the flour production. Sorghum grains was washed with clean water to remove dirt, stones and other foreign materials. Grain was solar dried by spreading it on a clean black polythene sheet till moisture content was 11 %. Grains was dehulled and solar dried for the second time to remove most of the moisture content. It was milled using the hammer mill into flour and sifted using sieve of mesh size of 250 microns to remove lumps.

Chemical analyses
Trace metal content determination
The method used by (Ofori et al. 2016) was used for the trace metal analyses. The dry ashing method was used for the Atomic Absorption Spectrometry (AAS) analysis (AOAC 2005). All glass ware was washed with 1 % nitric acid followed by demineralized water. Three grams of each sample was weighed into a platinum crucible. The crucible and test sample was placed in muffle furnace at a temperature of 550 °C for 8 hours. The crucible with ash

was put in desiccator to cool. Five (5) mL of nitric acid of mass fraction not less than 65 %, having a density of approximately ρ (HNO_3) = 1400 $mg.mL^{-1}$ was added, ensuring that all the ash came into contact with the acid and the resultant solution heated on hot plate until the ash dissolved. Ten (10) mL of 0.1 $mol.L^{-1}$ nitric acid was added and filtered into 50 mL volumetric flask. The resultant solution was topped up to the mark with 0.1 $mol.L^{-1}$ nitric acid. Blank solution was treated similar as the sample. Buck Scientific 210VGP Flame Atomic Absorption Spectrophotometer (Buck Scientific, Inc. East Norwalk, USA) was used to read the absorbance values at appropriate wavelength of the interested metal in sample solution. Cathode lamps used were As (wavelength 193.7 nm, lamp current 4.0 mA), Cu (wavelength 324.8 nm, lamp current 1.5 mA), Fe (wavelength 248.3 nm, lamp current 7.0 mA), Hg (wavelength 253.7 nm, lamp current 0.7 mA), Pb (wavelength 217.0 nm, lamp current 3.0 mA), and Zn (wavelength 213.9 nm, lamp current 2.0 mA). The metal content of the sample was derived from calibration graph made up of a minimum of three standards.

Aflatoxins determination
The extraction procedure used for the determination of aflatoxins was by the Stroka and Anklam (1991). A test portion (50 g) was extracted with 200 ml methanol/water solvent solution containing 5 g of sodium chloride. The sample extract was filtered, diluted with phosphate buffered saline to a specified solvent concentration and applied to the immunoaffinity column (R-Biopharm Rhone Ltd. Easi-Extract Aflatoxin) containing antibodies specific for aflatoxins B_1, B_2, G_1 and G_2. Aflatoxins were eluted from the immunoaffinity columns with neat methanol. The Aflatoxin level was quantified by reverse-phase high performance liquid chromatography (RP-HPLC) with post column derivatisation (PCD) involving bromination. The PCD was achieved with pyrimidinum hydrobromide perbromide (PBPB) followed by fluorescence detection. HPLC system used for analyses was from Waters Associates (Milford, MA, USA) and included Waters 1525 Binary HPLC pump, Waters 2707 Autosampler, Waters Model 1500 Column Heater, Waters 2475 Multi λ Fluorescence Detector and Breeze 2 software. Separation of the aflatoxin was carried out on a Spherisorb S5 ODS-1 column of dimensions 25x4.6 mm packed with 5 μm particles (phase separation In., Norwalk, USA) maintained at 35 °C. The HPLC mobile-phase flow rate was 10 ml/min and post column bromine derivatisation of Aflatoxin B_1, and G_1 was achieved by PBPB dissolved in 500 ml of demineralised water pumped at a flow rate of 1.0 ml/min using Elder precision metering pump (Elder laboratories Inc., Sam Carlos, USA). The excitation and emission wave length used were 360 nm and 440 nm respectively. The Aflatoxin were identified by means of their retention times, and quantification

was performed by comparing the peak areas of the samples to those of the standards prepared from pure aflatoxins standard (obtained from R. Biopharm) solutions under identified conditions.

Quality control of results

Samples were handled carefully to avoid contamination as part of measure to ensure reliability of results. The recovery test of the total analytical procedures was also carried out for the metals analyzed in the selected samples by spiking analyzed samples with aliquots of metal standards and then reanalyzed the samples. Acceptable recovery ranges of 94 ± 1 to 95 ± 1 %, 96 ± 1 to 97 ± 1 %, 93 ± 1 to 95 ± 1 %, 93 ± 1 to 94 ± 1 %, 95 ± 1 to 96 ± 1 %, 95 ± 1 to 97 ± 1 % were obtained for As, Cu, Fe, Hg, Pb and Zn respectively.

Data analysis

The standard deviations on mean values of duplicate samples were analyzed using Statistical Package for Social Scientist (SPSS 2013), version 21. Analyses of Variance (ANOVA), Duncan test was used to compare the means.

Results and discussion

Trace metal analyses

Trace metal variation of root and tuber flour is shown in Table 1. There was no significant differences in the concentration of As determined in all the three root and tuber flour samples at $p < 0.05$. WHO has recommended Provisional Tolerable Weekly Intake (PTWI) of As as 0.015 mg/Kg body weight (WHO 2010). The maximum concentration of As determined in the root and tuber flour was < 0.01 mg/Kg which is below the recommended PTWI by WHO. The minimum and maximum Cu concentration determined in the root and tuber flour was 0.69 ± 0.01 mg/Kg and 3.67 ± 0.02 mg/Kg, respectively which are far below the limit of 40 mg/Kg limit set by WHO as limit of Cu in foods (WHO 1982). Fe deficiency anemia affect one-third of the World population. However, excess intake of Fe causes colorectal cancer (Senesse et al. 2004). The concentration of Fe determined in root and tuber ranged from 10.97 ± 0.01 mg/Kg to 27.64 ± 0.16 mg/kg with the highest concentration above the 15 mg/Kg limit set by WHO as limit of Fe in food (WHO 1982). The maximum concentrations of Hg

and Pb determined in root and tuber flour was < 0.01 mg/Kg and 0.02 ± 0.01 mg/Kg respectively. Impairment related to Hg toxicity includes peripheral vision, disturbances in sensations, muscle weakness and lack of movement coordination (Xiong et al. 2013). WHO has recommended a Provisional Tolerable Weekly Intake (PTWI) of Hg as 1.6 µg/Kg body weight (WHO 2003). Pb is toxic even at trace levels (Dobaradaren et al. 2010). The maximum Pb concentration determined in root and tuber was below the 10 mg/Kg limit set by WHO as limit of Pb in raw plant material (WHO 1982). The concentration of Zn determined in root and tuber flour ranged from 6.79 ± 0.01 mg/Kg to 15.54 ± 0.52 mg/Kg, which are below the 60 mg/Kg limit that has been set by WHO (WHO 1982).

Table 2 shows the variation in trace metal concentration of cereal flour samples. The maximum concentration of As, Cu, Fe, Hg, Pb and Zn in the cereal flour samples were < 0.01, 3.07 ± 0.01, 201.40 ± 0.14, <0.01, 0.05 ± 0.01, 34.36 ± 0.06 mg/Kg respectively; of which the concentrations of As, Cu, Hg, Pb and Zn fell below the WHO set limit whilst that of Fe happens to be far above the WHO set limit (WHO 1982).

Table 3 compares trace metal concentrations in Sweet potato flour, cocoyam flour, water yam flour, sorghum flour, rice flour, maize flour and millet flour. The trace metal concentration of essential metals such as Cu, Fe and Zn in all the seven flour samples ranged from 0.03 ± 0.01 to 6.63 ± 0.02, 10.97 ± 0.01 to 201.40 ± 0.14, and 6.04 ± 0.03 to 34.36 ± 0.06 mg/Kg respectively. According to Silvestre et al. 2000 Cu is essential constituent of some metalloenzymes and is required for haemoglobin synthesis and in the catalysis of metabolic growth. The highest concentration of Cu determined in all the seven flour samples was 6.63 ± 0.02 mg/Kg which is far below the 40 mg/Kg limit that has been set by WHO (WHO 1982) as limit of Cu in foods. Statistically, there was no significant differences in the concentration of Cu in Sorghum and Millet flour which are cereals grown in the northern part of Ghana. Fe is also essential metal for the body but excess intake may lead to colorectal cancer (Senesse et al. 2004). The concentration of Fe determined in Cocoyam flour, Sorghum flour, Maize flour and Millet flour were above the 15 mg/Kg limit set by WHO as limit of Fe in Foods (WHO 1982). Ma and Betts (2000) stated that Zn constitutes about 33 ppm of

Table 1 Variations in trace metal concentration in mg/Kg dried weight of root and tuber flour

Sample	As	Cu	Fe	Hg	Pb	Zn
Sweet potato flour	<0.01[a]	0.69 ± 0.01[a]	10.97 ± 0.01[a]	<0.01[a]	<0.01[a]	6.79 ± 0.01[a]
Cocoyam flour	<0.01[a]	6.63 ± 0.02[b]	27.64 ± 0.16[b]	<0.01[a]	<0.01[a]	15.54 ± 0.52[b]
Water yam flour	<0.01[a]	3.67 ± 0.02[c]	15.11 ± 0.06[c]	<0.01[a]	0.02 ± 0.01[a]	9.19 ± 0.04[c]

Results are presented as means ± standard deviation. Superscript to figures in the same column implies significant or insignificant differences at $p < 0.05$ (ANOVA, Duncan test)

Table 2 Variations in trace metal concentration in mg/Kg dried weight of cereal flour

Sample	As	Cu	Fe	Hg	Pb	Zn
Sorghum flour	<0.01[a]	2.77 ± 0.33[a]	116.75 ± 0.07[a]	<0.01[a]	<0.01[a]	25.36 ± 0.06[a]
Rice flour	<0.01[a]	1.09 ± 0.05[b]	11.87 ± 0.06[b]	<0.01[a]	<0.01[a]	7.84 ± 0.01[b]
Maize four	<0.01[a]	0.03 ± 0.01[c]	20.44 ± 0.02[c]	<0.01[a]	<0.01[a]	6.04 ± 0.03[c]
Millet flour	<0.01[a]	3.07 ± 0.01[a]	201.40 ± 0.14[d]	<0.01[a]	0.05 ± 0.01[b]	34.36 ± 0.06[d]

Results are presented as means ± standard deviation. Superscript to figures in the same column implies significant or insignificant differences at $p < 0.05$ (ANOVA, Duncan test)

adult body weight and is essential as a constituent of many enzymes involved in a number of physiological functions such as protein synthesis and energy metabolism. Zinc deficiency has been linked to increased risk of stunting in children (Black et al. 2008). The minimum and maximum Zn concentrations determined in the cereal and root and tuber flour were below the limit of 60 mg/Kg set by WHO (WHO 1982). The highest concentrations of As, Hg and Pb determined in the seven flour samples were < 0.01, < 0.01, and 0.05 ± 0.01 mg/Kg respectively which were all below the WHO set limit of As, Hg and Pb in foods. The maximum concentration of Hg determined in the root and tuber and cereal flour samples was < 0.01 mg/Kg which is similar to that determined in cassava flour samples by Ofori et al. (2016). Arsenic (As) and Pb was not detected in processed cassava flour samples analyzed in Ghana by Ofori et al. (2016) but was detected in all the root and tuber and cereal flour sample analyzed presently even though their concentrations were within acceptable levels by WHO (WHO 1982).

Aflatoxin analyses
The recovery of the analytical method was 90.50, 76.56, 95.58, 91.76 % for aflatoxin B_1, B_2, G_1, and G_2 respectively and a linearity of $R^2 = 0.999$. The seven flour samples of Maize, Sorghum, Millet, Rice, water Yam, Cocoyam and Potato did not contain aflatoxin B_1, B_2, G_1 and G_2 at detectable levels, which may be attributed to good processing techniques used. The limit of detection (LOD) for aflatoxin B_1 and B_2 was 0.15 µg/kg and

aflatoxin G_1 and G_2 was 0.13 µg/kg whiles the limit of quantification (LOQ) for aflatoxin B_1, B_2, G_1, and G_2 was 0.16, 0.30, 0.28 and 1.08 µg/kg, respectively. Limit of detection (LOD) and LOQ for aflatoxin was obtained using the formula; LOD = standard concentration at which no peak was observed (3*baseline noise/peak height) and LOQ = 2*LOD. The findings from the present work conforms to a similar work on aflatoxins content in High Quality Cassava Flour (HQCF) done by Ofori et al. (2016) where aflatoxin B_1, B_2, G_1, and G_2 was not detected in all HQCF samples analyzed but contradict previous studies on aflatoxin determination in Maize, Sorghum, Millet and Water yam flour in some West African countries such as Ghana, Gambia and Nigeria. For example Kpodo (2001) conducted aflatoxin determination in maize from Ghana which revealed that eighty-four (84) out of one hundred and twenty-eight (128) maize kernel sampled from markets and maize processing sites were contaminated with aflatoxin at level up to 200 ng/g whiles a similar studies on aflatoxin determination in cereal-based food products intended for infants and young children done by Blankson and Mill-Robertson (2016) also showed aflatoxin contamination at levels of 0.18 ± 0.01 to 36.10 ± 0.32 µgkg^{-1}. Hudson et al. (1992)) analyzed nine millet samples for aflatoxin in Gambia a West African country and all the nine sample were contaminated with aflatoxin to level ranging from 1-27 ng/g. Gbolagade et al. (2011) in Nigeria analyzed six yam flour samples which were all contaminated with aflatoxins to a level ranging from 25.17 µg/kg to 32.33 µg/kg.

Table 3 Variations in trace metal concentration in mg/Kg dried weight of cereal and root and tuber flour

Sample	As	Cu	Fe	Hg	Pb	Zn
Sweet potato flour	<0.01[a]	0.69 ± 0.01[a]	10.97 ± 0.01[a]	<0.01[a]	<0.01[a]	6.79 ± 0.01[a]
Cocoyam flour	<0.01[a]	6.63 ± 0.02[b]	27.64 ± 0.16[b]	<0.01[a]	<0.01[a]	15.54 ± 0.52[b]
Water yam flour	<0.01[a]	3.67 ± 0.02[c]	15.11 ± 0.06[c]	<0.01[a]	0.02 ± 0.01[a]	9.19 ± 0.04[c]
Sorghum flour	<0.01[a]	2.77 ± 0.33[d]	116.75 ± 0.07[d]	<0.01[a]	<0.01[a]	25.36 ± 0.06[d]
Rice flour	<0.01[a]	1.09 ± 0.05[e]	11.87 ± 0.06[e]	<0.01[a]	<0.01[a]	7.84 ± 0.01[e]
Maize flour	<0.01[a]	0.03 ± 0.01[f]	20.44 ± 0.02[f]	<0.01[a]	<0.01[a]	6.04 ± 0.03[f]
Millet flour	<0.01[a]	3.07 ± 0.01[d]	201.40 ± 0.14[g]	<0.01[a]	0.05 ± 0.01[b]	34.36 ± 0.06[g]

Results are presented as means ± standard deviation. Superscript to figures in the same column implies significant or insignificant differences at $p < 0.05$ (ANOVA, Duncan test)

Conclusion

The maximum concentration of essential metal, Cu and Zn determined in the cereal and root and tuber flour samples was 6.63 ± 0.02 mg/Kg and 34.36 ± 0.06 mg/Kg respectively which are below the WHO set limits and therefore pose no health threat. The highest Fe concentration determined in the flour samples was 201.40 ± 0.14 mg/Kg which is far above the WHO set limit and therefore pose a health threat. The toxic metals: As, Hg and Pb had their concentrations below the WHO set limit and therefore pose no health threat. Aflatoxin B_1, B_2, G_1 and G_2 were not detected in all the cereal and root and tuber flour samples analyzed and hence pose no health threat.

Acknowledgement

This is an output from the West Africa Agriculture Productivity Program (WAAPP 2A)-Ghana, funded by the World Bank and Government of Ghana. The views expressed are not necessarily those of World Bank.

Authors' contributions

The work was done in collaboration between all authors. Authors HO and CT designed the experiment. Author HO conducted the experiments. Authors HO, CT, PTA and JA conducted the literature search, performed the statistical analyses and wrote the first draft of the manuscript. All authors read and approved the final draft of the manuscript.

Competing interests

The authors declare that they have no competing interests.

References

Addo P, Tortoe C, Hagan L, Buckman ES, Akonor PT, Padi A, Addy P, Dawson AE, Wayo TC. Indigenous Cereal Composite Flour Processing and Recipe Training Manual. Accra: Council for Scientific and Industrial Research-Food Research Institute; 2015. p. 15–50.

AOAC. Official methods of analysis of AOAC International. 18th ed. Gaithersburg: AOAC International; 2005.

Aradhna G, Devendra KR, Ravi SP, Bechan S. Analysis of some heavy metals in the riverine water, sediments and fish from Ganges at Allahabad. Environ Monit Assess. 2009;157:449–58.

Bandara JMRS, Senevirathna DMAN, Dasanayake DMRS, Herath V, Bandara JMRP, Abeysekara T, et al. Chronic renal failure among farm families in cascade irrigation systems in Sri Lanka associated with elevated dietary cadmium levels in rice and freshwater fish (Tilapia). Environ Geochem Health. 2008;30:465–78.

Black, Caulfield LE, De Onis M, Ezzati M, Mathers C, Rivera J. Maternal and child undernutrition: global and regional exposures to health consequences. Lancet. 2008;371:243–60.

Blankson GK, Mill-Robertson FC. Aflatoxin contamination and exposure in processed cereal-based complementary foods for infants and young children in greater Accra, Ghana. Food Control. 2016;64:212–7.

Celik U, Oehlenschlager J. High of cadmium, lead, zinc and copper in popular fishery products sold in Turkish supermarkets. Food Control. 2007;18(3):258–61.

Dobaradaren S, Kaddafi K, Nazmara S, Ghaedi H. Heavy metals (Cd, Cu, Ni, and Pb) content in fish species of Persian Gulf in Bushehr Port, Iran. AJ Biotech. 2010;32:6191–3.

Essono G, Ayodele M, Akoa A, Foko J, Filtengborg O, Olembo S. Aflatoxin-producing Aspergillus spp and aflatoxin levels in stored cassava chips as affected by processing practices. Food Control. 2009;20:648–54.

FAO. FAOSTAT Agriculture data. Food and Agriculture Organisation of the United Nations. 2005.

FAO. Food and Agricultural Organization (FAO) production statistics. 2012.

Gbolagade J, Ibironka A, Yetunde O. Nutritional composition, fungi and aflatoxins detection in stored "gbodo" (fermented Dioscorea rotundata) and "elubo ogede" (fermented Musa parasidiaca) from South-western Nigeria. Afr J Food Sci. 2011;5(2):105–10.

Hudson GJ, Wild CP, Zarba A, Groopman JD. Aflatoxin isolated by immunoaffinity chromatography from foods consumed in The Gambia, West Africa. Nat Toxins. 1992;1:100–5.

IITA. International Institute of Tropical Agriculture. Nigerian's Cassava Industry: Statistical Handbook. 2004.

Kpodo KA. Fusaria and Fumonisins in maize and fermented maize products in Ghana. PhD. Thesis. University of Ghana, Legon, Ghana. 2001.

Kudadjie CY, Struik PC, Richards P, Offei SK. Assessing production constraints, management and use of sorghum diversity in North-East Ghana: A diagnostic study. University of Ghana, Legon Accra, University of Wageningen; 2004.

Ma J, Betts NM. Zinc and Copper intakes and their major food sources for older adults in the 1994-96 continuing survey of food intakes by individual 9CSF-II). J Nutr. 2000;130:2838–43.

MoFA. Agriculture in Ghana, Facts and Figures, 2011. Statistical Research and Information Directorate (SRID). Ghana Ministry of Food and Agriculture; 2012.

Ofori H, Akonor PT, Dziedzoave NT. Variation in trace metal and aflatoxin content during processing of High Quality Cassava Flour (HQCF). Int J Food Contam. 2016;3:1.

Onyeka J. Status of cocoyam (Colocasia esculenta and Xanthosoma spp) in West and Central Africa: Production, Household Importance and Threat from Leaf Blight. Lima (Peru).CGIAR Research Program on Roots, Tubers and Bananas (RTB). 2014. www.rtb.cgiar.org.

Senesse P, Meance S, Cottet V, Faivre J, Boutron-Ruault MC. High dietary iron and copper and risk of colorectal cancer: a case –control study in Burgundy, France. Nutr Cancer. 2004;49:66–71.

Silvestre MD, Lagarda MJ, Farra R, Martineze-Costa C, Brines J. Copper, iron and zinc determination in human milk using FAAS with microwave digestion. Food Chem. 2000;68:95–9.

SPSS 21 for Windows (2013). SPSS 21 for Windows. Chicago, Illinois, USA.

Stroka J, Anklam E. Quantitative analysis for aflatoxins. JAOAC. 1991;74:81–4.

Tortoe C, Akonor PT, Padi A, Boateng C, Opoku Asiama M, Addy P, Dawson AE, Wayo TCA. Root and Tuber Composite Flour Processing and Recipe Manual. Accra: Council for Scientific and Industrial Research-Food Research Institute; 2014. p. 3–24.

Tuzen M. Toxic and essential trace elemental contents in fish species from Black Sea, Turkey. Food Chem Toxicol. 2009;47(8):1785–90.

Vernier P. Yam chips production in West Africa. The newsletter of post-harvest system in Africa. International Institute of Tropical Agriculture. No. 2, 1998.

World Health Organisation (WHO). Evaluation of Certain Foods Additives and Contaminants (Twenty-Six Report of the Joint FAO/WHO Expert Committee on Food Additives). WHO Technical Report series, No. 683, Geneva; 1982.

World Health Organisation (WHO). Evaluation of certain food additives and contaminants (sixty-first report of the joint FAO/WHO expert committee on food additives). WHO technical report series, No. JECFA/16/SC, Rome; 2003.

World Health Organisation (WHO). Evaluation of certain food additives and contaminants (sixty-first report of the joint FAO/WHO expert committee on food additives). WHO technical report series, No. JECFA/72/SC, Rome; 2010.

Xiong C, Zhang Y, Xu X, Lu Y, Ouyang B, Ye Z, et al. Lotus roots accumulate heavy metals independently from soil in main production regions of China. Scientia Horticulturea. 2013;164:295–302.

Effectiveness of the sanitation regimes used by dairy actors to control microbial contamination of plastic jerry cans' surfaces

Wanjala Nobert Wafula[1*], Wafula Joseph Matofari[1], Masani John Nduko[1] and Peter Lamuka[2]

Abstract

Background: The most common milk handling containers used by dairy actors along the informal milk value chain in developing countries are plastics jerry cans which are difficult to effectively be cleaned thus contributing immensely to milk contamination and consequently post-harvest losses. The aim of this study was to determine the effectiveness of some common cleaning regimes used by the dairy actors in Kenya against reduction of surface microbial load on jerry cans. Milk handling plastic jerry can containers ($n = 16$) were obtained from dairy actors and then subjected to four different commonly used cleaning regimes alongside a control experiment of aluminium cans ($n = 4$). These containers were aseptically swabbed in three replicates before and after the application of a cleaning regime and the swabs ($n = 120$) analyzed for Total Viable Count (TVC), Total Coliform Count (TCC) and Lactic Acid Bacteria (LAB). The quantitative mean difference of the bacterial load reduction between before and after the application of a cleaning regime was used as the measure of its effectiveness.

Results: The study found out that irrespective of the cleaning, the type of container was significant ($P < 0.001$) in the reduction of microbial contaminants, whereby the aluminium cans had the highest microbial load reduction of 86, 85 and 96 % for TVC, TCC and LAB respectively as compared to 40, 28 and 42 % for TVC, TCC and LAB respectively for plastic jerry cans. The use of a commercial scourer in the cleaning was found to significantly reduce ($P < 0.05$) only TVC and TCC.

Conclusions: The results from this study explains the unsuitability of plastic jerry cans in handling of milk and a risk factor for milk post-harvest losses in Kenya through microbial contamination.

Keywords: Plastic jerry cans, Contamination, Post-harvest losses, Nakuru

Background

Milk from the mammary glands of healthy animals is initially sterile, but post-harvest handling like the milking personnel and milk handling containers; remain to be the major sources of bacterial contamination of raw milk (Coorevits et al. 2008; Reta et al. 2016). Therefore, milk should be produced under hygienic conditions so as to meet set standards (Ahmad et al. 2015) which are $<10^6$ colony forming units/ml in the case Kenya (KEBS 2010). However, the procedures used in cleaning and sanitizing the milk handling equipments are also key factors in influencing the level of microbial contamination of raw milk in terms of counts and the types of bacterial (Kelly et al. 2009).

Milk should be handled in hygienically designed equipment i.e. one that has no dead spaces and crevices, the major control method of surface route of milk contamination, is the use of an effective cleaning and disinfection programme. Failures in the cleaning and disinfection regimes will causes bacterial deposits on the container surfaces thus incubation site for them (Reinemann et al. 2003). In particular, dead ends, corners, joints, valves and the hard-to-reach places of milk handling equipment are the most appropriate regions for the existence of microbial contaminants. Bacteria attach on milk handling equipment surfaces either as single cells or in binary biofilms, which may become difficult to remove (Lindsay et al. 2002). The presence of crevices and scratches on equipment surfaces causes accumulation of organic debris that

* Correspondence: blessednobert@yahoo.com
[1]Department of Dairy and Food Science and Technology, Egerton University, P.O. Box 536-20115, Egerton, Kenya
Full list of author information is available at the end of the article

offers good condition for bacterial growth thus high concentration of microbial load whereby some withstand the cleaning and disinfection (Murphy and Boor 2000). Residual bacteria on surfaces that remain after cleaning and disinfection have the potential to proliferate and cause problems in the dairy value chain. Therefore the hygiene of equipment surfaces definitely affects the quality and safety of the milk and milk products to the public (Olivier and Moshoeshoe 2012).

A very wide range of plastics are available but it's only a few of them that are food grade approved such as polypropylene (PP), polycarbonate (PC), high-density polyethylene (HDPE), unplasticized polyvinyl chloride (PVC), and fluoropolymers such as polytetrafluoroethylene (PTFE, Teflon®). Some of these plastics are porous and lack resilience and must thus be used with carefully (Faille and Carpentier 2009). Common for most dairy actors' milk handling containers in many developing countries are the plastic jerry cans and plastic buckets.

These plastic jerry cans and buckets have been reported to be used in many areas including Kenya (Omore et al. 2005), Burkina Faso (Millogo et al. 2010), Ethopia (Welearegay et al. 2012; Worku et al. 2014), Mali (Bonfoh et al. 2006), Tanzania (Kivaria et al. 2006), Turkey (Tasci 2011), Peru (Fuentes et al. 2014) Iran (Fadaei 2014) and Zimbabwe (Gran et al. 2002) among many more countries. Plastic jerry cans are more complicated to clean than aluminium cans because of the small opening (Kivaria et al. 2006). The common methods of cleaning milk handling equipments throughout the dairy value chain are use of a bar soap either with hot water, warm water or cold water and sometimes use of a commercial scourer (Orregard 2013). However there are no guidelines especially in Kenya on the use of this bar soaps and commercial scourers in the sanitation of the milk handling containers. Also, no studies have carried out in these areas where the plastic jerry cans are used to handle milk against how effective they are the sanitation practices. This study focused on the evaluating the effectiveness of the cleaning regimes commonly used by dairy actors in Kenya in sanitation of plastic jerry cans against reduction of surface microbial contaminants.

Methods
Study area
This cross-sectional study was conducted between March to October 2015 in two different areas of Nakuru County. The two areas were Olenguruone in Kuresoi sub-county (0°34'60"S, 35°40'60"E) and Bahati/Wanyororo in Bahati sub-county (0°9'0"S, 36°7'0"E). Nakuru county has an altitude of 2490 m above sea level, rainfall of 600 mm/year and temperature range of 7–25 °C. Analysis of the samples was done at Egerton University, Food Microbiology laboratory.

Sampling and sample preparation
A total of sixteen plastic jerry can milk handling containers were randomly obtained from dairy actors: dairy farmers, milk transporters and milk vendors and four aluminium cans from a dairy plant (Guildford Dairy) and transported to the laboratory under aseptic conditions for microbial load analysis at 4 °C. The plastic jerry were cut open and then surface swabs for collecting microorganisms were done using a sterile cotton swab buds pre-wetted in peptone water at an area of 5 × 5 cm in three replicates. The replicates were from the same obtained milk container (aluminum cans and plastic jerry cans) at randomly selected different places. Swabs were taken before and after a cleaning regime by rotating the cotton end in contact with the prepared milk handling container surfaces before and after a cleaning regime. The swabbed samples were then transferred to the 9 ml 0.1 % (w/v) buffered peptone water and shaken using a vortex for 2 min to dislodge the bacteria.

Experimental design of the cleaning regimes on the plastic jerry cans
This study was conducted in a completely randomized design in a 5 × 2 factorial arrangement where the plastic jerry cans and the control were subjected to different treatments with or without interaction with the use of a commercial scourer as shown in Table 1. The commercial scourer was placed in the milk containers during the washing together with the bar soap and water (cold, warm or hot), closed with the lid and shaken vigorously for 2 min. After the cleaning the milk containers were either only rinsed or disinfected. The containers were then inverted on a rack in the sun to dry. Each treatment is one of the common sanitation practice (Wafula et al. 2016) used by the dairy actors in Kenya for the sanitation of milk containers and the surface swabs were taken before and after application of the treatment A total of sixty surface swabs were obtained from the ten treatments used on the twenty milk handling containers (plastic jerry cans (*n* = 16) and aluminium cans (*n* = 4)) and in three replicates for microbial analysis.

The control milk container were aluminium cans were washed, rinsed and disinfected with a chlorine based disinfectant at a concentration of 300 ppm as shown in Table 1. For the treatment with disinfection, it was applied after the cleaning of the aluminium cans and two plastic jerry cans with bar soap and commercial scourer. The contact time between the rinsed container and the disinfectant was 3 min. After drying, swabbing was done using sterile pre-wetted swabs to collect microorganisms on the surfaces. The swabbed samples were serially diluted in buffered peptone water and cultured in the same regime as plastic jerry can containers surface swabs.

Table 1 The treatments used in the sanitation of milk handling containers

Treatment (Factor A)						Factor (B)		
Regime		Container	Bar soap	Disinfectant	Water temperature	Commercial scourer	No. of containers	Reps
1	a	Aluminium	Yes	Yes	45 °C	Yes	2	3
	b	Aluminium	Yes	Yes	45 °C	No	2	3
2	a	Plastic	Yes	Yes	45 °C	Yes	2	3
	b	Plastic	Yes	Yes	45 °C	No	2	3
3	a	Plastic	Yes	No	22 °C	Yes	2	3
	b	Plastic	Yes	No	22 °C	No	2	3
4	a	Plastic	Yes	No	45 °C	Yes	2	3
	b	Plastic	Yes	No	45 °C	No	2	3
5	a	Plastic	Yes	No	85 °C	Yes	2	3
	b	Plastic	Yes	No	85 °C	No	2	3

Microbial analysis

For total plate count, swabbed samples were serially diluted appropriately in buffered peptone water. The dilutions were then plated using pour plate method on Plate Count Agar (PCA) (Oxoid, UK) at 37 °C for 48 h. The dilutions giving the expected total number of colonies (30–300) on a plate were selected (Richardson 1985) and the colony counting was made using Dr. N. Gerber digital colony counter (Schneider and Co., Zurich) and counts recorded. For Total Coliform Count (TCC) were surface swabs pour plated on Violet Red Bile Agar (Oxoid, UK) and incubated at 36 °C for 24 h on and typical dark red colonies on the plates was considered as coliforms and counted. For lactic acid bacteria (LABs) surface swabs were also serially diluted following similar methods as for total bacterial count but the dilutions were pour plated on MRS (Oxoid, UK) agar then incubated at 37 ± 2 °C/48 h.

The colonies were further isolated and identified according to their morphological, physiological and biochemical tests characteristics. The tests that were carried out were Gram reaction test, catalase test, oxidase test, and methyl-red test (MR), Voges-proskauer test (VP), Indole test and sugar fermentations. The sugars used for testing of the fermentation of the isolates were sucrose, galactose, glucose, lactose and mannitol (Grainger et al. 2001). Sugar fermentation and gas production was considered as sufficient evidence for the presence of coliforms (Ombarak and Elbagory 2014). For the TCC, a confirmatory test by transferring 1 ml of the aliquots from each dilution into three tubes of Lauryl sulfate tryptose (LST) broth and incubating at 35 °C for 48 h was also done. While for the LABs, colonies were further determined by their ability to grow at 15 °C, 35 °C and 45 °C for 5 days and in NaCl at 2, 4 and 6.5 % strength in MRS broth. The growth of LAB at the different temperatures and salt concentrations was visually confirmed by turbidity changes in the MRS broth after 24, 48 and 72 h (Azadnia and Khan 2009).

Statistical analysis

Data obtained from the difference in microbial counts (TBC, TCC and LAB) between before and after application of the treatment were transformed to logarithmic values (log_{10}) of colony forming units per cm^2 (cfu/cm^2) before statistical analysis. Logarithmic transformations were applied to the data to meet the assumptions of analysis of variance (ANOVA) using Komolgorov–Smirnoff's test was used to test the normality and Levene's test to test the homogeneity of variances (Goberna et al. 2005). ANOVA was analyzed using General Linear Model (PROC GLM) procedure, Komolgorov–Smirnoff's test was done using PROC NPAR1WAY procedure and Levene's test done using PROC GLM with LEVENE,s option in SAS software version 9.1. Treatments means separations were done using Least Significant Difference (LSD) at $P \leq 0.05$. Also planned orthogonal contrast method was done for the comparisons of means among the treatments.

Results

The effect of milk container type on the surface microbial load reduction

The effect of milk handling container type on the surface microbial load reduction for TVC, TCC and LAB is shown in Table 2.

The study found out that irrespective of the treatment, the overall microbial reduction was very high on aluminium milk handling containers by 86, 85 and 96 % for TVC, TCC and LAB respectively, as compared to plastic milk handling containers that was 36, 28 and 42 % for TVC, TCC and LAB respectively. After the cleaning, aluminium container surfaces had a mean microbial residual load of 0.82 ± 0.88, 0.70 ± 0.25 and 0.13 ± 0.08 log_{10} cfu/cm^2 while the plastic jerry cans had a

Table 2 Comparison of mean reduction in microbial load between the aluminium milk handling cans (Control) and plastic milk handling jerry cans

Type of the equipment	N	Type of microorganism (mean log$_{10}$ cfu/cm^2)	Initial microbial load	Final microbial load	Change in microbial load	% Microbial load reduction
Aluminium	12	TVC	5.86 ± 0.92	0.82 ± 0.34	5.06 ± 0.88	86.3
	12	TCC	4.53 ± 1.17	0.70 ± 0.25	3.83 ± 1.02	84.5
	12	LAB	3.77 ± 0.74	0.13 ± 0.08	3.60 ± 0.77	95.5
Plastic	48	TVC	5.99 ± 1.03	3.84 ± 0.92	2.50 ± 0.66	35.9
	48	TCC	5.07 ± 0.91	3.64 ± 0.80	1.43 ± 0.68	28.2
	48	LAB	4.81 ± 0.81	2.75 ± 1.07	2.03 ± 0.87	42.2

Key: *TVC* total viable count, *TCC* total coliform count and *LAB* lactic acid bacteria

mean microbial residual load of 3.84 ± 0.92, 3.64 ± 0.80 and 2.75 ± 1.07 log$_{10}$ cfu/cm^2. The mean microbial load reduction on the aluminium type of milk handling containers was significantly higher ($P < 0.001$) than on plastic jerry cans and therefore the type of the container had an influence on the effectiveness of the sanitation process.

The effect of the treatments on the reduction of surface microbial load

The effect of different treatments applied on the milk handling containers on the reduction of surface microbial load for TVC, TCC and LAB is shown in Table 3.

The study found out a significant difference ($P < 0.05$) among the treatments used in the experiment for reduction of all microbial types (TVC, TCC and LAB). There was only a significant difference ($P < 0.05$) in the use of a commercial scourer for reduction of TVC and TCC in Table 3. Also the interaction effect of the use of commercial scourer with different treatments was found to be significant ($P < 0.001$) in the reduction of TVC and TCC. Though not significant ($P > 0.05$), the overall mean reduction of the LAB was higher where commercial scourer was not used i.e. 2.39 ± 1.06 log$_{10}$ cfu/cm^2 when compared where it was used i.e. 2.31 ± 1.07 log$_{10}$ cfu/cm^2 as shown in Fig. 1 and Table 3. The study found out that though there was no significant difference ($P < 0.05$) in LAB reduction among the regimes, but the interaction effect of the treatment and use of a commercial scourer had the highest microbial load reduction on the

Table 3 The analysis of variance's mean of squares table for the reduction of microbial load from surfaces of the milk containers

Source of variation	DF	TVC	TCC	LAB
Treatment	4	20.319***	14.299***	12.520***
Commercial scourer	1	0.699*	0.432*	0.103^{ns}
Treatment* Commercial scourer	4	3.575***	3.405***	0.635^{ns}
Replication	2	0.061	0.405	0.247
Error	50	0.269	0.349	0.267

Key: *C.V* coefficient of variation, *S.D* standard deviation and *R^2* coefficient of determination *ns* not significant at $P = 0.05$, *significant at $P < 0.05$, **significant at $P < 0.01$ and ***significant at $P < 0.001$

aluminium cans and regime where the disinfectant was used by 98 and 65 % respectively. However, in the regime that used the disinfectant, there was a marginally high reduction of bacteria where the commercial scourer was not used than where it was used. There are very high chances that the commercial scourer used in the cleaning is the one introducing the microorganisms.

The effect of the use of a commercial scourer during cleaning of the milk handling containers on the reduction of surface microbial load for TVC, TCC and LAB are shown in Fig. 1. The study found out that the treatments that used a commercial scourer in the cleaning had significantly ($P < 0.05$) less microbial reduction as compared to treatments that did not use the commercial scourer for the TVC and TCC but did not differ significantly ($P > 0.05$) for the reduction of the LAB. The use of a commercial scourer in the cleaning of the aluminium containers had the highest reduction of the LAB and lowest for TVC whereas TCC remained almost the same to when compared to the cleaning of the aluminium cans without a commercial scourer. However on plastic jerry cans, use of the commercial scourer had a least microbial reduction of TVC, TCC and LAB when compared to where it was not used. Though the LAB were the type of bacteria had the highest reduction on plastic jerry cans, the plastic jerry cans that were cleaned without the commercial scourer had a reduction of 44.9 % for the LAB when compared to the jerry cans that were cleaned with a commercial scourer that had a reduction of 39.5 %.

The effect of cold, warm and hot water in the cleaning of the aluminium cans and plastic Jerry cans on the reduction of surface microbial load for TVC, TCC and LAB are shown in Table 4.

The study found out that the irrespective the temperature of water for sanitation, aluminium cans had significantly ($P < 0.001$) much higher microbial load reduction from the surfaces when compared to the plastic jerry cans for TVC, TCC and LAB. The mean reduction of the LAB on the plastic jerry cans treated with a disinfectant was not significantly different ($P > 0.05$) from the control but significantly different ($P < 0.01$) from

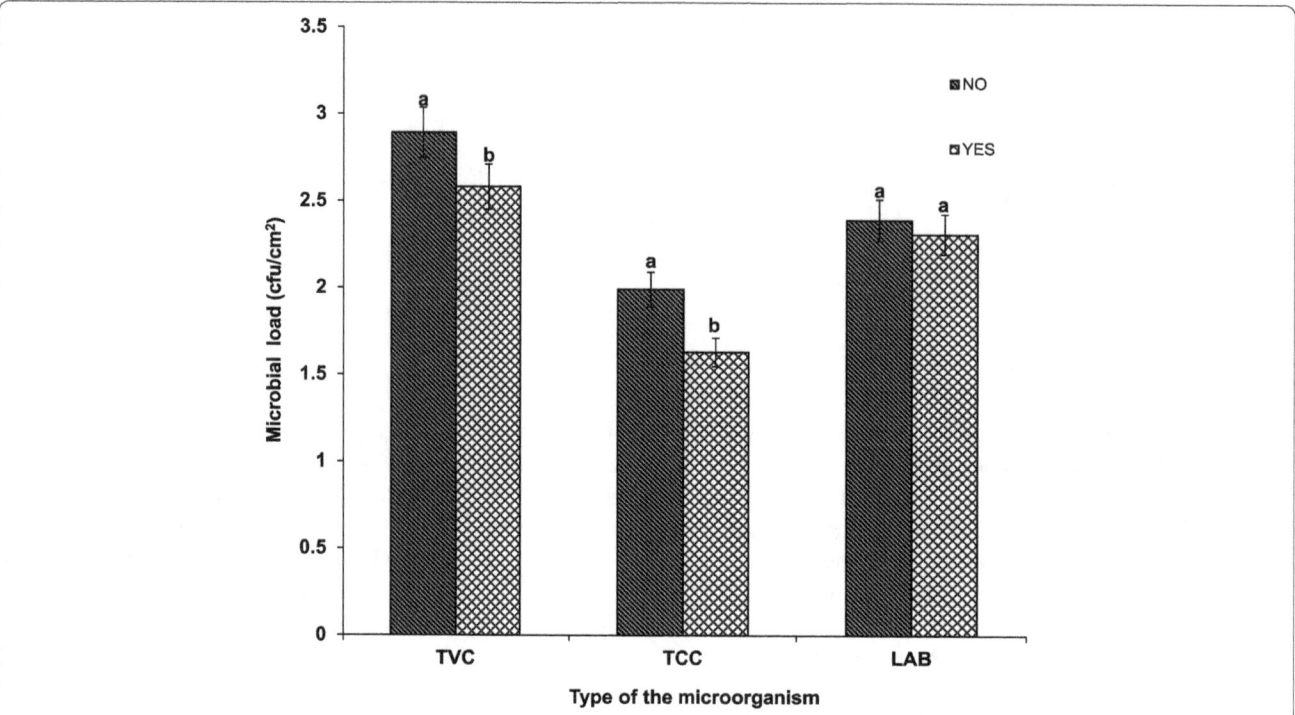

Fig. 1 Effect of the use of a commercial scourer on the reduction of surface microbial loads. Key: TVC- Total Viable Count, TCC- Total Coliform Count and LAB- Lactic Acid Bacteria. Error bars with same letter are not significantly different at $P = 0.05$ using Least Significant Difference (LSD) for mean ± standard deviation of sixty swabs taken from twenty milk container samples in three replicates

treatments with hot water, warm water and cold water. The treatments with warm water had the least reduction of the LAB while the treatments with hot water had the least reduction of the TCC.

It was found out in this study (Table 5) that the lethality of the disinfectant was very high on aluminium cans by 59, 69 and 66 % for TVC, TCC and LAB respectively than on the jerry can containers given the same conditions of cleaning. This can be due to presence of more biofilms on jerry cans than on aluminum cans that shield the disinfectant from accessing the microorganisms. It was also found out that the use of hot water was highly lethal on LAB, which is gram positive on both aluminium and jerry cans. This can be due to the susceptibility nature

Table 4 The mean comparison of microbial reduction from milk container surfaces between the control treatment and the treatments that used cold water, warm water, hot water and use of a disinfectant

Treatments of sanitation regimes	TVC	TCC	LAB
The control treatment	5.06ª	3.83ª	3.60ª
The treatments having Hot water	2.30ᵇ	1.18ᶜ	1.87ᵇ
The treatments having Cold water	2.14ᵇ	1.74ᵇ	1.79ᵇ
The treatments having Warm water	2.09ᵇ	1.49ᵇᶜ	1.22ᶜ
The treatments having Disinfectant	2.08ᵇ	1.30ᵇᶜ	3.24ª

Key: Means with the same letter (along the column) is not significantly different at $P < 0.05$ using *LSD* least significant difference

of gram positives as a result of having a single cell membrane as compared to gram negatives that have an extra cellular membrane that offers resistance. When milk is put in these containers, the microorganisms in the biofilm move out so as to access the nutrients. This results in the higher milk contamination of the plastic jerry cans when compared to the aluminium cans.

Discussion

The study found out that the microbial load reduction on the aluminium cans was significantly higher than on plastic jerry cans and therefore the type of the container had an impact on the effectiveness of the sanitation process. In addition, it was found out that irrespective of the temperature of water used for sanitation, aluminium cans still had significantly much higher microbial load reduction from the surfaces when compared to the plastic jerry cans. The lethality of the disinfectant was also higher than the on aluminium cans. Similar studies had found out that aluminium cans are more hygienic in handling milk when compared to the plastic jerry cans (Omore et al. 2005). The efficacy of the sanitation programme is assessed by the reduction of microbial load on a surface, as taken before and after cleaning and disinfection (Gibson et al. 1999), and the one with high death rate or high percent reduction of microbial load is considered to be of high efficiency (Salustiano et al.,

Table 5 The effect of using a commercial scourer on the reduction of microbial load from the aluminium cans and plastic jerry can surfaces

Milk container type	Scourer	N	Type of microorganism (mean $\log_{10}$ cfu/cm^2)	Initial microbial load	Final microbial load	Microbial load change	% Reduction in microbial load
Aluminium	NO	6	TVC	5.60 ± 1.16	0.58 ± 0.13	5.02 ± 1.12	89.6
			TCC	3.96 ± 1.14	0.63 ± 0.34	3.33 ± 0.99	84.1
			LAB	3.57 ± 0.78	0.18 ± 0.09	3.33 ± 0.82	93.3
	YES	6	TVC	6.15 ± 0.59	1.06 ± 0.33	5.10 ± 0.60	82.9
			TCC	5.10 ± 0.97	0.78 ± 0.11	4.32 ± 0.86	84.7
			LAB	3.96 ± 0.70	0.08 ± 0.04	3.88 ± 0.67	98.0
Plastic	NO	24	TVC	5.97 ± 1.11	3.67 ± 0.81	2.30 ± 0.61	38.5
			TCC	5.03 ± 0.99	3.37 ± 0.81	1.66 ± 0.73	33.0
			LAB	4.79 ± 0.85	2.63 ± 1.08	2.15 ± 0.98	44.9
	YES	24	TVC	6.02 ± 0.98	4.01 ± 1.01	2.01 ± 0.68	33.4
			TCC	5.12 ± 0.86	3.92 ± 0.71	1.20 ± 0.55	23.4
			LAB	4.83 ± 0.79	2.86 ± 1.06	1.91 ± 0.75	39.5

2004). Studies have shown that handling small quantities of milk is subject to a high rate of contamination with a small ratio of milk volume to container volume (Bonfoh et al., 2003). Despite plastic jerry cans being of greater risk to milk contamination, many dairy actors in Kenya prefer them for milk handling because they are cheaper than the aluminium cans as shown in Fig. 2 (Omore et al. 2005).

These are the difficult areas to clean in plastic containers

Fig. 2 a: Milk transportation in plastic jerry can containers using a donkey **b**: Cross-sectional view of inside the plastic container showing difficult areas to be cleaned

The temperature of the water to be used in the sanitation plays a big role in the efficacy of the process. Hot water was found to be the more effective against TVC and LAB reduction followed by cold water and the least was warm water. The TVC and LAB was reduced as a result of high temperatures of hot water that denatured the microbial cells and the antimicrobial activity of the soap. However, use of cold water with soap was more effective against TCC than the use of warm and cold water. Hence, increase in the temperature of water in the sanitation process causes reduction of the soap antimicrobial lethality against TCC. Also, irrespective of the water temperature used in the sanitation, the use of a scourer had significantly less reduction of TVC and TCC but did not differ significantly for the reduction of the LAB as compared to where it was not used. Generally the LAB, which are the Gram positive bacteria are the most susceptible to reduction than the TCC which are the Gram negative bacteria. Several reasons can explain the high prevalence of Gram negatives. First, the Gram-negative bacteria are innately resistant by virtue of their double membrane structure that prevents the disinfection agents from accessing the cell wall target or enzymatic inactivation of the cleaners (Russell 2001). Secondly is that the bar soap and the commercial scourer used in the cleaning are contaminated and introduce bacteria in the process. Earlier studies had found out bar soaps used in the sanitation were excessively colonized with bacteria which were predominantly Gram-negative bacteria (Zeiny 2009). The total viable count is for the overview of microbial contamination and total coliform count is for the evaluation of hygiene (Tamime 2009).

The scourer is used to offer mechanical abrasive energy to aid in detachment of bacteria from equipment surfaces, but some studies have shown that this mechanical scouring is then often insufficient (Bylund 1995). The scourer's abrasive forces cause scratches on the surface of the milk equipments thus facilitating the attachment and colonization of bacteria because roughness (Latorre et al. 2010). The abrasive force also causes wear of materials in the milking handling equipment that causes the appearance of cracks and crevices (Czechowski 1990). The energy required to remove deposit decreases with distance from the surface, suggesting that the cohesive forces between elements of the deposit are weaker than those of adhesion (Fryer et al. 2006).

From the study, the reduction of the microbial load was very low on the plastic jerry cans and the main reason was the nature of their material which was, first, hydrophobic thus exhibiting greater microbial surface adherence when compared with hydrophilic materials such as aluminium cans, glass and stainless steel (Sinde and Carballo 2000). Secondly, the surface texture and shape of equipments also determine the cleaning efficiency such as smooth surfaces are easier to clean than rough surfaces (Wirtanen et al. 1995).

Conclusion

Irrespective of the type of the water used (cold, warm or hot) and use or no use of commercial scourer in the sanitation process, microbial load reduction from plastic containers was very minimal compared to the aluminium cans. As a result, the uses of these plastic containers have high levels of milk contamination. Therefore, dairy actors should be encouraged to use food grade plastic containers, Mazican, in handling milk.

Abbreviations

ANOVA, analysis of variance; BMBF, German ministry of education and research; GLM, general linear model; HDPE, high-density polyethylene; LAB, lactic acid bacteria; LSD, least significant difference; LST, Lauryl sulfate tryptose; MR, methyl-red; MRS, de Man Rogosa and Sharpe agar; PC, polycarbonate; PCA, plate count agar; PP, polypropylene; PVC, polyvinyl chloride; ReLOAD, reduction of post-harvest losses and value addition in east african food value chains; SAS, statistical analysis system; TCC, total coliform count; TVC, total viable count; UK, violet red bile agar; VP, voges-proskauer

Acknowledgement

We also thank ReLOAD team members; Caroline M. Makau, Joy D. Orwa, Johnson K. Mwove, Samuel M. Nato, Olivier B. Kashongwe, Linnet W. Mwangi and Faith N. Ndunge for their assistance in field samples collection, laboratory samples analysis and data analysis.

Funding

This research work was supported financially by German Ministry of Education and Research (BMBF) through ReLOAD Project RELOAD: Reduction of Post-Harvest Losses and Value Addition in East African Food Value Chains.

Authors' contributions

This research work was part of NW Thesis research for the award of MSc. Food Science degree of Egerton University and supervised by JM, JN, PL and WN was in charge of field samples collection and laboratory samples analysis under the directorship of the three supervisors. The supervisors were also involved in the designing of the experiment, data analysis, interpretation of the results and development of this manuscript. All authors read and approved the final manuscript.

Authors' information

Mr. Nobert Wafula is a graduate Student in the Department of Dairy and Food Science and Technology in Egerton University, Kenya, undertaking an MSc. Food Science. Mr. Wafula also holds a BSc. Food Science and Technology and a Diploma in Dairy Technology from the same university. He also has working experience in the dairy and food manufacturing industries in Kenya. Prof. Joseph Matofari and Dr. John Nduko are senior lecturers in the Department of Dairy and Food Science and Technology in Egerton University while Mr. Peter Lamuka is a lecturer at the Department of Department of Food Science, Nutrition and Technology, University of Nairobi, Kenya.

Competing interests

The authors declare that they have no competing interests.

Author details

[1]Department of Dairy and Food Science and Technology, Egerton University, P.O. Box 536-20115, Egerton, Kenya. [2]Department of Food Science, Nutrition and Technology, University of Nairobi, P.O. Box 29053, Nairobi, Kenya.

References

Ahmad MM, Owni E, and Osman AO. Assessment of Microbial Loads and Antibiotic Residues in Milk Supply in Khartoum State, Sudan; 2015.

Azadnia PK, Khan Nazer AH. Identification of lactic acid bacteria isolated from traditional drinking yoghurt in tribes of Fars province. Iran J Vet Res. 2009; 10(3):235–40.

Bonfoh B, Wasem A, Traore AN, Fane A, Spillmann H, Simbé CF, Alfaroukh IO, Nicolet J, Farah Z, Zinsstag J. Microbiological quality of cows' milk taken at different intervals from the udder to the selling point in Bamako (Mali). Food Control. 2003;14(7):495–500.

Bonfoh B, Roth C, Traore AN, Fane A, Simbe CF, Alfaroukh IO. Effect of washing and disinfecting containers on the microbiological quality of fresh milksold in Barnako (Mali). Food Control. 2006;17(2):153–61.

Bylund G. Dairy Processing Handbook: Tetra Pak Processing Systems AB S 221 86 Lund, Sweden; 1995.

Coorevits AN, Valerie DJ, Vandroemme J, Reekmans R, Heyrman J, Messens R, De Vos P, Heyndrickx L. Comparative analysis of the diversity of aerobic spore-forming bacteria in raw milk from organic and convectional dairy farms. Syst Appl Microbiol. 2008;31(2):126–40.

Czechowski MH. Bacterial attachment to Buna-a-gaskets in milk processing equipments. Aust J Dairy Tech. 1990;45(2):113–4.

Fadaei A. Bacteriological quality of raw cow milk in Shahrekord, Iran. Vet World. 2014;7(4):240–3.

Faille C, Carpentier B. Food contact surfaces, surface soiling and biofilm formation. In: Fratamico PM, Annous BA, Gunther IV NW, editors. Biofilms in the food and beverage industries. Oxford, Cambridge, New Delhi: Wood head Publishing Limited; 2009. p. 304–30.

Fryer PJ, Christian GK, Liu W. How hygiene happens: Physics and chemistry of cleaning. May Int J Dairy Technol. 2006;59:76–84.

Fuentes E, Bogue J, Gómez C, Vargas J, Le Gal PY. Effects of dairy husbandry practices and farm types on raw milk quality collected by different categories of dairy processors in the Peruvian Andes. Trop Anim Health Prod. 2014; 46(8):1419–26.

Gibson H, Taylor H, Hall K, Holah J. Effectiveness of cleaning techniques used in the food industry in terms of the removal of bacterial biofilms. J Applied Microbiol. 1999;87:41–8.

Goberna M, Insam H, Klammer S, Pascual JA, Sanchez J. Microbial community structure at different depths in disturbed and undisturbed semiarid Mediterranean forest soils. Microb Ecol. 2005;50(3):315–26.

Grainger J, Hurst J, Burdass D. Basic Practical Microbiology: A Manual. The Society for General Microbiology; 2001.

Gran HM, Mutukumira AN, Wetlesen A, Narvhus JA. Smallholder dairy processing in Zimbabwe: the production of fermented milk products with particular emphasis on sanitation and microbiological quality. Food Control. 2002;13(3):161–8.

KEBS (Kenya bureau of standards). Raw Milk –specification KS EAS 67: 2007. Third Edition. Nairobi 2010.

Kelly PT, O'Sullivan K, Berry DP, More SJ, Meaney WJ, O'Callaghan EJ, O'Brien B. Farm management factors associated with bulk tank total bacterial count in Irish dairy herds during 2006/2007. Ir Vet J. 2009;62:36–42.

Kivaria FM, Noordhuizen JPTM, Kapaga AM. Evaluation of the hygienic quality and associated public health hazards of raw milk marketed by smallholder dairy producers in the Dar es Salaam region, Tanzania. Trop Anim Health Prod. 2006;38(3):185–94.

Latorre AA, Van Kessel JS, Karns JS, Zurakowski MJ, Pradhan AK, Boor KJ, Jayarao BM, Houser BA, Daugherty CS, Schukken YH. Biofilm in milking equipment on a dairy farm as a potential source of bulk tank milk contamination. J Dairy Sci. 2010;93:2792–802.

Lindsay D, Brozel V, Mostert J, Holy A. Differential efficacy of a chlorine dioxide-containing sanitizer against single species and binary biofillms of a dairy-associated Bacillus cereus and a Pseudomonas fluorescens isolate. J Appl Microbiol. 2002;92:352–61.

Millogo V, Sjaunja KS, Ouedrago GA, Agenas S. Raw milk hygiene at farms, processing units andlocal markets in Burkina. Faso. Food Control. 2010;21(7):1070–4.

Murphy SC, Boor KJ. Trouble-shooting sources and causes of high bacteria counts in raw milk. Dairy Food Environ Sanit. 2000;20(8):606–11.

Olivier D, Moshoeshoe SL. Incidence of aerobic spoilage-and psychrotrophic bacteria in non-pasteurised and pasteurised bovine milk from Maseru. Medical Technology SA. 2012;26(2):22–7.

Ombarak RA, Elbagory AM. Bacteriological Quality and Safety of Raw Camel Milk in Egypt. Egyptian J Dairy Sci. 2014;42:95–103.

Omore A, Lore T, Staal S, Kutwa J, Ouma R, Arimi S, Kang'ethe E. Addressing the public health and quality concerns towards marketed milk in Kenya. Kenya: SDP Research and Development Report No. 3. Smallholder Dairy (R&D) Project; 2005.

Orregard M. Quality analysis of raw milk along the value chain of the informal milk market in Kiambu County, Kenya. Kenya: Thesis, Swedish University of Agricultural Sciences, Uppsala; 2013.

Reinemann DJ, Wolters GMVH, Billon P, Lind O. and Rasmussen MD. Review of practices for cleaning and sanitation of milking machines. Bulletin-International Dairy Federation. 2003: 3–18.

Reta MA, Bereda TW, Alemu AN. Bacterial contaminations of raw cow's milk consumed at Jigjiga City of Somali Regional State, Eastern Ethiopia. Int J Food Contam. 2016;3(1):1–9. doi:10.1186/s40550-016-0027-5.

Richardson GH. Standard Methods for the Examination of Dairy Products American Public Health Association. Washington, DC. 1985.

Russell AD. Mechanisms of bacterial insusceptibility to biocides. Am J Infect Control. 2001;29:259–61.

Salustiano VC, Nelio JA, Sebastiao CB, Willian M, Gabriela P. An assessment of chemical sanitizers on the microbiological profile of air in a milk processing plant. J Food Saf. 2004;24:159–67.

Sinde E, Carballo J. Attachment of Salmonella spp. and Listeria monocytogenes to stainless steel, rubber and polytetrafluorethylene: the influence of free energy and the effect of commercial sanitizers. Food Microbiol. 2000;17:439–47.

Tamime AY. Milk Processing and Quality Management: Quality Control. Belleque J., Chicon R. and Recio I., Eds. West Sussex, UK: John Wiley and sons publishers; 2009. Pg 75

Tasci F. Microbiological and chemical properties of raw milk consumed in Burdur. J Anim Vet Adv. 2011;10(5):635–41.

Wafula NW, Matofari JW, Nduko JM and Lamuka PO. Sanitation practices used by dairy actors along the smallholder raw milk value chain and its effect on milk post-harvest losses in Kenya. African Journal of Food Science and Technology. InPress. 2016.

Welearegay H, Yilma Z, Tekle-Giorgis Y. Hygienic practices and microbiological quality of raw milk produced under different farm size in Hawassa, southern Ethiopia. Agric Res Rev. 2012:132-42.

Wirtanen G, Ahola H, Mattila Sandholm T. Evaluation of cleaning procedures in elimination of biofilms from stainless steel surfaces in open process equipment. Trans I Chem E. 1995;73:9–16.

Worku T, Negera E, Nurfeta A, Welearegay H. Milk handling practices and its challenges in Borana Pastoral Community, Ethiopia. Afr J Agric Res. 2014; 9(15):1192–9.

Zeiny SMH. Isolation of Some Microorganisms from Bar Soaps and Liquid Soaps in Hospital Environments Iraqi. J Pharm Sci. 2009;18(1):28.

15

Assessment of antibiotic residues in commercial and farm milk collected in the region of Guelma (Algeria)

Samiha Layada[1*], Djemel-Eddine Benouareth[1*], Wim Coucke[2*] and Mirjana Andjelkovic[2*] ⓘⒹ

Abstract

Background: In an attempt to enhance the quality and quantity of food production (especially milk) and in order to prevent, or treat,animal diseases, the use of antibiotics in Algeria follows an increasing trend. The increased use evidently contributes to the emergence of increased contamination levels of antibiotic residues.

Results: In this work, two methods were used to detect presence of antibiotic residues in raw and fermented cow's milk collected in Guelma's farms (in Algeria). The screening comprised different points of sale in Guelma province. In a first step a widely used prescreening method based on microbial inhibition assay; Delvotest SP-NT; was used to analyze 131 milk samples. In a second step a liquid chromatography coupled to mass spectrometry (LC-MS/MS) was used. The latter was first optimized for extraction of 36 veterinary drugs of penicillins, quinolones, macrolides, tetracyclines, sulfonamides, and trimethoprim from the collected milk. After simple extraction and dilution, the 194 samples, including those previously tested by the Delvotest SP-NT, were analyzed by LC-MS/MS. Results obtained by both methods were compared. Among the LC-MS/MS findings, 65.46 % of non-conform samples contained authorized residues at levels higher than the MRL, residues without set MRL, or non-authorized residues.

Conclusion: The comparison of both methods showed that Delvotest SP-NT is less trustworthy due to number of false negative results. This was further confirmed by LC-MS/MS pointing out the traces of antibiotics in numerous samples. In 65.46 % of milk samples residues of antibiotics were found suggesting a lack of public health controls as well as an evidence of the negligent use of antibiotics in the livestock industry, which both form a risk to public health. This indication should be confirmed by a nationwide study with in-depth analyses of antibiotic's presence in food chain originated from animals. The study offered an LC-MS/MS based analytical method ready to be used in Algerian National Residues Control Plan as a versatile analytical tool to monitor and determine the occurrence of antibiotic multi-residues in milk.

Keywords: Antibiotic residues, Delvotest SP-NT, LC-MS/MS, False positive, False negative, Fermented milk (Iben)

Background

Antibiotics are widely used in livestock production for many purposes, such as: animal disease treatment (therapeutic application), animal disease prevention (prophylactic application), and feed efficiency (as growing promoters) (Jank et al. 2015). Their presence as residues

* Correspondence: layada.samiha@yahoo.fr; Benouareth_dje@yahoo.fr; wim.coucke@wiv-isp.be; mirjana.andjelkovic@wiv-isp.be
[1]Research Laboratory of Biology, Water and Environment, Biology Department, Faculty of Natural and Life Sciences, Earth and Universe Sciences, University 8 Mai 1945-Guelma, BP 401, Guelma 24000, Algeria
[2]Scientific Institute of Public Health, Juliette Wytsman Street 14, 1050 Brussels, Belgium

in food products especially milk implicates certain damages in the public health like: the development of allergic reactions in some hypersensitive individuals, increased risk of carcinogenicity (Petrović et al. 2008; Hou et al. 2014), the growth of resistant bacterial strains, and imbalances in intestinal microflora (Wang et al. 2006; Borràs et al. 2011). Low concentrations of antimicrobial drug residues create problems in the production of milk products by inhibiting the starter cultures (Petrović et al. 2008; Stead et al. 2008). Moreover, the risk of contamination with antibiotic residues of farms milk is higher if inappropriate practices are applied. For these reasons, control measures must be implemented to prevent drug residues from entering into the food chain. In this regard, and to ensure the safety of the consumer, worldwide regulatory authorities have set Maximum Residue Limits (MRLs) for several veterinary drugs (Kassaify et al. 2013). Since 2006, these substances are forbidden by the European Union (EU) to be used either in sub-therapeutic doses for prophylaxis or as growth promoters in veterinary medicine (Council Regulation No 1831/2003). Consequently, the presence of antibiotic residues in products that are targeted for food consumption has to be controlled. Similarly, various analytical methods have been described to analyze milk; especially microbiological and immune assays which are widely used because of their low cost and short time of analysis (Ramirez et al. 2003; Beltran et al. 2015). However, most current microbial screening tests have been initially developed to detect β-lactams in cow's milk. They are based on the inhibition of *Geobacillus stearothermophilus var. calidolactis* which is highly sensitive to these substances (Beltran et al. 2015). Delvotest SP-NT is considered as one of the most commonly used microbial inhibition tests (Stead et al. 2008). However, the limited sensitivity and selectivity of the method demand further confirmation of the results obtained using more sensitive technique (Ferrini et al. 2015).

Instrumental techniques such as liquid chromatography (LC) coupled to UV- VIS spectroscopy or mass spectrometry (MS), are widely used in food control for either screening or confirmation of positive findings of less specific test within the following studies: Hermo et al. (2008), Li et al. (2012), Boix et al. (2014), Hou et al. (2014), Martins et al. (2014), and Cepurnieks et al. (2015). Furthermore, LC-MS/MS is nowadays the most frequently used analytical tool for detecting a large number of multiclass veterinary drug residues in food and decreasing the rate of false negative and false positive results with high selectivity and sensitivity (Martins et al. 2014). Many papers have tackled the analysis of different classes of veterinary antibiotics in milk using LC-MS/MS (Bogialli et al. 2008; Hermo et al. 2008; Han et al. 2015, and Meng et al. 2015). But recent research themes have enlarged their interests to the development of multiclass veterinary drug residues methods which facilitate the discrimination of antibiotics in milk and other matrices; such as, the study of Martins et al. (2014), which has established two screening methods that can analyze 24 antibacterial and one metabolite residue in milk and liver using LC-MS/MS.

Very few studies in Algeria have been conducted on the potential presence of antibiotic residues in raw cow's milk samples and that was by using less specific microbiological methods. Of the few researches that are available, studies made by Tarzaali et al. (2008), Aggad et al. (2009), and Titouche et al. (2013) have described high levels of milk contamination by using Delvotest SP-NT.

They indicated that among all tested milk samples 89, 29, and 47 % were found to contain residues, respectively. This appears significantly high prevalence of positive samples in comparison to results from any European Union country where a regular system for antibiotic residues control in milk is many decades-old. On contrary in Algeria similar milk control appears to be immature. Consequently, there are no data available on occurrence of antibiotic residues in milk produced in Guelma province located in North East of Algeria. Moreover, in any of the published reports contamination of fermented milk was not considered. The fermented milk - called lben- and cow's milk are important components of Algerian's diet (Belhadia et al. 2011; Zoubeidi and Gharabi 2013). Lben is traditional cultured milk widely consumed in North Africa and in Middle Eastern countries. It is produced by spontaneous souring of cow's milk followed by churning in order to separate lben from butter. Whereas only one report about a similar milk type (raibi milk) is available (Zinedine et al. 2007), no other data on the specific products are available or known.

In this work, the analyses of two types of milk (raw and lben) collected in Guelma province (in the north of Algeria) is presented. The samples were collected directly from the farms, and further in the value chain from dairy markets. Two methods were used to detect the presence of antibiotic residues in those samples. The first one was Delvotest SP-NT, which is most commonly used test for this purposes in Algeria. The second one was an optimized LC-MS/MS multi-residues screening method. The results obtained by both tests were compared and correlated to evaluate the feasibility of the use of Delvotest SP-NT as a prescreening test in Algerian setting. The optimization and validation of multi-residue LC-MS/MS screening method was performed following the international norms.

Methods
Samples and accompanying data
A total of 194 cow's milk samples were collected. Among them, 156 samples were collected from two farms (A and B), whereas 38 (including fermented and raw cow's milk) were purchased from 16 various points of sale in Guelma province during the period of January 2013 to July 2014. Of 156 farm samples 53 were collected from farm A located in the centrum of Guelma province. The samples were gathered as such: 8 samples were taken from individual cows which have been treated less than a month prior to the collection, 28 from individual untreated cows, and 17 samples were collected from bulk tank milk at the farm. Similarly, 103 milk samples were collected from farm B located more than 7 Km far from Guelma center-. Their collection and number were as follows: milk from individual cows

treated less than a month prior to the collection were 18, milk from individual untreated cows were 70, and bulk tank milk were 15. In addition to sampling for the assessment of antibiotic residues, questionnaires were made to collect data on the cow's treatment like: the kind of antibiotic administered to the cow, the day of treatment, and the withdrawal time of each antibiotic used. The latter may vary from 2 to 7 days depending on the administered antibiotics (penicillins, tetracyclines, macrolides and sulfonamides) and this either intermammary or intramuscular.

Out of 38 milk samples purchased at various points of sale in Guelma province, 22 were fermented milk "lben" and 16 were raw cow's milk. Seven points of sale were situated in the province's capital whereas the others were situated in separate regions surrounding the state. For this part of the value chain a short questionnaire was used in order to collect more specific information about the milk origin. Those data specified milk as collected either from a farmer, private collection, milk factories, or from farms in Guelma municipalities and Souk Ahras state. Due to missing data, other collections were indicated as unknown origin.

Approximately a volume of 140 mL of each milk sample was collected and conserved in sterile flask and transported to the laboratory at 4 °C. Ten millilitre were used to be analyzed by a microbial test Delvotest SP-NT for the prescreening of antibiotics in milk samples. The remaining 130 mL were frozen and kept at –20 °C prior to further analysis by LC/MS-MS. Two blank milk samples were assigned in the collection after a negative test for antibiotics by Delvotest SP-NT and one purchased from Belgian supermarket for LC-MS/MS analysis.

Chemicals

The following reference standards were from Sigma-Aldrich (Bornem, Belgium). The sulfonamides (sulfapyridine, sulfamethoxypyridazine, sulfadoxine, sulfadimidine, sulfamonomethoxine, sulfaquinoxaline, sulfamoxole, sulfachloropyridazine, sulfadimethoxine, sulfathiazole, sulfaguanidine, sulfamerazine, sulfamethoxazole, sulfadiazine, sulfacetamide, sulfisoxazole, sulfamethizole), trimethoprim, penicillins (amoxicillin, ampicillin, penicillin V, penicillin G, dicloxacillin, cloxacillin, oxacillin, nafcillin), quinolones (flumequine, difloxacin, sarafloxacin, ciprofloxacin, enrofloxacin, danofloxacin, nalidixic acid, marbofloxacin, norfloxacin, ofloxacin, oxolinic acid), macrolides (erythromycin, spiramycin, josamycin, clindamycin, lincomycin, neospyramycin, tilmicosin, tylosin, tylvalosin, tulatrhomycin), and tetracyclines (doxycycline, oxytetracycline, 4-epi- chlortetracycline, 4-epi-oxytetracycline, 4-epi- tetracycline, tetracycline, chlortetracycline) were with a purity of 95 to 100 %. These standards were used to prepare stock standards solutions. Similarly, the internal

standards sulfadimidine C^{13}, flucloxacilline, norfloxacin_ D_5, roxythromycine, and demeclocycline, were from Sigma-Aldrich (Bornem, Belgium). Besides, stock standards solutions of sulphonamides, penicillins, macrolides, and tetracyclines were prepared by dissolving 10 mg of each substance into 10 mL of methanol, except for penicillins which were dissolved in 10 mL of Milli-Q water (Millipore corp., Bedford, MA, USA). For the quinolones, 5 mg of each substance were dissolved into 10 mL of methanol. In addition, mixed intermediate standard solutions at 5MRL (Commission Regulation 2377/1990) were prepared from diluting stock standard solutions to obtain a specific final concentration for each substance then conserved them at –20 °C. Actually, solvents acetonitrile and methanol used for mobile phase and extraction were of UPLC-MS grade. They were purchased from Biosolve (Valkenswaard, the Netherlands). Yet, Formic acid, oxalic acid 10 mM, Ethylene-diaminetetraacetic acid (EDTA) 100 mM, and sodium sulfate anhydride were purchased from Merck (Darmstadt, Germany). The analytical liquid chromatography column (Waters, Millford, MA, USA) C18 2.1 × 100 mm, 1.7 μm was used.

Microbial inhibitor test (Delvotest SP-NT)

Delvotest SP-NT, which is a non-specific microbial inhibitor test, was used to detect the presence of antibiotics in milk samples. In short this is an agar diffusion test that contains a standardized number of *Bacillus stearothermophilus* spores, selected nutrients, and pH indicator bromocresol purple. Four kits of Delvotest SP-NT (DSM, Netherlands) were used whereas three were provided by DSM Food Specialties located in Spain, and the fourth was purchased. After adding milk sample directly to the agar bed (ampoules), an incubation step was conducted for 3 h at 64 °C. During incubation, microbial metabolism resulted in a change in pH, and hence in a change of color from purple to yellow. By contrast, if the sample contained sufficiently high concentrations of inhibitory substances, the color would remain purple (Stead et al. 2008). Except fermented milk, 10 mL of each raw milk sample was heated at 80 °C for 10 min to destroy natural inhibitors lysozyme and lactoferrin (Hillerton et al. 1999). In parallel, 0.1 mL was decanted into Delvotest SP-NT ampoules using a specific pipette for each sample. The ampoules were incubated in water bath at 64 °C ± 2 °C within 3 h. Test and data interpretation were performed according to the manufacturer's instructions.

Extraction optimization

The extraction procedure was based on an existent method for screening of antibiotics in meat. This method was previously developed and validated in accordance with the European commission (Commission Decision 2002). In this study, only 36 antibiotics were selected to be followed in milk samples during method optimization instead of 59 ones targeted in the initial method. The selected antibiotics had either the same level of MRL both in milk and meat, or lower MRL in milk than in meat as prescribed in European commission (Commission Regulation 37/2010). With required scrutiny, the selection of the appropriate method for the extraction of antibiotics residues in milk was based on the comparison between four modified extraction protocols as briefly presented in Table 1. All protocols were tested on a set of a blank milk sample and two control samples spiked with mixture of antibiotic standards. Out of the comparative overview of the average recoveries obtained by four extraction protocols (Table 2), method 3 was selected as the optimal. Further details are explained in the section of Optimization of the LC-MS/MS screening method results. The following list includes the steps of the extraction procedure. Ten grams of one blank and two control milk samples were weighed in 50 mL falcon. To these 500 µL of mixed internal standard solution at 1MRL, 667 µL of EDTA 0.1 M, 23 mL of the mixture (methanol/acetonitrile) and 3 g of sodium sulfate

anhydride were added. The spiked samples were fortified with 250 µL of each mixed standard solutions at 1MRL as above in order to obtain levels corresponding to 0.5 MRL. Samples were vortexed for 1 min, and centrifuged at 10,000 rpm for 10 min at 4 °C. Subsequently, 5 mL of the supernatant was decanted into 10 mL tubes and evaporated to dryness at 40 °C. Three-hundred microlitre of Milli-Q water was added into tubes after evaporation, vortexed for a few seconds and transferred to eppendorfs tubes to be centrifuged at 12,000 rpm for 30 min. Supernatant was filtered through a 0.2 µm filter directly to injection vial prior to LC-MS/MS analysis.

LC-MS/MS analysis

UPLC analysis was performed using an Acquity sample and solvent manager (Waters, Milford, MA, USA). Chromatographic separation was achieved using an Acquity UPLC column C18 2.1 × 100 mm 1.7 µ (Waters, Millford, MA, USA) at 30 °C with the mobile phase composed by 0.1 mM oxalic acid, 0.1 % formic acid (solution A) and acetonitrile 0.1 % formic acid (solution B) at a constant flow of 0.4 mL min-[1]. The gradient elution program used

Table 1 Comparative overview of four extraction methods parameters

Extraction steps	Method 1	Method 2	Method 3	Method 4 (Martins et al. (2014))
1) Sample amount	3g	10g	10g	0.5g
2) Type of container	50mL Falcon tube			2mL Eppendorf tube
3) Mixed internal standard solution, 1MRL (EU2377/90)	150µL	150µL	500µL	12.5µL
4) EDTA, 0.1M	200µL	200µL	667µL	30µL
5) Mixture methanol/acetonitrile	7mL	7mL	23mL	/
6) Sodium sulfate anhydride	3g	3g	3g	/
7) Mixed standard solution*	75µL	75µL	250µL	12.5µL
8) Vortexing	1min			10sec and equilibration for 10min from light
9) Centrifugation	10000rpm, 10min at 4°C			
10) Deprotonization				0.6mL Ethanol/acetic acid (96/4) Followed by short vortexing and 30min at -18°C incubation
11) Supernatant	5mL evaporated to dryness at 40°C followed by addition of 300 µL of water MilliQ			/
12) Vortex and centrifugation	Few seconds 12000rpm during 30min			
13) Supernatant	All supernatant			0.75mL with addition of 0.25mL Formic acid 0.1% in water/ formic acid in acetonitrile (98:2)
14) Filtration	0.2µm filter			/

*mixed standard solution contained: Sulfonamides (sulfapyridine, sulfamethoxypyridazine, sulfadoxine, sulfadimidine, sulfamonomethoxine, sulfaquinoxaline, sulfaquinoxaline, sulfamoxole, sulfachlorpyridazine, sulfadimethoxine, sulfaquinoxaline, sulfathiazole, sulfaguanidine, trimethoprim, sulfamerazine, sulfamethoxazole, Sulfadiazine, sulfacetamide, sulfisoxazole), penicillins (amoxicillin, ampicillin, penicillin V, penicillin G, dicloxacillin, cloxacillin, oxacillin, nafcillin) quinolones (flumequine, difloxacin, sarafloxacin, ciprofloxacin, enrofloxacin, danofloxacin, nalidixic acid, marbofloxacin, norfloxacin, ofloxacin , oxolinic acid), macrolides (erythromycin, spiramycin, josamycin, clindamycin, lincomycin, neospyramycin, tilmicosin, tylosin, tylvalosin, tulatrhomycin), tetracyclines (doxycycline , oxytetracycline ,4-epi- chlortetracycline, 4-epi-oxytetracycline, 4-epi- tetracycline, tetracycline, chlortetracycline)

Table 2 All parameters used for the comparison between four methods (M1, M2, M3 and M4) and the final selection of the test suitable method

Group and type of antibiotic	Spiked Concentration (µg/L)	Average Rec% ± STDEV M1	Average Rec% ± STDEV M2	Average Rec% ± STDEV M3	Average Rec% ± STDEV M4	MRL in milk (µg/L)	MRL in meat (µg/L)	MRL milk/MRL meat
Macrolides		35.65 ± 2.72	24.50 ± 8.54	73.66 ± 8.33	79.20 ± 6.51			
Erythromycin	2000	3.40^{ab} ± 0.98	0.67^a ± 0.35	20.10^b ± 3.77	0.00^a ± 0.00	40	200	0.2
Neospyramycin	2500	18.78 ± 1.33	22.14 ± 1.10	70.31 ± 11.95	102.99 ± 3.17	200	200	1
Spiramycin	2500	25.96 ± 3.06	29.07 ± 1.18	53.17 ± 4.80	37.45 ± 2.02	200	200	1
Tilmicosin	500	96.42 ± 2.34	63.54 ± 39.90	171.49 ± 11.93	245.02 ± 27.02	50	50	1
Tylosin	1000	33.69^a ± 5.89	7.08^b ± 0.16	53.23^a ± 9.22	10.55^b ± 0.35	50	100	0.50
Penicillins		34.36 ± 2.10	30.36 ± 1.87	72.33 ± 9.17	75.42 ± 7.91			
Amoxicillin	500	57.08^a ± 4.54	69.32^a ± 0.38	143.26^b ± 25.38	107.28^c ± 35.83	4	50	0.08
Ampicillin	500	78.19^a ± 7.09	104.03^a ± 7.99	204.73^b ± 26.17	273.58^c ± 7.65	4	50	0.08
Cloxacillin	30,000	1.43^a ± 0.12	1.31^a ± 0.03	3.05^a ± 0.21	5.76^a ± 0.49	30	300	0.1
Dicloxacillin	3000	9.13^a ± 0.44	6.96^a ± 0.09	21.14^{ab} ± 2.80	60.03^b ± 5.54	30	300	0.1
Nafcillin	3000	8.23^a ± 0.20	10.69^a ± 0.12	18.07^a ± 0.39	10.23^a ± 0.31	30	300	0.1
Oxacillin	3000	10.99^a ± 0.86	11.72^a ± 0.27	28.44^{ab} ± 1.39	53.30^b ± 2.03	30	300	0.1
Penicillin G	500	75.48^a ± 1.44	8.46^b ± 4.25	87.64^a ± 7.85	17.76^b ± 3.53	4	50	0.08
Quinolones		37.05 ± 2.20	26.58 ± 1.62	76.52 ± 1.74	83.07 ± 13.94			
Enrofloxacin	500	53.28 ± 3.60	40.40 ± 1.90	110.21 ± 1.22	95.99 ± 19.05	100	100	1
Flumequine	1000	9.05^a ± 1.45	0.00^a ± 0.00	24.33^a ± 0.70	52.72^c ± 9.48	50	200	0.25
Marbofloxacin	750	48.80^a ± 1.55	39.32^a ± 2.94	95.02^b ± 3.29	100.49^b ± 13.30	75	150	0.5
Sulfonamides		34.80 ± 1.32	29.49 ± 2.52	94.58 ± 6.43	99.73 ± 9.82			
Sulfacetamide	1000	18.54 ± 1.36	14.02 ± 1.57	35.42 ± 8.03	20.53 ± 2.80	100	100	1
Sulfachloropyridazine	1000	38.38 ± 0.48	26.78 ± 3.87	108.43 ± 0.11	132.90 ± 3.87	100	100	1
Sulfadiazine	1000	25.49 ± 2.58	24.09 ± 0.89	58.88 ± 0.63	65.94 ± 3.54	100	100	1
Sulfadimethoxine	1000	14.31 ± 1.46	7.10 ± 0.40	41.56 ± 1.33	85.52 ± 2.47	100	100	1
Sulfadimidine	1000	39.33 ± 0.21	38.29 ± 1.55	114.46 ± 7.31	123.00 ± 9.00	100	100	1
Sulfadoxine	1000	30.31 ± 0.74	16.97 ± 2.60	82.33 ± 0.96	153.84 ± 1.11	100	100	1
Sulfaguanidine	1000	18.64 ± 0.12	14.25 ± 0.14	39.51 ± 7.71	26.60 ± 10.07	100	100	1
Sulfamerazine	1000	9.70 ± 1.57	7.65 ± 0.11	39.23 ± 1.08	89.31 ± 11.19	100	100	1
Sulfamethizole	1000	42.44 ± 0.80	55.31 ± 7.74	92.82 ± 1.55	84.97 ± 4.78	100	100	1
Sulfamethoxazole	1000	43.95 ± 6.02	18.70 ± 3.82	127.40 ± 4.92	161.97 ± 15.02	100	100	1
Sulfamethoxypyridazine	1000	43.94 ± 0.02	51.72 ± 1.76	140.44 ± 5.21	165.70 ± 12.54	100	100	1
Sulfamonomethoxine	1000	42.22 ± 1.56	50.56 ± 0.46	148.67 ± 9.67	174.34 ± 7.26	100	100	1
Sulfamoxole	1000	38.61 ± 1.46	19.51 ± 2.34	80.30 ± 10.71	100.31 ± 23.38	100	100	1

Table 2 All parameters used for the comparison between four methods (M1, M2, M3 and M4) and the final selection of the test suitable method (*Continued*)

Sulfapyridine	1000	29.02 ± 0.68	31.62 ± 2.64	89.32 ± 6.45	101.35 ± 18.40	100	100	1
Sulfaquinoxaline	1000	13.64 ± 0.32	10.28 ± 2.07	28.72 ± 5.89	79.34 ± 14.84	100	100	1
Sulfathiazole	1000	36.39 ± 1.95	39.05 ± 0.88	88.64 ± 5.73	41.14 ± 21.28	100	100	1
Sulfisoxazole	1000	55.57 ± 0.80	29.05 ± 2.59	168.29 ± 12.57	62.20 ± 4.43	100	100	1
Trimethoprim	1000	86.00 ± 1.57	75.98 ± 9.94	217.95 ± 25.88	126.27 ± 10.81	50	50	1
Tetracyclines		25.73 ± 0.54	21.89 ± 17.13	73.50 ± 12.79	52.61 ± 7.69	50	50	
Chlortetracycline	1000	18.87 ± 0.49	21.77 ± 9.36	30.30 ± 6.29	35.67 ± 6.17	100	100	1
Oxytetracycline+ 4-Epi-Oxytetracycline	1000	32.79 ± 0.77	22.24 ± 23.07	108.49 ± 15.90	48.16 ± 2.89	100	100	1
Tetracycline+ 4-Epi-Tetracycline	1000	25.53 ± 0.36	21.67 ± 18.97	81.72 ± 16.19	74.02 ± 14.00	100	100	1

Means with the same letter do not differ significantly at the level of 0.05

STDEV Standard Deviation, Rec Recovery

*the estimated concentration was based on the calibration curve produced with the five different standard concentrations (250, 500, 1000, 1500 and 2000 µg/L) after screening the quantifier ion in the MS spectra.

Four methods used (n = 2) in milk samples spiked each antibiotic of 500 µg/L as a final concentration was initially 98 % of A decreasing to 2 % within 10.5 min (0–10.5 min). After that, the composition was set back to the initial levels A: B (98:2) (10.5–12.5 min). The total run time was 12.5 min. XEVO TQ MS triple quadrupole mass spectrometer (Waters, Millford MA, USA), operating in positive Electrospray ionization (ESI) MS/MS mode was used for detection. Data was controlled and evaluated by MassLynx software (version 4.1). The selected reaction monitoring (MRM) mode was used and the following tune parameters were applied: capillary, 3 kV; cone 15 V; extractor, 3.00 V; source temperature, 150 °C; desolvation temperature, 500 °C, cone gas flow, 80Lh^{-1}; desolvation gas flow, 1000Lh^{-1}; collision gas flow, 0.16 mL min^{-1}, resolution (LM1, HM1, LM2, HM2 where LM is low mass and HM is high mass), 2.7, 15, 2.8, 14.8; ion energy (1 and 2), 0.3, 0.6; multiplier 546.52 V. Cone voltage (V), collision energy (eV) and transition mass parameters for all antibiotic residues analyzed in milk samples are presented in Table 3.

Validation of the LC-MS/MS screening method

Validation of the selected method was performed on all antibiotic substances cited in Table 4 and performed following the international norm (Commission Decision 2002) through determining: specificity/selectivity, detection capability CCβ, linearity, and applicability. Milk samples bought from a Belgian supermarket were analyzed, and after confirming the absence of antibiotics were used as blank samples in the validation studies.

To determine the specificity of the proposed method, a set of extract of blank milk samples ($n = 20$) were injected into the chromatographic system on the same day. The process included also an analysis of the spiked samples on three levels. This permits the specificity to be evaluated through the average and standard deviation of the noise amplitude (S/N), expressing relative to the internal standard signal amplitude. When a ratio of signal to noise (S/N) of blank sample is higher than 3, its relative retention time (RRT) is equivalent to RRT of the spiked milk sample and its response area is higher than 1 % of that of the spiked milk sample, then, the result is deemed false positive. Moreover, the selectivity was guaranteed by following up the multi-reaction monitoring (MRM) transitions per substance on LC-MS/MS and the relative retention time. The detection capability (CCβ) which is the smallest content of the substance that may be detected, identified and/or quantified in a sample with an error probability of β = 5 %. Probability of false non-compliance ≤5 % was tested with 20 milk samples. The latter, had been spiked at 0.5 MRL and analyzed according to two criteria. Firstly, relative retention time (RRT) of the suspect sample had to be in a range of ±2.5 % around the RRT of the standard solution. Secondly, S/N ratio of the daughter ion

Table 3 Mass spectrometry parameters used for the screening of antibiotic residues in milk

Analyte	Transition	Cone voltage (V)	Collision energy eV
Amoxicillin	366 > 114	10	20
Ampicillin	350 > 106.1	30	20
Chlortetracycline	479.1 > 302.8	30	40
Cloxacillin	436.24 > 160.0	30	10
Dicloxacillin	469.9 > 160.1	30	10
Enrofloxacin	360.4 > 341.9	30	20
Erythromycin	734.4 > 576.6	30	20
Flumequine	262.1 > 244	30	20
Marbofloxacin	363.26 > 72	30	20
Nafcillin	415 > 199	30	10
Neospiramycin	699.4 > 174.2	30	20
Oxytetracycline + 4-epi-Oxytetracycline	461.1 > 443.1	30	10
Oxacillin	402.2 > 243.3	15	10
Penicillin G	335.1 > 176.1	30	10
Spiramycin	843.6 > 174.4	30	40
Sulfacetamide	215.2 > 156	30	10
Sulfachloropyridazine	285 > 155.8	25	15
Sulfadiazine	251 > 108	30	20
Sulfadimethoxine	311.1 > 156	30	20
Sulfadimidine	279.2 > 155.9	30	20
Sulfadoxine	311.2 > 155.9	30	20
Sulfaguanidine	215.1 > 156	25	15
Sulfamerazine	265.1 > 155.9	30	20
Sulfamethizole	271.1 > 107.8	30	20
sulfamethoxazole	254.1 > 108.1	30	20
sulfamethoxypyridazine	281 > 155.7	30	15
Sulfamonomethoxine	281.2 > 155.9	30	20
Sulfamoxole	268.2 > 155.9	30	10
Sulfapyridine	250.1 > 184	30	20
Sulfaquinoxaline	301.2 > 107.7	30	20
Sulfathiazole	256.1 > 107.7	30	30
Sulfisoxazole	268.1 > 156.2	30	10
Tetracycline + 4-epi-tetracycline	445.1 > 410.1	30	20
Tilmicosin	869.8 > 174.1	65	45
Tylosin	916.6 > 127.2	30	40
Trimethoprim	291.3 > 230.3	60	20

had to be equal or higher than 10. In order to get a linear calibration curve at 5 points, standard solutions series of 0.25, 0.5, 1, 1.5 and 2 MRL were injected with the series of the samples extracted with 4 different methods (Table 1). In particular, these solutions were prepared using three mixtures of antibiotic standards at

Table 4 Maximum residues limits (MRL*), CCβ and number of samples analyzed for each antibiotic used for the validation of the selected method

Family of antibiotics	Antibiotics	MRL in milk (µg/L)	CCβ (µg/L)	Detection	
				CCβ spiked	CCβ Blank samples
Penicillins	Amoxicillin	4	2	20/20	20/20
	Penicillin G	4	2	20/20	20/20
	Ampicillin	4	2	20/20	19/20
	Cloxacillin	30	15	20/20	20/20
	Dicloxacillin	30	15	20/20	19/20
	Nafcillin	30	15	20/20	20/20
	Oxacillin	30	15	20/20	20/20
Quinolones	Enrofloxacin	100	50	20/20	20/20
	Marbofloxacin	75	37.5	20/20	20/20
	Flumequine	50	25	20/20	19/20
Macrolides	Erythromycin	40	20	20/20	19/20
	Spiramycin	200	100	20/20	20/20
	Tilmicosin	50	25	20/20	20/20
	Tylosin	50	25	20/20	19/20
	Neospiramycin	200	100	20/20	20/20
Tetracyclines	Chlortetracycline	100	50	20/20	20/20
	Oxytetracycline+ 4-epi-oxytetracycline	100	50	20/20	20/20
	Tetracycline+ 4-epi-Tetracycline	100	50	20/20	20/20
Sulfonamides	Sulfadiazine	100	50	20/20	20/20
	Sulfapyridine	100	50	20/20	20/20
	Sulfamethoxypyridazine	100	50	20/20	20/20
	Sulfadoxine	100	50	20/20	20/20
	Sulfadimethoxine	100	50	20/20	20/20
	Sulfadimidine	100	50	20/20	20/20
	Sulfamonomethoxine	100	50	20/20	20/20
	Sulfamoxole	100	50	20/20	20/20
	Sulfaquinoxaline	100	50	20/20	20/20
	Sulfachloropyridazine	100	50	20/20	20/20
	Sulfathiazole	100	50	20/20	20/20
	Sulfamerazine	100	50	20/20	20/20
	Sulfamethoxazole	100	50	20/20	20/20
	Sulfacetamide	100	50	20/20	20/20
	Sulfisoxazole	100	50	20/20	19/20
	Trimethoprim	50	25	20/20	20/20
	Sulfaguanidine	100	50	20/20	20/20
	Sulfamethizole	100	50	20/20	20/20

*MRL (EU 37/2010)

1MRL: mix 1 (macrolides – tetracyclines), mix 2 (sulfonamides - quinolones), and mix 3 (penicillins). After adding specific volume of each internal standard to the mix, the intermediate solutions were evaporated to dryness at 40 °C. The specific volumes of penicillins and Milli-Q water were added in order to obtain final solutions at 300 µL with final concentrations of 250, 500, 1000, 1500, and 2000 µg/L, respectively. In order to evaluate applicability of the present multi-residue method, the collected milk samples (194) were analyzed. Additionally, the interpretation of results was based on five criteria (Commission Decision 2002) which were: a S/N ratio of the ionic transitions greater than ten, the difference of the chromatographic retention time was within 2.5 % range of the retention time of the same peak in standard solution, area of the sample was higher than area of blank, area of sample was higher than area of spiked sample, and concentration of the sample analyzed was higher than MRL and LOD.

Stability of antibiotics in milk

Taking in consideration that lben is produced within 1 day of fermentation at temperature room; the stability of antibiotic residues was evaluated. Consequently, milk purchased in a Belgian supermarket was used either as a negative control ($n = 2$), or enriched ($n = 3$) as a test material. The enrichment was done at 0.5 MRL as prescribed for meat matrices (Commission Regulation 2377/1990) with mixtures of antibiotic standards of sulfonamides, penicillins, quinolones, macrolides, and tetracyclines. The samples were stored at different storage temperature (4 and 21 °C) and tested at various time periods (day 0, 2 and 7). The analysis was performed using LC-MS/MS after extraction (method 3). First and for most, the initial values were determined on the day of extraction (day 0). After that, samples stored at 4 and 21 °C were analyzed at day 2 and day 7 of storage. Some of these abused storage conditions also mimicked or overestimated the transport of the samples from Algeria to Belgium as well as possible practices in Algeria. These data also served for antibiotic stability evaluation.

Statistical analysis

In order to select the most suitable method extraction; the analytical methods were compared on the basis of the recoveries sufficient for the screening purposes, and only for the antibiotics having lower MRL in milk than in meat. The latter was defined as the ratio of MRL in milk to MRL in meat by being lower or equal to 0.5. In addition, Sstatistical analysis of the recovery results obtained with the four methods was performed using two-factor weighted ANOVA test with replication ($n = 2$) using S-Plus 8.0 for Linux (Tibco, Palo Alto, CA, US). The first factor was the method used and the second was the type of antibiotic. At the same time, the interaction between these two factors was also studied. Since the latter was significant, the comparison between the four methods was done per each antibiotic and corrected for simultaneous hypothesis testing according to Sidak (1967). Similar was done for comparison of storage effect on antibiotics. Statistical significance was tested at significance level of 0.05 and 0.01.

A generalized linear model for binomially distributed data was fit to model the frequency of positive LC-MS/

MS and Delvotest SP-NT results using the logit link. These results were contrasted between different types of milk collected at the farms and points of sale. In particular, milk type was considered as a fixed factor, whereas, farm and points of sale were random ones. Also, the comparison between the different milk types was evaluated and P-values were corrected for simultaneous hypothesis testing, according to Tukey. For correlation purposes, results of the examined residues (without MRL, and non-authorized) using LC-MS/MS were considered as positive.

The agreement between the two tests was calculated using Cohen's kappa and its associated P-value. In addition, since the Delvotest SP-NT appeared to lack specificity, a list of antibiotics was set up containing the positive LC-MS/MS results and those for which there were more negative than positive Delvotest SP-NT.

Results
Optimization of the LC-MS/MS screening method

The main requirement for a reliable screening method is to detect authorized substances above the regulatory limits MRL and unauthorized at a level of MRPL; which is the minimum required performance limit, minimizing false negative results (Freitas et al. 2013). Before the final method validation, four methods were compared and the results presented in Table 2 were statistically evaluated to select the appropriate one. Comparison using ANOVA was limited to antibiotics (11 antibiotics) with lower MRL in milk than in meat (MRL ratio equal or lower than 0.5). Since highly significant interactions were observed between methods and antibiotics, methods were compared with each other for each of those selected antibiotics separately. Among the comparative results, the main differences were marked between two method groups. Particularly, out of 11 compared antibiotics, the recoveries of eight (method 1 as opposed to method 4) and seven (method 2 as opposed to method 3) were significantly different ($p < 0.05$). This result may be explained by the analytical differences of the methods. Method 3 presented recoveries that were higher than other methods and closer to 100 %. In a second step the lowest recoveries were compared among the methods to ensure that final method may detect all selected antibiotics. It was observed that only for erythromycin and nafcilin higher recoveries were obtained when applying method 3. Taking this in account and the fact that other compounds were acceptable for screening purposes by method 3, it was decided to use this method further in the study.

The optimized extraction method was validated for macrolides, penicillins, quinolones, sulfonamides, and tetracyclines. Except, ampicillin, dicloxacillin, flumequine, erythromycin, tylosin, and sulfisoxazole which were characterized with some interference peaked within retention time range resulting in 5 % of false positive, the rest of the specificity compounds analyzed in 20 blank milk samples were negative. The detection capability levels were 50 % of MRL for all compounds in accordance to the guidelines for the validation of screening methods for residues of veterinary medicines (Table 4) (Community Reference Laboratories 2010). A linear calibration curve was obtained according to concentrations used. The correlation coefficient was ≥0.80 for all analyzed substances. One hundred ninety- four (194) milk samples were analyzed using this screening method. Within these, one hundred twenty-seven (127) were found non-conform. More details are presented in LC-MS/MS results section.

Delvotest
Out of 154 samples analyzed in this study 39 were positive forming 25.3 % of the total samples. According to the results presented in Table 5, the highest frequencies of positive results by Delvotest SP-NT prescreening were obtained for treated cow's milk (61.5 %), followed by untreated cow's milk (20.8 %), and market fermented cow's milk (20 %). Yet, market raw cow's milk (16.7 %) and bulk tank milk (10.3 %) were tested less positive than treated cow's milk. Correspondingly, a high percentage of positive results in the sample category of treated cow's milk were significantly different from untreated cow's milk and bulk tank milk ($p \leq 0.01$).

LC-MS/MS results
The comparison between different types of milk collected from distinct sources did not reveal any significant differences. As presented in Table 5, 127 out of 194 samples analyzed were found positive (65.5 %). The highest frequencies of positive results with LC-MS/MS analysis were obtained for milk collected from the market where each sample of collected raw milk was contaminated (100 % samples positive) and followed by fermented milk (85 % samples positive). High percentage of positive results was similarly found in samples collected from the farms. Hereby, the milk collected from treated cows contained the most antibiotic residues

(68 %), followed by milk samples collected from untreated cows (58 %), and lastly bulk tank milk (54 %).

Correlation between LCMS/MS and Delvotest SP-NT results and occurrence of specific antibiotics
Only 131 milk samples were used for the comparative assessment of results obtained by Delvotest SP-NT and LC-MS/MS. Specifically, the comparison of the total numbers of results using Cohen's Kappa illustrated that there was no evidence of agreement between results obtained by LC-MS/MS and Delvotest SP-NT (Table 6). Furthermore, from 131 milk samples only 18 samples were negative by both LC-MS/MS and Delvotest SP-NT methods. In this study, 15 samples showed doubtful results and 13 samples presented positive results with Delvotest SP-NT. Both groups of samples (doubtful and positive results) were later confirmed negative by LC-MS/MS. Furthermore, results presented in Tables 6 and 7 also show that in 52 samples initially found negative by Delvotest SP-NT mutltiply antibiotics were detected by LC-MS/MS. The most abundant residues (β-lactams) were followed by macrolides. Sulfonamides, quinolones, and tetracyclines consecutively were present at similar low frequencies. Finally, 20 milk samples were confirmed positive by both methods and additionally 13 samples initially found questionable by Delvotest SP-NT were confirmed positive by LC-MS/MS. This all implies that LC-MS/MS method was more sensitive than Delvotest SP-NT.

Stability of antibiotics in milk during storage
The noticeable increase in measurable antibiotics was observed after storage. The followed antibiotics were spiked to blank milk samples which were stored during short (2 or 7 days) period. After 2 days there was no change of antibiotics independently of the storage conditions, but a significant increase in measurable antibiotics was noticed after 7 days both at 4° and 21 °C. These differences are shown in Fig. 1. For most of the antibiotics this difference was significant independently of the temperature at which milk was stored whereas for certain antibiotics it was insignificant ($p > 0.05$). This was for neospyramycine, sulfamethoxypyridazine, sulfaquinoxaline,

Table 5 Comparison of the results for different milk types obtained by Delvotest and LC-MS/MS

	Delvotest negative	Delvotest positive	%Delvotest positive	LC-MS/MS conform	LC-MS/MS non-conform	% LC-MS/MS positive
Bulk tank milk	26	3	10.3[a]	16	19	54[a]
Individual untreated cow's milk	57	15	20.8[a]	41	58	59[a]
Individual treated cow's milk	10	16	61.5[b]	7	15	68[a]
Market raw cow's milk	10	2	16.7[ab]	0	18	100[a]
Market fermented cow's milk	12	3	20[ab]	3	17	85[a]

Groups with the same letter behind the percentage of % Delvotest positive are not significantly different from each other at level of 0.05
Groups with the same letter behind the percentage of % LC-MS/MS positive are not significantly different from each other at level of 0.05

Table 6 Statistical agreement between results obtained by two techniques Delvotest SP-NT and LC-MS/MS

	Delvotest negative	Delvotest positive	Delvotest doubtful
LC-MS/MS conform	18	13	15
LC-MS/MS non-conform	52	20	13

Total number of results for each results category was compared and Cohen's kappa (−0.1042, P value: 0.1578) was calculated

OTC+epi-OTC2 and TC+epiTC1 at 4 °C; for penicilline G, trimetoprime at 21 °C and for: tylosine, dicloxacilline, sulfapyridine, sulfisoxazole at both temperatures. The results also shown lack of significant difference in antiobitic concentration in milk kept 2 days on either temperatures. However, the significant difference among some antibiotics was noticed after analyzing samples kept 7 days. At 21 °C after 7 days it was possible to measure significantly more antiobitics, in particular neospyramycine, spyramicine, cloxacilline, dicloxacilline, penicilline G, sulfacetamide, sulfadoxine, sulfamethoxazole, sulfamoxole.

Discussion

Monitoring large numbers of milk samples for the presence of residues in excess of the levels laid down under community legislation requires low cost screening methods. In practice, it is primarily performed using microbiological screening methods, because of their high

Table 7 Comparison of the results obtained by LC-MS/MS and Delvotest SP-NT given as frequency (%) of the measurement appearances

Antibiotic	(Frequency of results, %)		
	LC-MS/MS positive & Delvotest negative	LC-MS/MS positive & Delvotest positive	LC-MS/MS positive & Delvotest doubtful
Ampicillin	26.9	10	38.5
Oxacillin	9.6	/	/
Tylvalosin	11.5	5	/
Tilmicosin	13.5	10	/
Amoxicillin	42.3	40	/
Erythromycin	1.9	/	/
Tilosyn	1.9	/	/
Sulfathiazole	1.9	/	/
Sulfamerazine	1.9	/	/
Sulfamethizole	1.9	/	/
Enrofloxacin	1.9	/	/
Spiramicin	1.9	/	/
Nafcillin	1.9	/	/
Josamycin	1.9	/	7.7
Tulathromycin	1.9	/	/
Doxycyclin	1.9		/

cost-effectiveness compared to physical–chemical detection. In general, these assays can be operated: without special training, do not depend upon specialized equipment, and target a broad spectrum of antimicrobial residues within one test (Pikkemaata et al. 2009). The most widely used tests which are commercially available are microbial inhibitor tests with spores of *Bacillus stearothermophilus var. calidolactis* –Delvotest SP, Copan Test, Charm Farm-960 Test, and others (Žvirdauskienė and Šalomskienė. 2007). Within this study, Delvotest SP-NT was the selected one to be used for the assessment of antibiotic residues in milk samples collected in Guelma (Algeria). As the main limitation of this and other similar microbial assays non-specificity is usually assumed (Bilandzic et al. 2011). For a visual reading of in particular Delvotest SP-NT, clear yellow and clear purple colors are easy to determine. However, visual assessment of samples containing intermediate concentrations of antimicrobials is more difficult, even for experienced technicians which render the visual judgment of the colored reaction as in Delvotest SP-NT subjective (Suhren and Luitz. 1995; Stead et al. 2008). In such samples, the test medium often shows a cloud of purple in a yellow background indicating a suspect positive result. In addition, different types of milk and the different modes of action of antimicrobial compounds can lead to different colors in the test, making the interpretation more difficult (Stead et al. 2008). In turn, this makes Delvotest SP-NT less suitable for decisive analyses leaving spaces for disputable results (false positives, eg). Another explanation could be that false positive Delvotest SP-NT reactions may occur in samples from freshly calved cows due to the fact that natural inhibitors and incomplete milking could be responsible for positive reaction (Hillerton et al. 1999).

To confirm the results obtained by those screening tests more specific, fast and sensitive techniques could be used. One such a method is LC-MS/MS technique which was employed in this study to verify the results of the Delvotest SP-NT. The comparative analyses pointed out several discrepancies between the methods, a number of false negative results using Delvotest SP-NT were identified and the sensitivity of Delvotest SP-NT for different groups of antibiotics as well as milk type was evaluated.

Whereas according to Delvotest SP-NT the most contaminated samples were those obtained from treated cows, LC-MS/MS revealed that the samples collected from the markets were mostly contaminated. Antibiotics in milk of treated cows could arise from a collection of milk shortly after administration of antibiotics. Subsequently, it may also reflect a certain disrespect of a prescribed antibiotic withdrawal time due to overdose usage, failure to observe withdrawal time, drug misuse or the bad hygiene (Zinedine et al. 2007; Petrović et al. 2008; Mensah

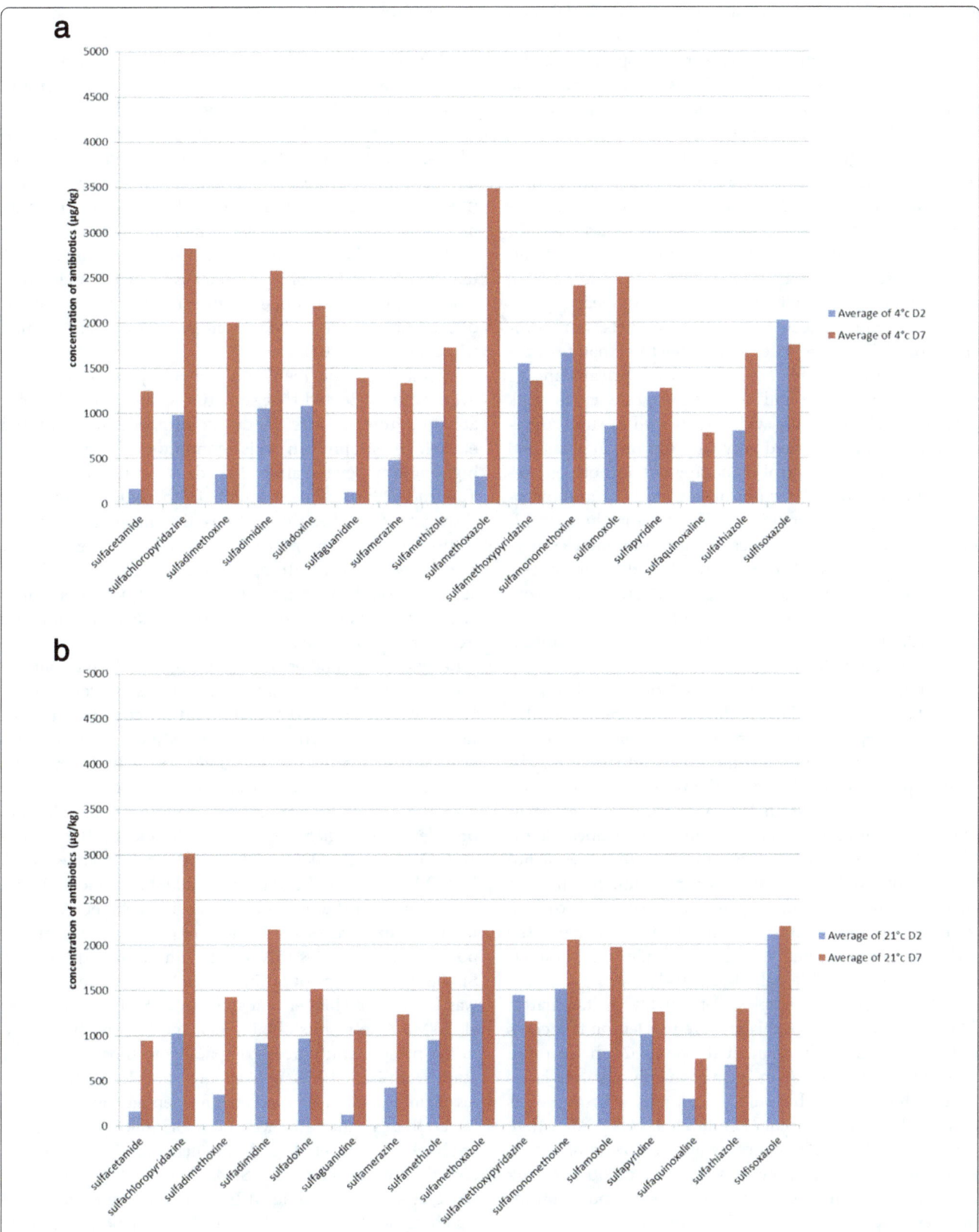

Fig. 1 Estimation of the antibiotic concentration in samples stored at two temperatures (4° and 21 °C) and during two periods (2 and 7 days). *The results present the estimated concentration of antibiotics measured in samples stored at 4 °C (**a**) and at 21 °C (**b**). The results are average of three measurements

et al. 2014b; Boultif. 2015). Establishment of a withdrawal time depends on different criteria (type of used antibiotic, quantity of given antibiotic, the way of applications, and as well age, health status, lactation stage and individual features of dairy animals) (Nikolić et al. 2011). Certain studies indicated that some African breeds may differ in terms of their genetic heritage making them more suited to local climatic conditions (water consumption, volume of distribution, and renal clearance) but non directly comparable to antibiotics withdrawal time as determined for breeds representative of a large-scale production in developed countries (Mensah et al. 2014a). Therefore, certain adaptation to local conditions could be envisaged.

Milk samples collected form the markets were provided by different private milk collectors that brought milk in isotherm tanks collected from private and/or governmental farms, and from small-scale cow's milk producers. The latter are widely dispersed in rural Algerian territories and located very far from urban areas and far away from any control by competent authorities. Actually, the accurate detection of low levels of antimicrobial drug residues in milk, as it was done in the study, is not only of great importance for governmental control laboratories and the dairy industry, but also for farmers to enable them to ensure that contaminated milk from individual cows is not consigned to the bulk tank (Stead et al. 2008) and prevent the further transfer of antibiotics in the food chain.

Algerian executive decree N°14-366 of December 15th, 2014 that regulates the conditions and the applicable modalities regarding contaminants tolerated in foodstuffs, does not mention sufficient information on the control plan and the dosage of the residues of antibiotics (Décret exécutif 2014). In fact, auto-control in dairy industries is commonly done following the European legislation. In contrast, quality assurance is rather low or not strictly followed in many African countries like in Algeria. Hence, the lack of statutory legal framework in Algeria may be a contributor to a high presence of the forbidden or regulated antibiotics residues in animal-derived foodstuffs (Mensah et al. 2014b).

In this study lben samples were highly contaminated. The fermented milk 'lben' is prepared by letting raw cow's milk to be naturally fermented within 24 h at room temperature. Fermentation of milk mainly involves lactic acid bacteria (LAB), but micrococci, coryneforms, yeasts and moulds can also occur (Zamfir et al. 2006). After that, it must be subjected to churning and removing of butter. Literature suggests that the presence of antibiotics in milk may affect fermentation processes in food production industries (Hsieh et al. 2011). Therefore, the fermentation during lben production may be affected if residues of antibiotics are present by possibly inhibiting the growth of the starter cultures (Nikolić et al. 2011). For this reason,

the performance of Delovtest SP-NT was questioned. Whereas no specific limitations for analyzing antibiotic residues in fermented milk are neither mentioned by the manufacturer nor verified in some studies (Hennart and Faragher 2012), the fate of antibiotics in milk type as lben is not fully evaluated. Except a study on raibi milk from Morocco (Zinedine et al. 2007), there is a lack of scientific data on fate of antibiotic residues in fermented milk (lben). This milk type is sour milk and may be a product of mostly lactic acid bacteria fermentation (Ouadghiri et al. 2009) and therefore may contain some byproducts, certain natural antimicrobial substances of that fermentation type. In turn, results obtained may represent the reaction of the test to those antibiotics already present in raw milk (Zinedine et al. 2007).

The shelf life of 'lben' is about 3 days at 4 °C. However, it was reported that sometimes "lben" may be kept at room temperature in the countryside with limited electricity supply. In this conditions 'lben' reaches high acidity levels after 2–3 days (Benkerroum and Tamime 2004). The results from our study have shown that in these conditions the antibiotic concentration appears to increase. This was independent of the temperature conditions (both 4° and 21 °C). More studies would be necessary to reveal whether these antibiotics would be easier bioavailable increasing the exposure to the consumers.

Recently, few studies in Algeria were performed on the presence of antibiotic residues in milk. Using Delvotest SP NT, Hakem et al. (2012) had found no contamination of raw milk collected from two Dairies Mitidja's Farms. However, in the present study, around 10 % of bulk tank milk samples and 20 % of samples of untreated cow's milk were positive. The latter was higher than the contamination prevalence in Algiers (9. 87 %) indicated by Ben-Mahdi and Ouslimani (2009). Other authors, Zinedine et al. (2007) in Morocco Tarzaali et al. (2008) in Mitidja, Aggad et al. (2009) in the west of Algeria, Titouche et al. (2013) in Tizi-Ouzu, reported higher frequency of antibiotics positive milk samples (57, 89, 29, and 46 %, respectively). Similarly, frequency of positive farm bulk tank milk (40 %) was found to be higher in a study from Serbia (Petrović et al. 2008). Also results obtained for market raw cow's milk in the present study were comparable to those found in a study from Iran (19.78 %) (Aalipour et al. 2015). Nevertheless, contamination of marketed fermented cow's milk was lower than results attained for raibi milk (50 %) in the study of Zinedine et al. (2007). In general, 65.5 % of all analyzed samples using LC-MS/MS contained antibiotic residues at levels exceeding MRL. This positive frequency is much higher than 15 % (Li et al. 2012), 16.66 % (Meng et al. 2015) in China, 1.76 % (Martins et al. 2014) in Brazil, and 28 % (García et al. 2016) in Spain. Pereira et al. (2014) indicated that 47.17 % of analyzed samples were at detectable concentration whereas Han et al. (2015) found 12 % at levels lower than MRL. Delvotest SP-NT has varying sensitivity to different antibiotics groups. According to the manufacturer and several reports (Althaus et al. 2003; Stead et al. 2008; Sierra et al. 2009a; DSM 2012, and Beltran et al. 2015), the sensitivity of Delvotest SP NT to β-lactams, as a main group of veterinary drugs used in therapy of cows in many countries is high except for cloxacillin. This one may be detected at levels higher than the MRL (Petrović et al. 2008). Delvotest SP-NT sensitivity to macrolides is limited to tylosin but erythromycin and spiramycin are detected at levels higher than MRL (DSM 2012). Althaus et al. (2003); Stead et al. (2008); Sierra et al. (2009b) and Beltran et al. (2015) also confirmed that the detection of erythromycin was at levels higher than MRL whereas, there are no data linked to Delvotest SP-NT sensitivity for the other macrolides (tilmicosin, tulathromycin, tylvalosin and josamycin). Furthermore, Delvotest SP-NT sensitivity to sulfonamides is limited only to sulfathiazole (DSM 2012). Enrofloxacin and doxycycline detection levels of this test were described by Sierra et al. (2009b). The latter argued that when the milk sample contained residues below the detection limit (LOD), the spores germinated and grew, thus, their metabolic activity made the indicator change the color. Althaus et al. (2003), Sierra et al. (2009b), Le Breton et al. (2007), and Comunian et al. (2010) noticed that Delvotest SP-NT demonstrated a lower ability to detect some other tetracyclines.

Both methods were in accordance for some milk samples that were found positive. The most frequently detected antibiotics were β-lactams. Mainly, amoxicillin (the most abundant residue in the studied milk) and ampicillin were found in half of the samples; which was still less than penicillin (97 %) and/or tetracycline (88 %) as reported by Ben-Mahdi and Ouslimani (2009), and Titouche et al. (2013), respectively where standard microbiological methods were used. The low cost of β-lactams in Algeria makes them easily available. The latter facilitated the use of penicillin by private farmers without veterinarian supervision in isolated places. Moreover, the presence of macrolides with low frequencies of tilmicosin and tylvalosin as -a residue without MRL - could be explained by their sporadic use. Aminoglycosides were not analyzed by LC-MS/MS method due to their relatively high MRL in comparison to meat. Fundamentally, Delvotest SP-NT manufacturer reported that the sensitivity of Delvotest SP-NT for dihydrostreptomycin and streptomycin is higher than MRL and lower than MRL for neomycin. Neomycin, in its turn, could be present within the 20 positive results of both methods. The high levels of contamination of milk samples by antibiotic residues can mainly be explained by massive and uncontrolled intermammary pharmaceutical preparations used for the treatment and prevention of bovine

mastitis, while the withdrawal times after treatment were probably not correctly respected. Similarly, the voluntary addition of bacterial growth inhibitors (antibiotics, antiseptics) in order to stop microbial growth and stabilize the microbial quality of milk (Zinedine et al. 2007) may also be considered as a possible cause. Another argumentation is supported by the study of Reybroeck (2010) where it was stipulated that the main reason for antibiotic residues in milk was the accidental milking of treated cows that went unnoticed in 66 % of the cases and the non -compliance of withdrawal time deadlines in 41 % of the cases.

Conclusion

To conclude, the study findings preliminary revealed that the presence of antibiotics in raw and fermented cow's milk collected in Guelma region and intended for either direct consumption and/or fermentation was high. The results highlighted 65.5 % of non-conform samples contained authorized residues at levels higher than the MRL, residues without set MRL, or non- authorized residues. The occurrence of antibiotic starting from farm's milk and ending in milk purchased from markets in Guelma province indicated the necessity of further control of milk. Providing that there is a lack of data in this domain, the control and assessment might be considered on a national level. Additionally, to obtain results an extraction method with LC-MS/MS detection was validated following regulatory criteria and demonstrated satisfactory results for all parameters. The comparison of results of both methods showed that Delvotest SP-NT could not be accurately trusted under these circumstances. However, LC-MS/MS represented a better screening alternative with a possibility to further investigate each group of compounds applying the specific extraction to increase the sensitivity of the method and decrease the LOD. The present LC-MS/MS method could be used in Algerian National Residue Control Plan as a versatile analytical tool to monitor and determine the occurrence of antibiotic multi-residues in milk and other food matrices after optimization.

Acknowledgements
The authors are thankful to those farmers who kindly provided milk samples and cows treatment data. The authors are very grateful for the financial support of the Algerian Ministry of the Superior Education and Scientific Research and University of Guelma for the training. A great acknowledgement goes to the technical support for LC-MS/MS analysis given by Khariklia Tsilikas, Jean-Yves Michelet, and Tim Reyns from Scientific Institute of Public Health (ISP.WIV) Brussels, Belgium. Special acknowledgments goes to Radhia Layada for the considerable work on refining the language of the manuscript. As well as, appreciations are delivered to Mr. Joens Vinas (Spain) and DSM food specialties (Netherlands) for providing Delvotest SP-NT Kits.

Authors' contributions
SL and DEB conceived the study. SL performed the collection of the samples, al practical work, the interpretation of the results, and drafted the manuscript with supervision of MA. MA designed the stability study and the

LC-MS/MS analytical method approach. WC performed the statistical analysis of all the results. All authors took parts in drafting the manuscript. All authors read and approved the final manuscript.

Competing interests

The authors declare that they have no competing interests.

References

Aalipour F, Mirlohi M, Jalali M, Azadbakht L. Dietary exposure to tetracycline residues through milk consumption in Iran. J Environ Health Sci Eng. 2015;13: 80. doi:10.1186/s40201-015-0235-6.

Aggad H, Mahouz F, Ammar YA, Kihal M. Evaluation de la qualité hygiénique du lait dans l'ouest algérien. Revue de Méd Vét. 2009;160(12):590–5.

Althaus RL, Torres A, Montero A, Balasch S, Molina MP. Detection limits of antimicrobials in ewe milk by delvotest photometric measurements. J Dairy Sci. 2003;86:457–63.

Belhadia M, Yakhlef H, Kouache B (2011) Filière lait, performances et insuffisances de la production et de la collecte du lait cru, cas des élevages bovins laitiers dans la région du haut Chéliff. Dans : Résumés papier présenté dans le Premier Séminaire National sur le lait et ses dérivés : « entre réalité de Production et réalités de transformation et de consommation », Université de Guelma, Algérie, 4–5 Octobre 2011

Beltran MC, Berruga MI, Molina A, Althaus RL, Molina MP. Performance of current microbial tests for screening antibiotics in sheep and goat milk. Int Dairy J. 2015;41:13–5.

Benkerroum N, Tamime AY. Technology transfer of some Moroccan traditional dairy 28 products (lben, jben, smen) to small industrial scale. Food Microbiol. 2004;21:399–413.

Ben-Mahdi M, Ouslimani S. Mise en évidence des résidus d'antibiotiques dans le lait de vache produit dans l'Algérois. Eur J Sci Res. 2009;36(3):357–62.

Bilandzic N, Kolanovic BS, Varenina I, Scortichini G, Annunziata L, Brstilo M, Rudan N. Veterinary drug residues determination in raw milk in Croatia. Food Control. 2011;22:1941–8.

Bogialli S, D'Ascenzo G, Di Corcia A, Laganà A, Nicolardi S. A simple and rapid assay based on hot water extraction and liquid chromatography- tandem mass spectrometry for monitoring quinolone residues in bovine milk. Food Chem. 2008;108:354–60.

Boix C, Ibanez M, Sancho JV, Leon N, Yusa V, Hernandez F. Qualitative screening of 116 veterinary drugs in feed by liquid chromatography- high resolution mass spectrometry: Potential application to quantitative analysis. Food Chem. 2014;160:313–20.

Borràs S, Companyó R, Granados M, Guiteras J. Analysis of antimicrobial agents in animal feed. TrAC Trends Anal Chem. 2011;30(7):1042–64.

Boultif L. Détection et quantification des résidus de terramycine et de pénicilline dans le lait de vache par chromatographie liquide haute performance (HPLC). Thèse de Doctorat: Université Mentouri Constantine, Algérie; 2015.

Cepurnieks G, Rjabova J, Zacs D, Bartkevics V. The development and validation of a rapid method for the determination of antimicrobial agent residues in milk and meat using ultra performance liquid chromatography coupled to quadrupole- Orbitrap mass spectrometry. J Pharm Biomed Anal. 2015;102:184–92.

Commission Decision (2002/657/EC) implementing Council Directive 96/23/EC concerning the performance of analytical methods and interpretation of results. Off. J. Eur. Commun L 221: 8–36

Commission Regulation (EU) N° 37/2010 text with EEA relevance of 22/12/2009 on pharmacologically active substances and their classification regarding maximum residue limits in foodstuffs of animal origin. Off. J. Eur. Commun L 15: 1–72

Commission Regulation (EU) N° 2377/1990 Laying down a Community procedure for the establishment of maximum residue limits of veterinary medicinal products in foodstuffs of animal origin. Off. J. Eur. Commun L224

Community Reference Laboratories Residues (CRL) (2010) Guidelines for the Validation of Screening Method for residues of Veterinary Medicines (Initial Validation and Transfer). Available on http://ec.europa.eu/food/safety/docs/cs_vet-med-residues_guideline_validation_screening_en.pdf

Comunian R, Paba A, Dupré I, Daga ES, Scintu MF. Evaluation of a microbiological indicator test for antibiotic detection in ewe and goat milk. J Dairy Sci. 2010; 93:5644–50.

Council Regulation No 1831/2003 of the European parliament and of the council of 22 September 2003 on additives for use in animal nutrition. Off. J. Eur. Union L268: 29–43

Décret exécutif n° 14–366 du 22 Safar 1436 correspondant au (15 décembre 2014) de la République Algérienne fixant les conditions et les modalités applicables en matière de contaminants tolérés dans les denrées alimentaires. J.O.R.A 74: 13–14

DSM food Specialties (Netherlands) (2012) Bulletin technique Delvotest® SP NT. Available on https://www.hygialim.com/Assets/Client/images/HYGIALIM/Schema/1008123.pdf.

Ferrini AM, Agrimi U, Appicciafuoco B, et al. Accred Qual Assur. 2015;20:267. doi:10.1007/s00769-015-1127-2.

Freitas A, Barbosa J, Ramos F. Development and validation of a multiresidue and multiclass ultra-high-pressure liquid chromatography-tandem mass spectrometry screening of antibiotics in milk. Int Dairy J. 2013;33(1):38–43.

García ND, Junza A, Zafra-Gomez A, Barron D, Navalon A. Simultaneous determination of quinolone and b-lactam residues in raw cow milk samples using ultrasound-assisted extraction and dispersive-SPE prior to UHPLC-MS/MS analysis. Food Control. 2016;60:382–93. doi:10.1016/j.foodcont.2015.08.008.

Hakem A, Yabrir B, Khelef D, Laoun A, Mouffok F, El-gallas N, Titouche Y, Ben-aissa R. Evaluation of microbial quality of raw milk into two dairies Mitidja's Farms (Algeria). Bulletin USAMV Veterinary Medicine. 2012;69(1–2). http://journals.usamvcluj.ro/index.php/veterinary/article/view/8850.

Han RW, Zheng N, Yu ZN, Wang J, Xu XM, Qu XY, Li SL, Zhang YD, Wang JQ. Simultaneous determination of 38 veterinary antibiotic residues in raw milk by UPLC–MS/MS. Food Chem. 2015;181:119–26.

Hennart SL, Faragher JJ. Validation of the Delvotest SP NT DA "Performance Tested Method^SM 011101". J AOAC Int. 2012;95(1):252–60. doi:10.5740/jaoacint.11-138.

Hermo MP, Nemutlu E, Kir S, Barron D, Barbosa J. Improved determination of quinolones in milk at their MRL levels using LC- UV, LC- FD, LC- MS and LC- MS/MS and validation in line with regulation 2002/657/EC. Anal Chim Acta. 2008;613:98–107.

Hillerton JE, Halley BI, Neaves P, Ros MD. Detection of antimicrobial substances in individual cow and quarter milk samples using delvotest microbial inhibitor tests. J Dairy Sci. 1999;82(4):704–11.

Hou XL, Chen G, Zhu L, Yang T, Zhao J, Wang L, Wu YL. Development and validation of an ultrahigh performance liquid chromatography tandem mass spectrometry method for simultaneous determination of sulfonamides, quinolones and benzimidazoles in bovine milk. J Chromatogr B Analyt Technol Biomed Life Sci. 2014;962:20–9.

Hsieh MK, Shyu CL, Liao JW, Franje CA, Huang YJ, Chang SK, Shih PY, Chou CC. Correlation analysis of heat stability of veterinary antibiotics by structural degradation, changes in antimicrobial activity and genotoxicity. Vet Med-Czech. 2011;56(6):274–85.

Jank L, Martins MM, Arsand JB, Motta TM, Hoff RB, Barreto F, Pizzolato TM. High-throughput method for macrolides and lincosamides antibiotics residues analysis in milk and muscle using a simple liquid–liquid ex-traction technique and liquid chromatography–electrospray–tandem mass spectrometry analysis (LC–MS/MS). Talanta. 2015;144:686–95. doi:10.1016/j.talanta.2015.06.078.

Kassaify Z, Abi Khalil P, Sleiman F. Quantification of antibiotic residues and determination of antimicrobial resistance profiles of microorganisms isolated from bovine milk in Lebanon. FNS. 2013;4:1–9.

Le Breton MH, Savoy-Perroud MC, Diserens JM. Validation and comparison of the Copan Milk Test and Delvotest SP-NT for the detection of antimicrobials in milk. Anal Chim Acta. 2007;586:280–3.

Li H, Xia X, Xue Y, Tang S, Xiao X, Li J, Shen J. Simultaneous determination of amoxicillin and prednisolone in bovine milk using ultra-high performance liquid chromatography tandem mass spectrometry. J Chromatogr B. 2012;900:59–63.

Martins MT, Melo J, Barreto F, Hoff RB, Jank L, Bittencourt MS, Arsand JB, Schapoval EES. A simple, fast and cheap non-SPE screening method for antibacterial residue analysis in milk and liver using liquid chromatography–tandem mass spectrometry. Talanta. 2014;129:374–83.

Meng Z, Shi Z, Liang S, Dong X, Li H, Sun H. Residues investigation of fluoroquinolones and sulfonamides and their metabolites in bovine milk by quantification and confirmation using ultra-performance liquid chromatography-tandem mass spectrometry. Food Chem. 2015;179:597–605. doi:10.1016/j.foodchem.2014.11.067.

Mensah SEP, Koudandé OD, Sanders P, Laurentie M, Mensah GA, Abiola FA. Antimicrobial residues in foods of animal origin in Africa: public health risks. Rev Sci Tech Off Int Epiz. 2014a;33(3):975–86.

Mensah SEP, Aboh AB, Salifou S, Mensah GA, Sanders P, Abiola FA, Koudandé OD. Risques dus aux résidus d'antibiotiques détectés dans le lait de vache produit dans le Centre Bénin J Appl Biosci. 2014b;80(1) doi:10.4314/jab.v80i1.9

Nikolić N, Mirecki S, Blagojević M. Inhibitory substances in raw milk. Mljekarstvo. 2011;61(2):182–7.

Ouadghiri M, Vancanneyt M, Vandamme P, Naser S, Gevers D, Lefebvre K, Swings J, Amar M. 14 Identification of lactic acid bacteria in Moroccan raw milk and traditionally fermented skimmed milk 15 'Iben'. J Appl Microbiology. 2009; 106:486–95.

Pereira RV, Siler JD, Bicalho RC, Warnick LD. Multiresidue screening of milk withheld for sale at dairy farms in central New York State. J Dairy Sci. 2014;97:1513–9.

Petrović JM, Katić VR, Bugarski DD. Comparative examination of the analysis of β-lactam antibiotic residues in milk by enzyme, receptor–enzyme, and inhibition procedures. Food Anal Methods. 2008;1:119–25. doi:10.1007/s12161-007-9007-y.

Pikkemaata MG, Rapallinib MLBA, Oostra-van Dijka S, Elferinka JWA. Comparison of three microbial screening methods for antibiotics using routine monitoring samples. Anal Chim Acta. 2009;637(1–2) doi:10.1016/j.aca.2008.08.023

Ramirez A, Gutiérrez R, Diaz G, Gonzalez C, Pérez N, Vega S, Noa M. High-performance thin- layer chromatography- bioautography for multiple antibiotic residues in cow's milk. J Chromatogr. 2003;B784:315–22.

Reybroeck W. Screening for residues of antibiotics and chemotherapeutics in milk and honey. Doctorat dessertation: Faculteit Diergeneeskunde Universiteit Gent; 2010.

Sidak Z. Rectangular confidence regions for the means of multivariate normal distributions. J Am Stat Assoc. 1967;62:626–33.

Sierra D, Sánchez A, Contreras A, Luengo C, Corrales JC, Morales CT, de la Fe CI, Guirao I, Gonzalo C. Detection limits of four antimicrobial residue screening tests for β-lactams in goat's milk. J Dairy Sci. 2009a;92:3585–91.

Sierra D, Contreras A, Sánchez A, Luengo C, Corrales JC, Morales CT, de la Fe CI, Guirao I, Gonzalo C. Short communication: detection limits of non-β-lactam antibiotics in goat's milk by microbiological residues screening tests. J Dairy Sci. 2009b;92:4200–6.

Stead SL, Ashwin H, Richmond SF, Sharman M, Langeveld PC, Barendse JP, Stark J, Keely BJ. Evaluation and validation according to international standards of the Delvotest® SP-NT screening assay for antimicrobial drugs in milk. Int Dairy J. 2008;18:3–11.

Suhren G, Luitz M. Evaluation of microbial inhibitor tests with indicator in micro-titer plates by photometric measurements. Milchwissenschaft. 1995;50:467–70.

Tarzaali D, Dechicha A, Gharbi S, Bouaissa MK, Yamnaine N, Guetarni D (2008) Recherche des résidus des tétracyclines et des bêta-lactamines dans le lait cru par le MRL Test (ROSA TEST) à Blida, Algérie. Papier présenté dans : 6èmes Journées Scientifiques Vétérinaires sur le médicament vétérinaire : nouvelles approches thérapeutiques et impact sur la santé publique. E.N.V, Algérie, 23–24 Avril 2008.

Titouche Y, Hakem A, Houali K, Yabrir B, Malki O, Chergui A, Chenouf N, Yahiaoui S, Labiad M, Ghenim H, Kechih-Bounar S, Chirilă F, Nadăş G, Fiţ NI. Detection of antibiotics residues in Raw Milk produced in freha area (Tizi-Ouzou), Algeria. Bulletin UASVM, Veterinary Medicine. 2013;70:1843–5378.

Wang J, Leung D, Lenz SP. Determination of five macrolide antibiotic residues in raw milk using liquid chromatography-electrospray ionization tandem mass spectrometry. J Agric Food Chem. 2006;54:2873–80.

Zamfir M, Vancanneyt M, Makras L, Vaningelgem F, Lefebvre K, Pot B, Swings J, De Vuyst L. Biodiversity of lactic acid bacteria in Romanian dairy products. Syst Appl Microbiol. 2006;29(23):487–95.

Zinedine A, Faid M, Benlemlih M. Détection des résidus d'antibiotiques dans le lait et les produits laitiers par méthode microbilogique. REMISE. 2007;1:1–9.

Zoubeidi M, Gharabi D. Impact du PNDA sur la performance économique des filières stratégiques en Algérie: cas de la filière lait dans la wilaya de Tiaret. Revue Ecologie-Environnement 9. 2013

Žvirdauskienė R, Šalomskienė J. An evaluation of different microbial and rapid tests for determining inhibitors in milk. Food Control. 2007;18:541–7.

16

Aflatoxin B_1 and Deoxynivalenol contamination of dairy feeds and presence of Aflatoxin M_1 contamination in milk from smallholder dairy systems in Nakuru, Kenya

Caroline Mwende Makau[1*] (iD), Joseph Wafula Matofari[1], Patrick Simiyu Muliro[1] and Bockline Omedo Bebe[2]

Abstract

Background: Mycotoxins are metabolites produced by phytopathogenic and spoilage fungi in animal feed as a result of poor storage. The mycotoxins can also originate in the field and are excreted in milk when dairy animals consume such feeds, posing a public health risk concern.

Methods: The aim of this study was to conduct a risk assessment in the informal sub-value chains of rural and peri-urban dairy systems in Nakuru County, by determining the prevalence and quantity levels of mycotoxins in animal feeds and milk. A total of 74 animal feed samples and 120 milk samples were simultaneously collected from individual cows and actors in the informal dairy value chain. Feed samples were analyzed for Aflatoxin B_1 (AFB_1) and Deoxynivalenol (DON) while milk samples were analyzed for Aflatoxin M_1 (AFM_1) using commercial Enzyme Linked Immune Sorbent Assay (ELISA) method.

Results: Aflatoxin B_1 contamination levels in 56 % (41/74) of the animal feeds exceeded the European Union (EU) limits of 5 μg/kg ranging between 0 and 147.86 μg/kg. Deoxynivalenol (DON) was identified in 63 % (27/43) of all the animal feeds ranging between 0 and 179.89 μg/kg. In the peri-urban dairy system, 48.5 % (33/68) of the milk samples were contaminated with the AFM_1 concentration above the EU regulation of 0.05 μg/L ranging between 0.017 and 0.083 μg/L. All milk samples from the rural dairy system had AFM_1 contamination levels below the EU limits of 0.05 μg/L ranging between 0 and 0.041 μg/L. Linear regression model showed significant association of abiotic factors; pH, water activity and moisture content of animal feed with AFB_1 and DON contamination of the animal feeds.

Conclusions: The results obtained from this study indicate that the peri-urban dairy farms, where intensive management predominate face the challenge of quality feeds, and one contributing factor is the on-farm production and handling of animal feeds.

Keywords: Animal feeds, Milk, Aflatoxin B_1, Deoxynivalenol and Aflatoxin M_1

Background

Mycotoxins are a diverse group of fungal secondary metabolites that are harmful to animals and humans. These toxins are produced by saprophytic fungi during storage or by pathogenic fungi during plant growth. Aflatoxin B_1 and B_2 are the main metabolites produced by fungi of the genus *Aspergillus* particularly *A. flavus*, *A. parasiticus* and *A. nomius* (Richard 2007; Reddy et al. 2010). Animals fed on

AFB_1 and B_2 contaminated feeds excrete into their milk the toxic AFM_1 and M_2, respectively which are metabolized in the liver. AFM_1 is of particular interest being the hydroxylated metabolite of the AFB_1 parent compound. AFs are highly carcinogenic causing liver cancer in humans (Zinedine et al. 2007). Deoxynivalenol (DON) is associated primarily with *Fusarium graminearum* and *F. culmorum*, both of which are important plant pathogens which cause fusarium head blight in wheat and fusarium ear blight in maize (Bottalico and Perrone 2002). DON is a mycotoxin belonging to the group of trichothecenes, which contaminates grains and cereal-based food and feed (Korosteleva

---
* Correspondence: makau.carol@gmail.com
[1]Egerton University, Faculty of Agriculture, Department of Dairy & Food Science and Technology, P.O. Box 536-20115, Egerton, Kenya
Full list of author information is available at the end of the article

et al. 2009). It is associated with acute gastrointestinal adverse effects such as vomiting (emesis) both in animals and humans (Vincelli et al. 2002).

The toxins can be passed down the food chain and contaminate milk and meat posing a greater danger to the health of humans (Flores-Flores et al. 2015; Leszczynska et al. 2001). Milk and milk products are traditionally the staple food commodities for the African communities. They are among the main entry routes of AFM_1 into the human dietary system in Africa (Hell and Mutegi 2011). Considering that milk and milk derivatives are consumed daily and, moreover, that they are of primary importance in the diet of children who are most vulnerable, many African countries have accepted the maximum admissible levels of 0.05 µg/L, set by the European Union (EC 2006a).

Worldwide, a high and increasing proportion of dairy cattle are kept in intensive systems making AFs be an increasing problem to dairy farmers (Unnevehr and Delia 2013). About 80 % of the milk in Kenya is produced by smallholder farmers, especially at rural and peri-urban dairy farming (Muriuki 2011). Kenyan small-scale dairy farmers practicing intensive dairy systems lack knowledge on the safe formulation of feed rations and as a result, they feed their dairy animals mostly on farm formulations from crop residues and cereals that are discarded due to mold spoilage. These farmers also feed their dairy animals on commercially prepared concentrates from uncertified agro-vet dealers. Rural and peri-urban dairy farmers lack knowledge on proper feed formulation and storage such as silage making and lack of properly constructed feed stores (Lukuyu et al. 2011).

Research studies reveal that urban dairy farmers in Kenya spend nine times more money to purchase commercial feeds than their rural counterparts (Thorpe et al. 2000) and are at a higher risk of feeding AFB_1-contaminated animal feeds (Kang'ethe and Lang'a 2009). There is also no monitoring and evaluation (M&E) system and inadequate enforcement of regulation in Kenya to evaluate the standards of market animal feeds (Nyaata et al. 2000). These factors contribute to the occurrence of mycotoxins in animal feeds and result in the carry over effect of AFM_1 in milk in the dairy value chain in Kenya. However, only very scarce data exists on the occurrence of mycotoxin contamination of animal feeds and AFM_1 in milk on rural and intensive small-scale dairy farms. Thus, this study aimed to provide information on the occurrence of common mycotoxins in the feeds and milk at rural and peri-urban dairy sub-value chains.

Methods
Study design and site description
A cross-sectional study was conducted in the Nakuru County dairy value chain between March 2015 and October 2015. The study was carried out in three divisions in Nakuru County, Kenya namely; Olenguruone, Wanyororo, and Bahati. Olenguruone division represented a rural dairy system which lies at 35° 40'60"E and 0° 34'60"S in DMS (degree minute seconds). Wanyororo and Bahati divisions represented the peri-urban dairy system as they surround Nakuru town and lie at 36° 16′ 12″ E and 0° 12′ 0″ S. Nakuru County has 52,670 small-scale farms with a population density of 35,500 dairy cows, 20,500 zebu (*Bos indicus*) and 15,000 exotic dairy cattle (*Bos taurus*). Both production systems in Nakuru County have high production capacities with 110,000 l of milk per day which translates to 40,150,000 l per year (MoALF 2012). This study implemented a value chain approach by investigating occurrence of mycotoxin contamination of the animal feeds and milk at the on-farm production stage, transportation, processing and marketing outlets from which milk is channeled to urban Nakuru consumers mostly through informal market agents.

Participant selection and sample collection
The approximate sample size for dairy farms was determined from the formula $n = Z^2 P_{exp} (1 - P_{exp})/L^2$, where Z is confidence level of 95 %, L is desired precision of 10 % and P_{exp} is expected prevalence of 50 % (Thursfield 1995). The calculated sample size was 78 farms in the rural dairy system and 42 farms in the peri-urban dairy system. A total of 97 animal feed samples were collected from randomly selected individual smallholder farms in rural and peri-urban dairy systems. A representative sample of 500 g was taken after mixing from storage bags into sterile plastic sampling bags and transported to the laboratory for analysis. Drying of samples was done by aseptically segregating 400 g of the sample and keeping it in an oven with the temperature set between 50 and 60 °C for 2 days to an average of 88 % dry matter content. The samples were then ground and stored at 20 °C under cool, dry conditions for analysis. The remainder of the sample was used for physico-chemical analysis. Milk samples from lactating cows on the same farm were collected. A total of 120 milk samples were collected from the individual lactating cows on small holder farms ($n = 69$), milk transporters bringing milk to cooperative dairy outlets ($n = 30$), cooperatives ($n = 12$) and milk bars ($n = 19$) in sterile 60 ml tubes. Samples were transported in cool boxes to the Egerton university laboratory under ice and frozen at −20 °C until analyzed within 3 months of collection.

Sample analysis
Determination of moisture content of animal feed samples
The moisture content was determined according the procedure provided by the Association of Official Analytical Chemists (AOAC) International (AOAC 2000). Samples weighing 2 g were dried in triplicates an oven at 105 °C for 3 h. Cooling of the dried samples was done in a desiccator for 10 min. Moisture content was calculated as the

loss in weight expressed as a percent of the original weight of the animal feed. The amount of moisture was reported in terms of loss in weight.

Determination of water activity in animal feed samples

A durotherm (Aw Messer- Germany) was calibrated using a saturated solution of barium chloride and left to stand for 3 h until water activity reading was at 0.900 in an incubator. This was done as described by manufacturer's instructions. Approximately 10 g of feed sample was finely chopped into small pieces and placed in triplicates in the durotherm. The water activity levels were recorded after 3 h at a temperature of 20 °C.

pH analysis of animal feed samples

The samples of animal feeds in triplicates were subjected to pH analysis of the glass electrode according to manufacturer's instructions. Approximately 20 g of air dried feed samples was transferred into 100 ml shaking bottle. 50 ml of distilled water was added and shaken for 2 h in the reciprocal shaker. The pH was determined by a precision pH meter PHS-3B (China) after a short but vigorous shaking. The pH meter was calibrated with buffers 4.0 and 7.0.

Enzyme immunoassay for aflatoxin B₁ totals in animal feeds

The quantitative analysis of AFB_1 in animal feed samples was performed by competitive ELISA (RIDASCREEN® Aflatoxin total, R-Biopharm) procedure as described by R-biopharm GmbH, Product code R4701. Prior to analysis of the samples, the ELISA method was validated to ensure data quality.

The sample preparation procedure was based on manufacturers of ELISA kit recommendations. The manufacturer's recommendations were followed except 20 g of the sample and 100 ml of methanol/water (70:30 v/v) was used instead of two grams of the sample and 10 mls of methanol/water (HPLC grade methanol was purchased from Fisher Scientific, USA) to extract the AFB_1. The entire extract was filtered. A Filtrate of 100 µL was diluted with 600 µL of the sample dilution buffer. A sample of 50 µL per well was employed in the assay. According to the manufacturer's instructions, the detection limit for feed samples was 1.75 µg/kg with the recovery rate of 85 %. AFB_1 in animal samples was measured according to the instructions of the manufacturer using standards (0, 0.5, 1.5, 4.5, 13.5, and 40.5 µg/kg). All samples were run in duplicates. The AFB_1 was measured photometrically at 450 nm (Readwell strip, ROBONIK, India).

Enzyme immunoassay for Deoxynivalenol in animal feeds

The quantitative analysis of DON in animal feed samples was performed by competitive ELISA (RIDASCREEN® DON, R-Biopharm) procedure as described by R-biopharm GmbH, Product code R5906. Prior to analysis of the

samples, the ELISA method was validated to ensure data quality.

The sample preparation procedure was based on manufacturers of ELISA kit recommendations. Five grams of the ground sample was weighed and added into a suitable container with 25 ml of distilled water and shaken vigorously for 3 min. The extract was filtered through Whatman paper No. 1. A sample of 50 µL per well was employed in the assay.

According to the manufacturer's instructions, the detection limit for feed samples was 18.5 µg/kg with the recovery rate of 85 %. DON in animal samples was measured according to the instructions of the manufacturer using standards (0, 3.7, 11.1, 33.3, and 100 µg/kg). All samples were run in duplicates. The DON was measured photometrically at 450 nm (Readwell strip, ROBONIK, India).

Enzyme immunoassay for aflatoxin M₁ in milk

The quantitative analysis of AFM_1 in pasteurized milk samples was performed by competitive ELISA (RIDASCREEN® AFM_1, R-Biopharm) procedure as described by R-biopharm GmbH, Product code R1121. Prior to analysis of the samples, the ELISA method was validated to ensure data quality.

The sample preparation procedure was based on manufacturers of ELISA kit recommendations. Milk samples (20 mL of milk) were thawed and centrifuged at 3500 g for 10 min at 10 °C before they were analyzed using an ELISA kit for M_1. The creamy upper layer was completely discarded, and the lower phase was used for ELISA quantitative analysis.

According to the manufacturer's instructions, the detection limit for milk samples was 5 µg/L with a recovery rate of 95 %. AFM_1 in skimmed milk samples was measured according to the instructions of the manufacturer using standards (0, 0.01, 0.02, 0.04, and 0.08 µg/L). All samples were run in duplicates. The AFM_1 was measured photometrically at 450 nm using ELISA reader (Readwell strip, ROBONIK, India).

ELISA methods validation

The analytical quality of the ELISA methods was assured by the Limit of Detection (LOD) which determined experimentally by measuring the concentration of 20 blank matrix samples and then calculated by the formula: Mean concentration of blank samples + 3-fold standard deviation of the concentrations of blank samples. The Limit of Quantification (LOQ) was determined experimentally by measuring the concentration of 20 blank matrix samples and then calculated by the formula: Mean concentration of blank samples + 9-fold standard deviation of the concentrations of blank samples (R-Biopharm 2015). The validation of the ELISA methods was carried out with the determination of the recoveries and the coefficient of variation (% CV). The mean absorbance values obtained for the standards and the samples were divided by the absorbance value of the zero standards and multiplied by 100. The zero standard was thus made equal to 100 %, and the absorbance values of other standards and samples were quoted in percentages of this value. The values calculated for the standards were entered in a system of coordinates semi- logarithmically and analyzed against the mycotoxin concentration using Excel (Microsoft, Inc. USA). The mycotoxin concentration in µg/L or µg/kg corresponding to the absorbance of each sample was read from the calibration curve. Calibration curves were prepared for each mycotoxin AFM_1, AFB_1, DON and coefficients of determination (r^2) were calculated respectively. In milk, AFM_1 curve was prepared from standard solutions in range 0.005–0.08 µg/L with $r^2 = 0.988$. In animal feeds, the AFB_1 curve was obtained from the standard solutions in range 0.05–40.5 µg/kg with $r^2 = 0.982$. In animal feeds, DON curve was prepared from standard solutions in the range of 3.7–100 µg/kg with $r^2 = 0.987$. In milk, recovery of the method was evaluated by analyzing spiked certified extracts in triplicates at the level of 0.01 and 0.05 µg/L corresponding to the maximum value allowed by the European Commission. Recoveries for AFB_1 and DON were calculated by comparing the response for each mycotoxin with that of known spiked mycotoxin levels expressed as a percentage. The validation experiments were performed as described for the samples above. The validation parameters (Table 1) were calculated and expressed using European Official Decision procedure for screening methods (EC 2002) and their values were in accordance with recommendations given in Commission Decision (EC 2006a). Both the recovery and % CV (Table 2) are in compliance with Commission Regulation (EC 2006a).

Statistical analysis

One regression model was fitted to determine the overall association between the predictors and outcomes. The predictors were moisture content, water activity and pH. The outcome were the concentration of AFB_1 and DON

Table 1 Validation of ELISA data for AFM_1 AFB_1 and DON

Parameter	Ridascreen® test kit		
	AFB_1 (µg/kg)	DON (µg/kg)	AFM_1 (ng/kg)
MC	0.82	16.94	3.92
SD	0.31	0.53	0.50
LOD	1.75	18.53	5.42
LOQ	3.61	21.68	8.42

LOD limit of detection
LOQ limit of quantification
MC mean concentration of the blanks
SD standard deviation

Table 2 Validation of ELISA data for AFM_1 AFB_1 and DON

Spiked sample	Spiked level[a]	Recovery (%)	Coefficient of variation
Milk (AFM_1)	0.01	101	1.1
Milk (AFM_1)	0.05	98	3.2
Animal feed (DON)	11.1	97	2.7
Animal feed (DON)	33.3	97	3.8
Animal feed (DON)	100	98	4.4
Animal feed (AFB_1)	4.5	98	2.1
Animal feed (AFB_1)	13.5	98	4.7
Animal feed (AFB_1)	40.5	98	3.6

[a]µg/L for AFM_1 and µg/kg for AFB_1 /DON with three replicates at each level

in animal feed. To improve model fit, outcome variables were natural log transformed and then the regression coefficients were back transformed to original scale to ease interpretation. Homoscedasticity was assessed by plotting standardized residuals against predicted values and by the use of Cooke-Weisberg test. Examination of residuals for normality was done using the normal probability plot and the Shapiro-Wilk test while the model fit was checked with Akaike's Information Criteria (AIC). Data obtained from mycotoxin analysis in feeds and milk was also tested for analysis of variance (ANOVA) and means comparisons was done using Tukey's Honestly Significant Difference (HSD) test at $P \leq 0.05$. STATA statistical software Version 9 (Statacorp, College TX, 2007) was used for analyses.

Results
AFB_1 contamination feeds

Animal feed contamination with AFB_1 was more frequent in rural (60 %) than in peri-urban (53 %) dairy system as illustrated in Fig. 1, but the median concentration was higher in peri-urban (60.43 µg/kg) than in the rural (12.25 µg/kg) system from the estimates in Table 3. The analysis of variance (ANOVA) showed concentrates had significantly higher AFB_1 levels compared to forage at $P < 0.001$ (Table 4). The lowest observed level of AFB_1 contamination of 2.31 µg/kg was from a hay feed sample from rural while the highest AFB_1 level of 147.86 µg/kg was observed in an on-farm formulated concentrate feed sample that contained maize germ from peri-urban as shown in Table 3. Association of water activity (a_w) and moisture content of the feeds with levels AFB_1 contamination of the feeds was significant at $P < 0.05$ and $P < 0.01$ respectively (Table 5).

DON contamination in animal feeds

Figure 2 illustrates that the contamination of feeds with DON was more frequent at 71 % with a higher concentration of median 60.61 µg/kg in the peri-urban than the rural dairy system at frequency of 53 % with concentration of median 21.62 µg/kg (Table 3). The analysis of

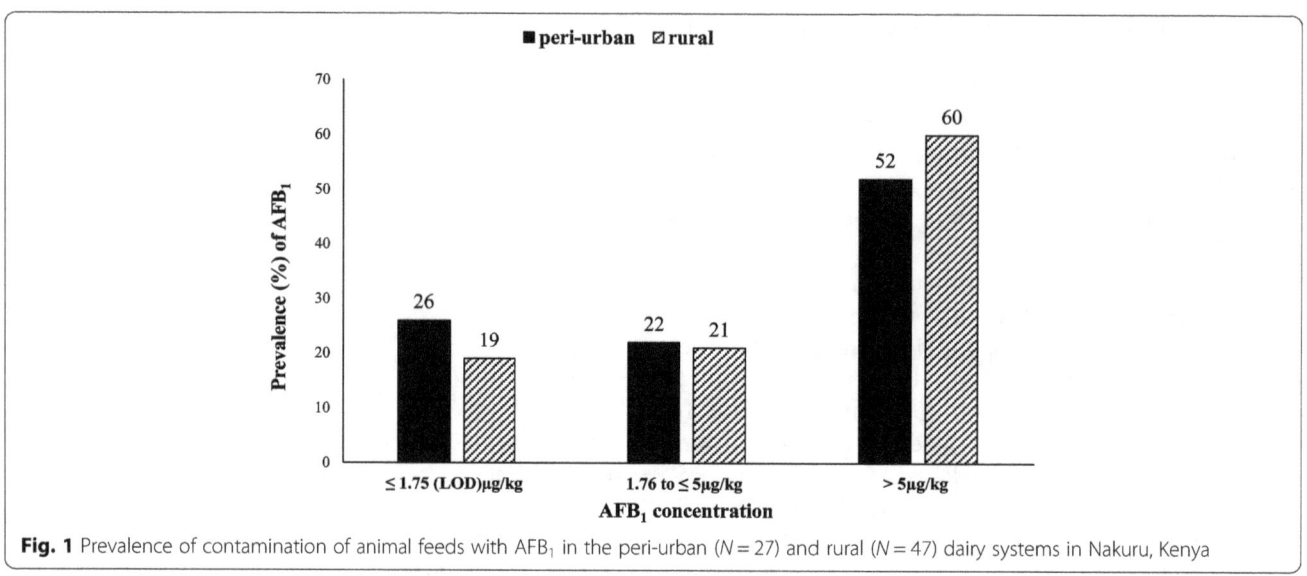

Fig. 1 Prevalence of contamination of animal feeds with AFB₁ in the peri-urban ($N = 27$) and rural ($N = 47$) dairy systems in Nakuru, Kenya

variance (ANOVA) (Table 4) showed the DON contamination was significantly higher in concentrates than in forage. The lowest level of DON contamination of 4.37 µg/kg was obtained from hay sample from rural dairy system while the highest level of DON contamination of 179.89 µg/kg was observed in silage feed sample from peri-urban system as shown in Table 3. Association of water activity, pH and moisture content of feeds with DON contamination levels was significant at $P < 0.01$, $P < 0.01$ and $P < 0.01$ respectively (Table 5).

AFM₁ contamination

In the peri-urban system, the prevalence of AFM₁ contamination ranged from 68 % at production, 29 % at transporters, 40 % at cooperatives and 17 % at milk bar outlets with a median value of 0.073 µg/L at production

level as shown in Fig. 3 and Table 6. All milk samples in the rural system were contaminated with AFM₁ concentration levels of less than 0.05 µg/L with a median value of 0.006 µg/L at production level as shown in Fig. 4 and Table 6. A majority of the samples along the rural value chain were below the limit of quantification of 0.005 µg/L as shown in Fig. 4.

The lowest level of AFM₁ contamination was 0.001 µg/L obtained from a milk sample from an individual cow at the rural dairy system. The highest AFM₁ level was 0.083 µg/L from a milk sample from an individual cow at peri-urban dairy system as shown in Table 6. The reducing trend of AFM₁ contamination along the value chain was observed with milk from cooperatives and milk bars having slighter range of contamination compared to milk from individual cows at production.

Table 3 The summary of mycotoxins in the animal feeds among different factors

Factor	Level	Statistic	AFB₁	DON
Dairy system	Rural	Mean ± SD	25.94 ± 28.71[a]	26.65 ± 28.00[a]
		Median	12.25	21.62
		Range	2.31–84.41	0.00–82.79
	Peri-urban	Mean ± SD	30.61 ± 43.33[b]	71.33 ± 62.29[b]
		Median	60.43	60.61
		Range	0.00–174.86	0.00–179.89
Type of animal feeds	Forage	Mean ± SD	5.14 ± 7.70[a]	17.88 ± 30.66[a]
		Median	7.52	1.33
		Range	2.31–29.52	0.00–96.20
	Concentrates	Mean ± SD	47.84 ± 36.81[b]	86.95 ± 51.70[b]
		Median	42.07	66.25
		Range	21.33–147.86	0.00–179.89

Means with same letter along the column are significantly different at $P \leq 0.05$ and SD standard deviation

Table 4 The analysis of variance of mycotoxins contamination in the animal feeds from the two dairy systems and type of the feeds

Source of variation	DF	MS for AFB_1	MS for DON
Dairy system	1	2029.670**	6362.286*
Type of Feed	1	33623.769***	51336.681***
Dairy system*Type of Feed	1	1354.295^{ns}	3275.537^{ns}
Error	70	730.152	1629.592

DF degree of freedom, *MS* mean squares, *AFB_1* aflatoxin B1, *DON* deoxynivalenol, *ns* not significant

* is significant at $P = 0.05$, ** is significant at $P = 0.01$ and *** is significant at $P = 0.001$

Discussion

Aflatoxin B_1 contamination in animal feeds

The analysis of variance (ANOVA) (Table 4) showed there was significant difference in aflatoxin B_1 contamination in animal feeds samples between the two dairy systems where the study was conducted. Feed samples from peri-urban had significantly higher levels of aflatoxin B_1 ranging between 0 and 147.86 µg/kg compared to rural dairy system which was ranging between 2.31 and 84.41 µg/kg (Table 3). This may be attributed to prolonged storage of animal feeds (hay, concentrates and silage) under precarious conditions in small stores by peri-urban dairy farmers, who practice stall feeding due to lack of grazing fields. The unsuitable storage conditions accompanied by the tropical climate in Kenya may provide the increasing fungi occurrence and mycotoxin production. In contrast, as indicated by previous studies (Baltenweck et al. 1998), rural farmers mainly practice free range grazing on fields directly with pasture grasses or using the cut and carry system without prior feed storage while supplementing the dairy cows with minimal proportions of commercial concentrates.

The analysis of variance (ANOVA) (Table 4) higher AFB_1 contamination of was observed in concentrates commonly utilized in the peri-urban ranging between 0 and 147.86 µg/kg than forages commonly used in the rural dairy system with AFB_1 levels of ranging between 0 and 29.52 µg/kg (Table 3). This could be attributed to lengthy storage of on-farm formulated concentrates by peri-urban farmers because of animal feed shortages observed in the dairy system due to lack of grazing fields

thus forcing them to formulate excess compounded concentrates. Prolonged storage conditions expose concentrates to the environmental conditions like humidity and temperatures that favour the growth of *Aspergillus spp.* (Soler et al. 2010). Besides, farmers lacked proper storage facilities for animal feeds with inadequate roofing leading to exposure of animal feeds to precipitation. These conditions contribute to mould growth leading to aflatoxin contamination.

Peri-urban farmers also used low-quality ingredients in the formulation of on-farm formulated concentrates leading to aflatoxin contamination of animal feeds. A study by (Richards and Godfrey 2003) in Nakuru County showed 42 % of urban and peri-urban farmers fed compounded concentrates to lactating cows. Other studies had shown that dairy farmers in the peri- urban areas of Kenya mostly use maize grains milled to make on-farm formulated concentrates to feed their cattle. The grains used are usually those that are contaminated with moulds at harvesting time and are separated from the healthy grains which are meant for human consumption. The mould invaded grains have been associated with aflatoxin contamination (Muture and Ogana 2005).

Extrinsic abiotic factors that affect growth of mycotoxin producing fungi measured in animal feeds included moisture content, water activity (a_w) and pH. Concentrates was identified moisture content, a_w and pH ranging between 11.20–71.30 %, 0.51–0.88 and 5.98–6.92 respectively that favour growth of mycotoxin producing fungi. Most storage fungi grow at a_w below 0.75. The required a_w for *Aspergillus spp.* growth is between 0.61 and 0.91 (Oviedo et al. 2011). Neutral pH ranging between 6 and 7 is also more suitable for mould growth which was exhibited in the study. From previous studies, optimum pH for aflatoxin production by *Aspergillus spp.* is between 3.5 and 8.0 (Oviedo et al. 2011). The toxin-producing fungi such as *Aspergillums flavus* and *A. parasiticus* species show enormous growth under environmental moisture of between 50–60 %, temperature conditions of 25 °C and 85–90 % relative humidity (Bakirci 2001).

Concentrates from both rural and peri-urban areas had high AFB_1 contamination above the European Directives (Directive 2002/32/EC (EC 2002) and amending

Table 5 Linear regression model showing association of abiotic factors with level of mycotoxins in animal feeds

Variable	AFB_1			DON		
	Coefficients	95 % confidence interval	*P*-value	Coefficients	95 % confidence interval	*P*-value
Constant	50.10	32.26–67.94	0.006	330.67	234.82–426.52	0.001
a_w	−3.04	−0.67– −5.41	0.024	125.06	46.33–203.79	0.012
pH	−1.39	−1.90– −0.88	0.610	−36.62	−46.66– −26.58	0.008
Moisture Content	−0.41	−0.26– −.056	0.007	0.92	0.57–1.27	0.004

AFB_1 aflatoxin B1, *DON* deoxynivalenol, *a_w* water activity

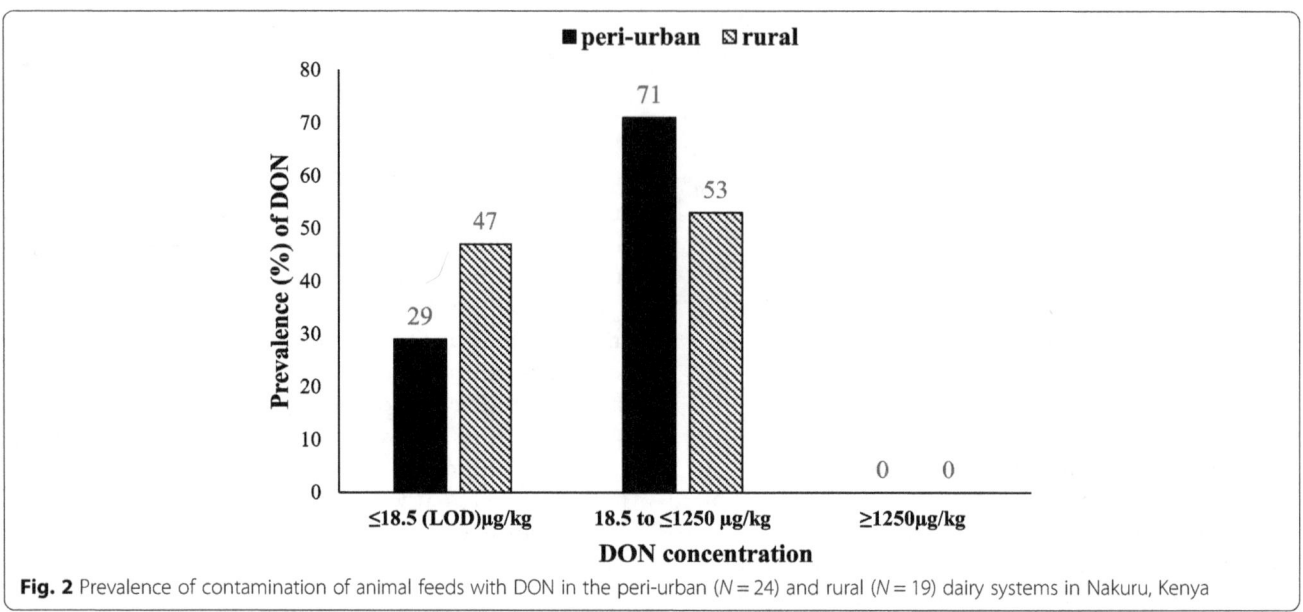

Fig. 2 Prevalence of contamination of animal feeds with DON in the peri-urban (N = 24) and rural (N = 19) dairy systems in Nakuru, Kenya

Directive 2003/100/EC (EC 2003) of 5 μg/kg. This could be attributed to lack of quality assurance system in the animal feeds value chain in Kenya. Animal feed ingredients used in formulations were not guaranteed of quality and safety while local agro-vets lack specified regulatory guidelines for animal feed distribution and proper storage.

Overall, the prevalence of AFB$_1$ contamination of animal feeds was above the EU maximum limit of 5 μg/kg in both rural dairy system and peri-urban system with 60 and 52 % respectively (Fig. 1). This condition presented a concern in the dairy industry in this region as the risk of AFB$_1$ toxicity in dairy cows was high in both dairy systems. This situation exposed cows to the risk of chronic intoxication with main target organ being the liver leading to hepatotoxicity, decreased weight gain, and decreased feed consumption, decreased reproductive

performance and abortions (Haschek et al. 2013). The reduced performance in dairy cows would cause farmers large milk and economic losses.

DON contamination in animal feeds

The analysis of variance (ANOVA) (Table 4) showed that there was significant difference in DON contamination in animal feeds samples between the two dairy systems where the study was conducted. Feed samples from peri-urban had significantly higher levels of DON contamination ranging between 0 and 179.89 μg/kg compared to rural dairy system ranging between 0 and 89.79 μg/kg (Table 3). This could be attributed to use of low-quality raw materials in feed formulation. This finding indicated that DON contamination may have occurred in the pre-storage period and probably the feed ingredients were contaminated before storage (Haschek et al. 2013).

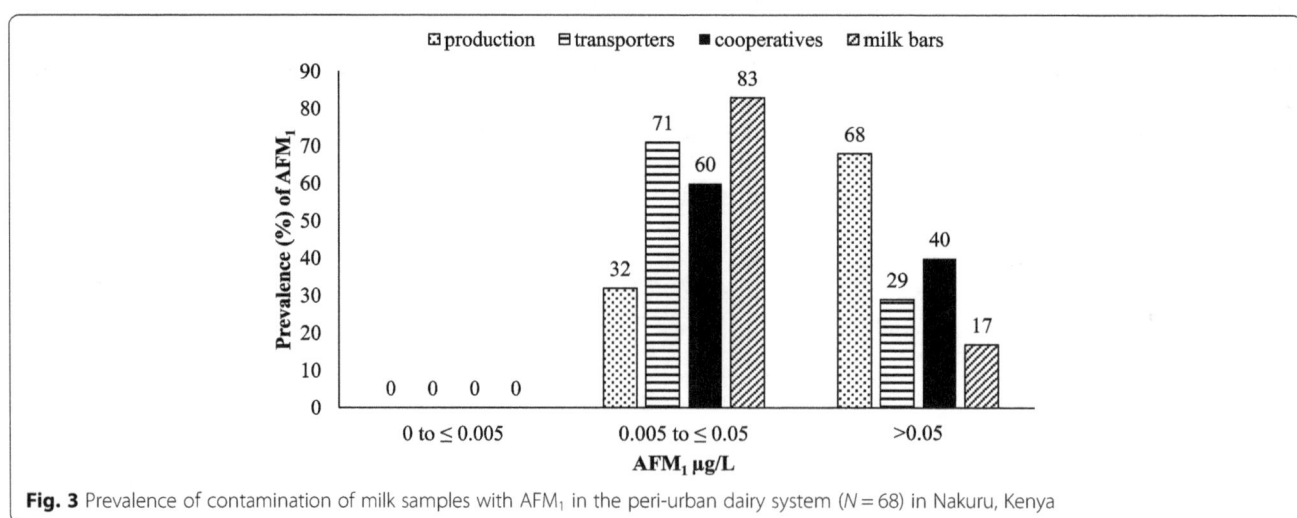

Fig. 3 Prevalence of contamination of milk samples with AFM$_1$ in the peri-urban dairy system (N = 68) in Nakuru, Kenya

Table 6 Aflatoxin M_1 contamination in milk in the rural dairy system and peri-urban dairy systems

Dairy system	Statistic	Production	Transporters	Cooperatives	Milk bars
Rural	Mean ± SD	0.011 ± 0.010[b]	0.007 ± 0.006[b]	0.005 ± 0.008[b]	0.006 ± 0.004[b]
	Median	0.006	0.006	0.00	0.00
	Range	0.00–0.041	0.00–0.019	0.00–0.022	0.00–0.034
Peri-urban	Mean ± SD	0.062 ± 0.019[a]	0.049 ± 0.021[a]	0.043 ± 0.025[a]	0.033 ± 0.015[a]
	Median	0.073	0.048	0.042	0.029
	Range	0.022–0.083	0.020–0.083	0.019–0.082	0.017–0.069

Means with same letter along the column are significantly different at $P \leq 0.05$ and *SD* standard deviation

The analysis of variance (ANOVA) (Table 4) showed high levels of DON contamination was observed in commercial and on-farm formulated concentrates. This is attributed to the fact that local feed processors and on-farm formulations contain a great proportion of on-farm produced cereals. In corn, *Fusarium* moulds are associated with ear rot and stalk rot, and in small grains, they are associated with diseases such as head blight (scab). In wheat, excessive moisture at flowering and afterward is associated with increased incidence of mycotoxin formation. In corn, *Fusarium* diseases are more commonly associated with insect damage, warm conditions at silking, and wet conditions late in the growing season (Placinta et al. 1999). The highest DON contamination of 179 μg/kg was observed in silage feed sample from the peri-urban system. This could have been caused by the silage being exposed to oxygen, causing yeast to utilize lactic acid in silage as a substrate causing an elevation of pH above 4.5 and the silage becoming conducive for mould growth.

Silage is green forage preserved by lactic acid fermentation under anaerobic conditions. Silage with a terminal pH of less than 4.5 is ideal since it prevents fungal growth (Liu et al. 2011). Neutral pH ranging between 6 and 7 is suitable for mould growth than a low pH level and for this reason well-prepared silage is less susceptible to fungal spoilage. Silos should be properly sealed to prevent aerobic conditions that favour mould growth and further mycotoxin production.

Linear regression model showed significant association of water activity, pH and moisture content of animal feeds with DON contamination of the animal feeds (Table 5). The maximum amount of DON is produced by *F. graminearum* at 0.98 a_w while Optimum DON production by *Fusarium spp.* is at pH of 7.5 (Comerio et al. 1999). The trichothecene DON persists in the animal feed at $\leq 0.90 a_w$ after it has already been produced (Hope et al. 2005).

All samples in this study were below the maximum limits for DON in the feed of 1250 μg/kg set in EU

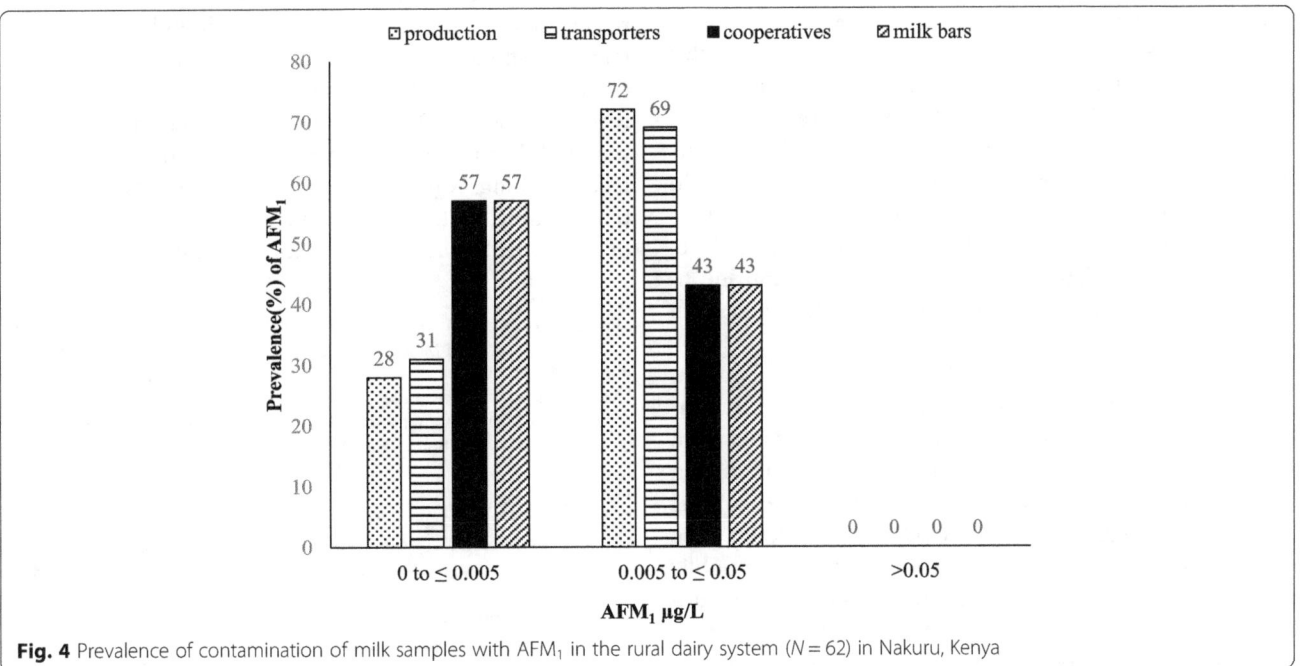

Fig. 4 Prevalence of contamination of milk samples with AFM_1 in the rural dairy system ($N = 62$) in Nakuru, Kenya

regulation 1881/2006 (EC 2006b). This implied that risk of DON toxicity in dairy cows was low in both dairy systems. The impact of DON on dairy cattle was not established, but clinical data shows an association between DON and poor performance in dairy herds (Côté et al. 1986). In previous studies, the DON-contaminated feed has caused a great economic loss in livestock, especially swine industry due to a well documented reduction in feed consumption and weight gain. High dose acute DON exposure resulted in emesis, abdominal distress, increased salivation and listlessness (Haschek et al. 2013).

The prevalence of AFM$_1$ in milk along the value chain of rural and peri-urban dairy systems

The study revealed that the peri-urban dairy system is contaminated with AFM$_1$ along the value chain ranging between 0.017 and 0.083 µg/L as shown in Table 6. The majority 48.5 % (33/68) of the milk samples from the peri-urban dairy system were above the EU regulation of 0.05 µg/L (EC 2006a) as shown in Fig. 3. The study also revealed that all milk samples in the rural dairy system were below the EU regulation of 0.05 µg/L (EC 2006a) as shown in Table 6. The cause for major differences in AFM$_1$ contamination levels of milk samples taken from rural and peri-urban farms can be explained by the different types of feeds that were provided to cows in these dairy systems. Peri-urban farms fed their cattle mainly on AFB$_1$ contaminated concentrates made of ingredients such as chicken feacal waste, maize germ, cotton and sunflower seed cake while most rural farms produced organic milk with lowest levels of AFM$_1$ by feeding their cows on a basic diet of pasture that comprised the tropical grass species *Pennisetum clandestinum* and *Pennisetum purpureum* also known as Napier grass. However, the concentrations of AFM$_1$ from the peri-urban dairy system in this study were lower compared to earlier studies in the urban Kenya reaching 0.68 µg/L (Kang'ethe and Lang'a 2009).

Milk samples from consumption nodes which comprise milk bars and processors level in the value chain had a narrower range of contamination compared to milk from farms in both systems as shown in Table 6. This could be explained by the effect of diluting due to bulking milk during transportation and at the collection centers prior to processing.

The study also showed a moderate correlation between AFB$_1$ contamination in feed samples and AFM$_1$ contamination in milk samples ($r = 0.46$ $P < 0.001$) collected from the same dairy farm. AFM$_1$ is excreted in milk within twelve hours of consumption of contaminated animal feeds (Fink-Gremmels 2008; Battacone et al. 2003). The occurrence of AFM$_1$ in milk and dairy products is a public health concern in the peri-urban dairy system which supplies milk to urban consumers. Milk is a primary part of the diet in Kenyan households and the effects of exposure to AFM$_1$ have been associated with poor growth in neonates and children (Haschek et al. 2013).

Recent studies in Ethiopia show that 91.8 % of milk samples exceeded the maximum level set by EU regulations (Gizachew et al. 2016). Serbia (76 %) of milk samples exceeded the maximum level set by EU regulations (Škrbić et al. 2014). In Brazil, 46 % of ultra-high temperature milk samples were AFM$_1$ positive with AFM$_1$ (Iha et al. 2013). In Pakistan, 71 % were positive with AFM$_1$ (Iqbal and Asi 2013). In this study, AFM$_1$ levels in milk from both dairy systems are lower than those found in some other studies in different countries published recently. The difficulty of comparing results among different countries in the world is attributed to different investigative procedures used, sources of feed AFB$_1$ contamination, different on-farm feeding practices, climatic situations, animal feed handling and storage conditions, the sampling time and procedures.

Milk consumption levels in Kenya are among the highest in the developing world (SDP 2004) with an average of 100 kg/year per capita. Keeping in view the high levels of AFM$_1$ in peri-urban dairy system production node, there is dire need to improve storage and handling conditions of animal feed. This will mitigate the AFB$_1$ levels in feed/feed ingredients and ultimately decrease the toxin in animal milk. The results showed that feeds used in peri-urban for the dairy animals are heavily contaminated with AFB$_1$.

Conclusion

Results of this study indicate that the level of mycotoxin contamination in commercial and on-farm formulated concentrates require attention in-order to put mitigation strategies in place to reduce AFB$_1$ exposure to dairy animals, especially in the peri-urban dairy system. The results suggest mitigation measures should focus on on-farm formulated concentrates with risk assessment evaluation of ingredients used in formulating them being recommended. DON contamination in animal feeds was minimal, but levels could vary year to year depending on feed handling conditions. Results of the survey indicate that organic milk produced in the rural dairy system is of high quality with low concentrations of AFM$_1$.

Abbreviations
AFB$_1$, aflatoxin B$_1$; AFM$_1$, aflatoxin M$_1$; ANOVA, analysis of variance; a$_w$, water activity; BMBF, German Ministry of Education and Research; DMS, degree minute seconds; DON, Deoxynivalenol; EC, European Commission; ELISA, Enzyme Linked Immune Sorbent Assay; EU, European Union; HPLC, High performance liquid chromatography; ReLOAD, Reduction of Post-Harvest Losses and Value Addition in East African Food Value Chains

Funding
This research work was supported financially by German Ministry of Education and Research (BMBF) through ReLOAD Project: Reduction of Post-Harvest Losses and Value Addition in East African Food Value Chains.

Authors' contributions
The experimental and analytical work was done by CMM, JWM, PSM, and BOB were involved in the designing of the experiment, data analysis,

interpretation of the results and manuscript development. All authors read and approved the final manuscript.

Authors' information

Ms. Caroline Makau is an MSc. student in the Department of Dairy and Food Science and Technology at Egerton University, Kenya. She also holds BSc. Food Science and Technology from the same university besides also having working experience in the food manufacturing industries in Kenya. Prof. Matofari and Dr. Muliro are senior lecturers in the Department of Dairy and Food Science and Technology in Egerton University while Prof. Bebe is a senior lecturer at the Department of Animal Science at Egerton University. All supervisors have many years of teaching experience at the University and have been involved in the supervision of many graduate students in their researches.

Competing interests

The authors of this article declare that they don't have any financial and non-financial competing interests.

Author details

[1]Egerton University, Faculty of Agriculture, Department of Dairy & Food Science and Technology, P.O. Box 536-20115, Egerton, Kenya. [2]Egerton University, Faculty of Agriculture, Department of Animal Science, P.O. Box 536-20115, Egerton, Kenya.

References

AOAC. Official methods of analysis of AOAC International. 17th ed. Gaithersburg: AOAC International; 2000.

Bakirci I. A study on the occurrence of aflatoxin M 1 in milk and milk products produced in Van province of Turkey. Food Control. 2001;12(1):47–51.

Baltenweck I, Staal SJ, Owango M, Muriuki H, Lukuyu B, Gichungu G, Kenyanjui M, Njubi D, Tanner J, Thorpe W. Intensification of dairying in the greater Nairobi Milk-shed: Spatial and housing analysis. Smallholder dairy (Research and Development) Project, MoA/KARI/ILRI. Collaborative Research Report. Nairobi: ILRI (International Livestock Research Institute); 1998. http://cgspace.cgiar.org/bitstream/handle/10568/1751/Baltenweck%20et%20al-1998-Dairy%20intensification%20greater%20Nairobi%20spatial&HH.pdf?sequence=1.

Battacone G, Nudda A, Cannas A, Borlino AC, Bomboi G, Pulina G. Excretion of aflatoxin M1 in milk of dairy ewes treated with different doses of aflatoxin B1. J Dairy Sci. 2003;86(8):2667–75.

Bottalico A, Perrone G. Toxigenic Fusarium species and mycotoxins associated with head blight in small-grain cereals in Europe. Eur J Plant Pathol. 2002;108:611–24.

Comerio RM, Pinto VF, Vaamonde G. Influence of water activity on deoxynivalenol accumulation in wheat. Mycotoxin Res. 1999;15(1):24–32.

Côté LM, Dahlem AM, Yoshizawa T, Swanson SP, Buck WB. Excretion of deoxynivalenol and its metabolite in milk, urine, and feces of lactating dairy cows. J Dairy Sci. 1986;69(9):2416–23.

EC Directive. Directive of The European Parliament and of the Council of 7 May 2002 on Undesirable Substances in Animal Feed 2002/32/EC. 2002.

EU Directive. Commission Directive 2003/100/EC of 31 October 2003 amending Annex I of Directive 2002/32/EC of the European Parliament and of the Council on undesirable substances in animal feed. L285/233–L285/237. Brussels: EU-commission; 2003.

European Commission. Commission Regulation 401/2006 of 23 February 2006 laying down the methods of sampling and analysis for the official control of the levels of mycotoxins in foodstuffs. Off J Eur Union. 2006a;L70:12–34.

European Commission. Commission Regulation (EC) No 1881/2006 of 19 December 2006 setting maximum levels for certain contaminants in foodstuff. 2006R1881-EN-01.09. 2014-014.001-1. 2006b.

Fink-Gremmels J. Mycotoxins in cattle feeds and carry-over to dairy milk: a review. Food Addit Contam. 2008;25(2):172–80.

Flores-Flores ME, Lizarraga E, de Cerain AL, González-Peñas E. Presence of mycotoxins in animal milk: a review. Food Control. 2015;53:163–76.

Gizachew D, Szonyi B, Tegegne A, Hanson J, Grace D. Aflatoxin contamination of milk and dairy feeds in the Greater Addis Ababa milk shed, Ethiopia. Food Control. 2016;59:773–9.

Haschek WM, Rousseaux CG, Wallig MA, editors. Haschek and Rousseaux's handbook of toxicologic pathology. New York: Academic; 2013. p. 1214–6.

Hell K, Mutegi C. Aflatoxin control and prevention strategies in key crops of Sub-Saharan Africa. Afr J Microbiol Res. 2011;5(5):459–66. Kor.

Hope R, Aldred D, Magan N. Comparison of environmental profiles for growth and deoxynivalenol production by Fusarium culmorum and F. graminearum on wheat grain. Lett Appl Microbiol. 2005;40(4):295–300.

Iha MH, Barbosa CB, Okada IA, Trucksess MW. Aflatoxin M 1 in milk and distribution and stability of aflatoxin M 1 during production and storage of yoghurt and cheese. Food Control. 2013;29(1):1–6.

Iqbal SZ, Asi MR. Assessment of aflatoxin M 1 in milk and milk products from Punjab, Pakistan. Food Control. 2013;30(1):235–9.

Kang'ethe EK, Lang'a KA. Aflatoxin B1 and M1 contamination of animal feeds and milk from urban centers in Kenya. Afr Health Sci. 2009;9(4):218–26.

Korosteleva SN, Smith TK, Boermans HJ. Effects of feed naturally contaminated with Fusarium mycotoxins on metabolism and immunity of dairy cows. J Dairy Sci. 2009;92(4):1585–93.

Leszczynska J, Maslowska J, Owczarek A, Kucharska U. Determination of aflatoxins in food products by the ELISA method. Czech J Food Sci. 2001;19(1):8–12.

Liu Q, Zhang J, Shi S, Sun Q. The effects of wilting and storage temperatures on the fermentation quality and aerobic stability of stylo silage. Anim Sci J. 2011;82(4):549–53.

Lukuyu B, Franzel S, Ongadi PM, Duncan AJ. Livestock feed resources: Current production and management practices in central and northern rift valley provinces of Kenya. Livest Res Rural Dev. 2011;23(5):112.

Ministry of Livestock Development Department and Fisheries -MoALF. District Livestock Production Annual Report, Nakuru North. Nairobi, Kenya. 2012.

Muriuki HG. Dairy development in Kenya. Rome: Food and Agricultural Organization; 2011.

Muture BN, Ogana G. Aflatoxin levels in maize and maize products during the 2004 food poisoning outbreak in Eastern Province of Kenya. East Afr Med J. 2005;82(6):275–9.

Nyaata OZ, Dorward PT, Keatinge JDH, O'Neill MK. Availability and use of dry season feed resources on smallholder dairy farms in central Kenya. Agrofor Syst. 2000;50(3):315–31.

Oviedo MS, Ramirez ML, Barros GG, Chulze SN. Influence of water activity and temperature on growth and mycotoxin production by Alternaria alternata on irradiated soya beans. Int J Food Microbiol. 2011;149(2):127–32.

Placinta CM, D'mello JPF, Macdonald AMC. A review of worldwide contamination of cereal grains and animal feed with Fusarium mycotoxins. Anim Feed Sci Technol. 1999;78(1):21–37.

R-Biopharm AG. Good ELISA Practice Manual. 2015. http://www.r-biopharm.com/wp-content/uploads/534/2015-09_Good_ELISA_Practice_Manual_EN_Web.pdf.

Reddy KRN, Raghavender CR, Reddy BN, Salleh B. Biological control of Aspergillus flavus growth and subsequent aflatoxin B 1 production in sorghum grains. Afr J Biotechnol. 2010;9(27):4247–50.

Richard JL. Some major mycotoxins and their mycotoxicoses — An overview. Int J Food Microbiol. 2007;119(1):3–10.

Richards JI, Godfrey SH. Urban livestock keeping in sub-Saharan Africa: Report of a workshop held on 3-5 March 2003 in Nairobi, Kenya. Aylesford, Kent: Natural Resources International Ltd; 2003. p. 118.

SDP. A series of policy briefs (demand for dairy products in Kenya; employment generation in the Kenya dairy industry; competitiveness of the smallholder dairy enterprise in Kenya; public health issues in Kenyan milk markets; improved child nutrition through cattle ownership in Kenya; and uncertainty on cattle numbers in Kenya) for the dairy industry policy reform forum held at Grand Regency Hotel, Nairobi. 2004.

Škrbić B, Živančev J, Antić I, Godula M. Levels of aflatoxin M1 in different types of milk collected in Serbia: assessment of human and animal exposure. Food Control. 2014;40(1):113–9.

Soler CM, Hoogenboom G, Olatinwo R, Diarra B, Waliyar F, Traore S. Peanut contamination by Aspergillus flavus and Aflatoxin B1 in granaries of villages and markets in Mali, West Africa. J Food Agric Environ. 2010;8:195–203.

Thorpe W, Muriuki HG, Omore A, Owango MO, Staal S. Development of smallholder dairying in Eastern Africa with particular reference to Kenya. In A paper prepared for the UZ/RVAU/DIAS/DANIDA-ENRECA Project Review Workshop 10–13 January 2000. 2000.

Thursfield M. Diagnostic testing in veterinary epidemiology. 2nd ed. Cambridge: Blackwell Science Ltd; 1995. p. 483.

Unnevehr L, Grace D (Eds.) Aflatoxins: finding solutions for improved food safety (vol. 20) Intl Food Policy Res Inst; Washington DC,USA 2013.

Vincelli P, Parker G. Fumonisin, vomitoxin, and other mycotoxins in corn produced by Fusarium fungi. University of Kentucky Cooperative Extension Service; 2002. ID 121(8).

Zinedine A, González-Osnaya L, Soriano JM, Moltó JC, Idrissi L, Mañes J. The presence of aflatoxin M1 in pasteurized milk from Morocco. Int J Food Microbiol. 2007;114(1):25–9.

17

A pilot study to assess lead exposure from routine consumption of coffee and tea from ceramic mugs: comparison to California Safe Harbor Levels

Grace L. Anderson[*], Lindsey Garnick, Mai S. Fung and Shannon H. Gaffney

Abstract

Background: Lead (Pb) is a pervasive metal that can be found in, and potentially leached from, ceramics, particularly into acidic foods and beverages. The purpose of this study was to investigate potential lead exposure from coffee and tea consumption, given that both are acidic and routinely consumed from ceramic mugs. We measured the concentration of lead in coffee and tea at two different time points brewed in five readily available mugs known to contain lead. Results were compared to EPA's action level for drinking water and FDA's allowable level for bottled water. The measured concentrations, along with consumption patterns, were also used to calculate potential daily lead doses, which were compared to California's Safe Harbor Levels under Proposition 65. Additionally, we estimated changes in adult and fetal blood lead levels using EPA's Adult Lead Methodology model.

Findings: The results of this pilot study suggest that lead in ceramic mugs can leach into coffee and tea. The measured lead concentrations ranged from 0.2 to 8.6 µg/L in coffee, and from <0.2 to 1.6 µg/L in tea. No statistical differences were found between the measured concentrations in coffee, tea, or water within each cup, or in the measured concentrations between retention times within each cup. However, a statistically significant difference was observed in the lead concentrations measured between cups, indicating that the lead concentrations were dependent on the cup used, rather than on the beverage or retention time. The estimated daily dose of lead exceeded the California Maximum Allowable Dose Level of 0.5 µg per day for one of the five mugs tested. Blood lead levels did not increase above regulatory or guidance values.

Conclusions: This preliminary investigation provides data on potential lead exposures from daily beverage consumption among typical consumers, relevant to a substantial portion of the population, with particular implications for pregnant women.

Keywords: Lead, Coffee, Tea, Ceramics, Proposition 65, Leaching

Introduction

Lead (Pb) is a naturally occurring metal pervasive in the environment that can cause well-known adverse health effects in humans upon sufficient exposure (Brown and Margolis 2012; ATSDR 2007b). The primary target of lead toxicity in both children and adults is the nervous system, although children are more sensitive to lead's neurotoxic effects. Moreover, children generally absorb more ingested lead into their blood than do adults.

Children absorb approximately 50% of ingested lead, while adults absorb approximately 10% (ATSDR 2007b, ATSDR 2007a; Philip and Gerson 1994). Lead exposure may begin *in utero,* as it can cross the placenta (Mason et al. 2014; Brown and Margolis 2012). Decreasing cognitive function has been observed with increasing lead exposure in both children and adults (Mason et al. 2014; Brown and Margolis 2012). IQ deficits of one to five points have been associated with blood lead level increases of 10 µg/dL or less in children (ATSDR 2007a). At high levels of exposure, lead can cause fatal damage to the brain and kidney

* Correspondence: gracekavanaugh@gmail.com
Cardno ChemRisk, 101 2nd Street, Suite 700, San Francisco, CA 94105, USA

in both adults and children (ATSDR 2007b). Moreover, the International Agency for Research on Cancer (IARC), the U.S. Department of Health and Human Services (DHHS), and the U.S. Environmental Protection Agency (EPA) have all considered lead or lead compounds as probably carcinogenic in humans (ATSDR 2007b).

Historically, lead was used in gasoline, paint, plumbing, and various industrial processes (Levin et al. 2008). Because of concerns over lead's human health effects, many of these uses have been limited or banned under regulatory actions in the U.S. For example, lead was phased out of gasoline beginning in 1973, residential lead-based paint was banned in 1978, and lead solder in food cans was banned in 1995 (Brown and Margolis 2012). While these efforts over the past several decades have significantly reduced the potential for lead exposure in the U.S., lead poisoning remains an appreciable health concern, especially for children.

In January 2012 *Consumer Reports* published an article indicating that 25% of 88 juices tested exceeded the Food and Drug Administration (FDA) bottled water standard for lead of 5 µg/L (CR 2012; USFDA 1995). More recently, the lead-contaminated drinking water in Flint, Michigan, has drawn the public eye to lead poisoning. Based on elevated blood lead levels, the President declared a state of emergency in Flint in early 2016 (The White House 2016). The Centers for Disease Control and Prevention (CDC) estimates that at least four million U.S. households contain children exposed to "high levels of lead" (CDC 2016b). In 2012, the CDC lowered the blood lead level at which it recommends public health action from 10 to 5 µg/dL for children. Though average blood lead levels have significantly declined in U.S. children since the 1970s, the CDC estimates that approximately half a million U.S. children between the ages of one and five have blood lead levels above 5 µg/dL (CDC 2016b; CDC 1997).

The most common route of lead exposure is through ingestion, and multiple regulatory and guidance values exist to limit lead ingestion in the U.S. (ATSDR 2007a). The Environmental Protection Agency (EPA) has set a lead action level of 15 ppb in drinking water, for example, and the FDA has set an allowable level for lead in bottled water of 5 ppb (EPA 1991; USFDA 1995). The World Health Organization (WHO) formerly set a provisional tolerable weekly intake (PTWI) of 25 µg/kg/week for lead from food and water, but withdrew it in 2010. The WHO stated that it did not issue an updated PTWI because it was "not possible" to establish a value that would be "health protective" (WHO 2010). In California, lead is listed as a chemical known to cause cancer and reproductive toxicity under the Safe Drinking Water and Toxic Enforcement Act (Proposition 65). Accordingly, the California Office of Environmental Health Hazard Assessment (OEHHA) has set Safe Harbor Levels

for lead, consisting of a Maximum Allowable Dose Level (MADL) of 0.5 µg/day, based on reproductive toxicity, and a No Significant Risk Level (NSRL) of 15 µg/day based on lead's potential carcinogenicity (OEHHA 2016b). There is currently a petition to lower the MADL to 0.2 µg/day (Cal/EPA 2015a; Cal/EPA 2015b).

Ceramics with lead-containing paint or glaze are one potential source of lead exposure (ATSDR 2007b). Several studies have suggested that lead may leach from such ceramics, particularly in acidic environments (Sheets 1997; Mohamed et al. 1995; Levin et al. 2008; Markowitz 2000; Gonzalez de Mejia and Craigmill 1996; Feldman et al. 1999). In fact, the FDA limits the amount of lead in cups and mugs to that which results in no more than 0.5 µg/mL lead in an acidic leaching solution (USFDA 2015). The purpose of the current pilot study, then, was to investigate potential lead exposure from coffee and tea consumption, given that both beverages are acidic and routinely consumed from ceramic mugs in the U.S. Specifically, we measured the concentration of lead in coffee and tea at two time points brewed in five commercial mugs known to contain lead.

Materials and methods

The five mugs chosen for this study were selected because they were found to contain lead in a screening-level assessment. Specifically, 24 mugs from the authors' office were tested using an Olympus Innov-X Delta handheld X-Ray fluorescent (XRF) analyzer. Each mug was measured once with the XRF gun at its highest sensitivity setting, which required the tester to hold the analyzer over the mug for 45 s. The three mugs with the highest resulting lead concentrations (1,223 to 7,034 mg/kg) were selected for the present study. These mugs each had decorative elements and will be referred to by their predominant colors: Green Decorative, Yellow Decorative, and Red Decorative. In addition, two representative mugs were selected from the batch of office mugs baring the authors' company's logo. These will be referred to as Black Logo1 and Black Logo2. All five mugs selected were in active use in the authors' San Francisco, California, office environment, and were typically washed daily in an automatic dish washer. Four of the five mugs were purchased in the U.S., and one was purchased in Europe (Red Decorative). The mugs all appeared to be in good condition, with no obvious signs of damage or wear.

Five beverage-making events were performed in the five mugs, as well as in a glass cup, in duplicate, allowing the collection of 60 total samples. The five mugs and one glass cup will hereafter be collectively referred to as 'cups'. The sampling protocol is described in greater detail below. The five beverages were hot water, coffee 10 and 60 min post-brewing, and tea 10 and 60 min post-brewing. Tap water was used for each scenario; the first tap water collection occurred at approximately 9:30 in the morning, after the tap had been used intermittently for several hours. Between each event, the cups and all utilized equipment were washed with dish soap and water and dried.

In each sampling situation, the "end temperature" of the beverage was measured immediately before pouring approximately 250 mL of the beverage into a plastic sampling container. Sampling containers are depicted in Fig.1a. Four control samples of tap water were also collected. All containers were shipped on ice to a laboratory accredited by the Environmental Laboratory Accreditation Program (ELAP), where they were analyzed for the presence of lead using a Perkin Elmer inductively coupled plasma mass spectrometer (IC-PMS), according to EPA Method 200.8. As specified in Method 200.8, with each set of samples, the laboratory ran method calibration blanks and multiple laboratory control samples and duplicates to verify the instrument performance and determine instrument precision. A matrix spike of 50 µg/L lead was utilized as a calibration procedure.

Water

Water was boiled in an electric kettle, and approximately 300 mL of water was poured into each of the six cups. After 30 min had elapsed, the water was stirred. After 60 min had elapsed, the water was stirred again, the temperature of the water was recorded, and the water was poured into the sample containers. The scenario was repeated for each of the six cups such that a total of 12 samples were collected.

Coffee

Coffee was brewed utilizing a single cup pour-over cone with a paper filter (Fig.1b, c). Approximately 300 mL

boiling water was poured over three teaspoons of a nationally available, freshly-ground, medium roast, 100% Arabica coffee for each of the six cups. Two scenarios were performed in duplicate: in the first, the coffee remained in the cup for 10 min, and in the second, the coffee remained in the cup for 60 min before being poured into the sample container. After 5 min had elapsed, the pour-over cone was removed, and the coffee was stirred in both scenarios. The 60-min samples were also stirred after 30 min, and all samples were stirred prior to transfer to the sampling containers.

Tea

Tea was prepared by pouring boiling water over one nationally available, 2 g black tea bag in each cup (Fig.1d). Water was boiled in an electric tea kettle, and 300 mL of water was poured into each cup. Two scenarios were performed in duplicate: in the first, the tea remained in the cup for 10 min, and in the second, the tea remained in the cup for 60 min before being poured into the sample container. The tea bag was dunked in the water several times within 5 min of steeping in both scenarios. After 5 min of steeping, the tea bag was removed, and the tea was stirred. The 60-min samples were also stirred after 30 min, and all samples were stirred prior to transfer to the sampling containers.

Tap water controls

On a separate day from the beverage making scenarios, two samples of tap water that had been boiled and retained in the electric tea kettle for 60 min were collected. After boiling was reached, the water was allowed

Fig. 1 Brewing and Sampling Equipment. **a** Sampling containers. **b**, **c** Coffee brewing. **d** Tea brewing

to cool, was stirred inside the kettle at the 30 and 60-min marks, and then was poured into the sampling containers. Two additional samples were collected directly from the tap.

Risk assessment: comparison to regulatory limits

For simple comparison purposes, and to put our results in the context of regulatory limits, our measured concentrations were compared to the EPA's lead action level of 15 ppb in drinking water and the FDA's allowable level for lead in bottled water of 5 ppb.

The sampling results were also utilized along with general assumptions to determine the level of lead exposure expected from daily consumption of coffee and tea from the cups. These estimated daily doses were compared to the current Safe Harbor Levels under Proposition 65 in California (MADL: 0.5 µg/day, NSRL: 15 µg/day). The EPA currently has no Reference Dose for lead.

The daily dose of lead was calculated simply as the amount of lead in a serving of the beverage from the study cups multiplied by the number of servings typically consumed in one day, or:

$$\frac{\text{Amount of Lead }(\mu g)}{\text{Beverage Serving }(L)} \times \frac{\text{Beverage Servings}(L)}{\text{Day}}$$
$$= \text{Amount of Lead Ingested per Day }(\mu g/\text{day})$$

Basing these calculations on typical coffee and tea consumption is appropriate, since Proposition 65 dictates that the MADL be calculated "using the reasonably anticipated rate of intake or exposure for average users of the consumer product" (OEHHA 2016a). Note that our estimate accounts only for lead exposure from drinking beverages from the subject cups, and ignores any other potential sources or routes of lead exposure.

Coffee

A 2009 survey by the National Coffee Association reported that the average consumption among coffee drinkers in the U.S. aged 18 and older was 3.3 eight-ounce cups per day (USFDA 2012). We thus assumed a daily consumption of 26.4 fluid ounces, or 0.78 L, of coffee per day in our risk assessment.

Tea

According to the most recent available data (2011–2012) from the National Health and Nutrition Examination Survey (NHANES), median tea consumption among U.S. tea drinkers age 18 to 80 is 355.2 g per day, or 12.01 fluid ounces per day (CDC 2014). We thus assumed a daily consumption of 12.01 fluid ounces, or 0.36 L, of tea per day in our risk assessment.

Estimation of blood lead levels following exposure to lead from mugs

As another point of comparison, we utilized our measured concentration data to estimate changes in adult blood lead levels (BLLs) using the EPA's Adult Lead Methodology (ALM) model. The ALM model has previously been used to estimate BLLs as a result of exposure to lead from beverages and consumer products (Monnot et al. 2015). The average baseline BLL for adults was assumed to be 1.0 µg/dL (ALM default value); however, we also ran the model assuming a baseline of 0 µg/dL in order to determine the contribution of lead from the beverages ingested in the lead-containing ceramic mugs. The gastrointestinal absorption for lead was assumed to be 12% in adults, the default for the model. Any ingestion from soil and dust was assumed to be zero because of their irrelevance to beverage exposure. The average consumption of coffee per day (in g/day) was used because average coffee consumption is higher than average tea consumption. We assumed an exposure frequency of 365 days/year, assuming that a person would drink the same amount of the beverage every day. The BLLs estimated from the model were compared to regulatory and guidance values for BLLs set by the CDC and EPA. Additionally, the model was used to estimate fetal blood lead concentrations in women exposed to lead from beverages contaminated by ceramic mugs.

Findings

Lead was measured at levels at or above the limit of detection (0.2 µg/L) in 56 out of 60 samples (93.33%) (Table 1, Fig. 2). The percent recovery from quality control samples ranged from 95.6 to 102%. The percent recovery from matrix spikes ranged from 98.3 to 102%. Out of the four samples for which the lead concentration was below the limit of detection, three results were from 60 min water samples and one was from a 10 min tea sample. The total range of lead levels measured was <0.2 µg/L to 8.6 µg/L. The highest concentration of 8.6 µg/L came from a 10 min coffee sample in the Green Decorative mug. The other lead concentrations measured from the Green Decorative mug, though consistently higher than the measures for all other cups, did not exceed 1.8 µg/L. All four tap water control samples resulted in measurements below the limit of detection (<0.2 µg/L) (Results not shown).

Statistical tests were performed in order to determine what factors contributed to the concentrations of lead measured in beverages from each cup. Non-parametric methods were used, as the data were found to be non-normally distributed. Additionally, a value of 0.1 µg/L (the limit of detection divided by the square root of 2) was substituted for the four samples that resulted in concentration measures below the limit of detection.

Table 1 Measured lead concentrations by cup, beverage, and time

Cup	Medium	Time in cup (min)	Result (µg/L)	End temperature (°C)
Black Logo1	Coffee	10	0.2	55
Black Logo1	Coffee	10	0.2	55
Black Logo1	Coffee	60	0.2	34
Black Logo1	Coffee	60	0.2	36
Black Logo1	Tea	10	<0.2	35
Black Logo1	Tea	10	0.2	35
Black Logo1	Tea	60	0.2	59
Black Logo1	Tea	60	0.2	60
Black Logo1	Water	60	0.2	36
Black Logo1	Water	60	<0.2	37
Black Logo2	Coffee	10	0.2	55
Black Logo2	Coffee	10	0.2	57
Black Logo2	Coffee	60	0.2	34
Black Logo2	Coffee	60	0.2	34
Black Logo2	Tea	10	0.2	35
Black Logo2	Tea	10	0.2	35
Black Logo2	Tea	60	0.2	60
Black Logo2	Tea	60	0.2	60
Black Logo2	Water	60	0.2	37
Black Logo2	Water	60	0.2	37
Green Decorative	Coffee	10	8.6	55
Green Decorative	Coffee	10	1.7	56
Green Decorative	Coffee	60	1.8	31
Green Decorative	Coffee	60	1.3	33
Green Decorative	Tea	10	1.6	33
Green Decorative	Tea	10	1.2	33
Green Decorative	Tea	60	1.0	57
Green Decorative	Tea	60	0.9	58
Green Decorative	Water	60	0.8	34
Green Decorative	Water	60	1.6	34
Yellow Decorative	Coffee	10	0.4	55
Yellow Decorative	Coffee	10	0.2	56
Yellow Decorative	Coffee	60	0.2	35
Yellow Decorative	Coffee	60	0.2	36
Yellow Decorative	Tea	10	0.2	36
Yellow Decorative	Tea	10	0.2	37
Yellow Decorative	Tea	60	0.2	60
Yellow Decorative	Tea	60	0.2	60
Yellow Decorative	Water	60	0.3	35
Yellow Decorative	Water	60	0.2	35
Red Decorative	Coffee	10	0.2	58
Red Decorative	Coffee	10	0.2	56
Red Decorative	Coffee	60	0.2	33

Table 1 Measured lead concentrations by cup, beverage, and time (*Continued*)

Cup	Medium	Time	Result	End temp
Red Decorative	Coffee	60	0.2	36
Red Decorative	Tea	10	0.2	35
Red Decorative	Tea	10	0.2	35
Red Decorative	Tea	60	0.3	60
Red Decorative	Tea	60	0.2	60
Red Decorative	Water	60	0.3	35
Red Decorative	Water	60	<0.2	37
Glass	Coffee	10	0.3	59
Glass	Coffee	10	0.2	58
Glass	Coffee	60	0.2	32
Glass	Coffee	60	0.2	35
Glass	Tea	10	0.2	35
Glass	Tea	10	0.2	35
Glass	Tea	60	0.2	57
Glass	Tea	60	0.2	60
Glass	Water	60	0.2	36
Glass	Water	60	<0.2	35

The Mann-Whitney Wilcoxon test was performed between the coffee samples retained in the cups for 10 and 60 min, and between the tea samples retained in the cups for 10 and 60 min, respectively. No statistical significance was found for either test, indicating that there was no difference in lead concentrations measured depending on the amount of time the beverages were retained in the cups (coffee samples: $p = 0.43$; tea samples: $p = 0.60$). As a result of this test, the 10 and 60 min samples for coffee and tea, respectively, were pooled from each cup to perform the Kruskal-Wallis test. This test was performed to determine if there was a difference in lead measured depending on the beverage in the cups. No statistical significance was found ($p = 0.64$). Finally, the Kruskal-Wallis test was performed on the samples from each cup to determine if there was a difference in lead concentrations measured depending on which cup was used. The Nemenyi multiple comparison test was utilized to determine statistical differences among the cups, and it found that samples from the Green Decorative mug were statistically significantly higher than samples from all other cups ($p = 0.0001–0.0125$).

Risk assessment: comparison to regulatory limits

Measured lead concentrations in the beverage samples from the six cups in this study ranged from <0.2 to 8.6 µg/L. Mean lead concentrations across all beverage scenarios were 0.18 µg/L in the Black Logo1 mug, 0.2 µg/L in the Black Logo2 mug, 2.05 µg/L in the Green Decorative mug, 0.23 µg/L in the Yellow Decorative

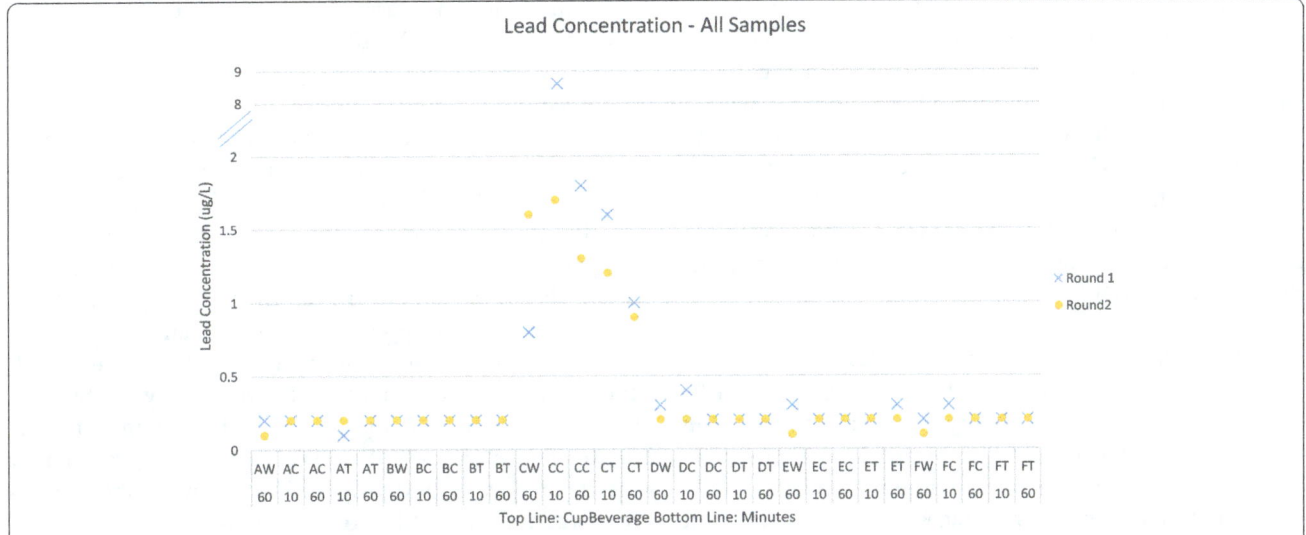

Fig. 2 Lead Concentration – All Samples. Measured lead concentrations by mug/cup, beverage and time. Cup A = Black Logo1 mug; Cup B = Black Logo2 mug; Cup C = Green Decorative mug; Cup D = Yellow Decorative mug; Cup E = Red Decorative mug; Cup F = Glass Cup. Beverage W = Water. Beverage C = Coffee. Beverage T = Tea. Samples below the limit of detection (<0.2 µg/L) graphed at the limit of detection divided by the square root of two (0.1 µg/L)

mug, 0.21 µg/L in the Red Decorative mug, and 0.2 µg/L in the glass cup (Table 2).

All measured lead concentrations were below the EPA's lead action level of 15 ppb in drinking water. However, the highest concentration measured from the Green Decorative mug (8.6 µg/L or 8.6 ppb) exceeded the FDA's allowable level for lead in bottled water of 5 ppb. This simple comparison, although interesting, does not have clear health implications, since these EPA and FDA health - based values use water consumption rates, which are much higher than coffee or tea rates of consumption.

For comparison the to the California Proposition 65 Safe Harbor Levels, given that the samples from the Green Decorative mug were statistically different from samples from all other cups, we calculated daily lead intake based on three values: 1) the mean lead concentration from all cups excluding the Green Decorative mug

(0.2 µg/L); 2) the mean lead concentration from the Green Decorative mug (2.05 µg/L); and 3) the maximum concentration measured from the Green Decorative mug (8.6 µg/L). These three values incorporate all beverage scenarios because, as described above, there was no statistical difference between beverages or retention time in the cups in this study.

Coffee

Using the mean lead concentration of 0.2 µg/L from all cups except the Green Decorative mug, and the average daily coffee intake among coffee drinkers of 0.78 L, the daily dose of lead from coffee consumption would be 0.156 µg. This dose is below both the current Proposition 65 NSRL of 15 µg/day and the MADL of 0.5 µg/day. Using the mean lead concentration of 2.05 µg/L from the Green Decorative mug, the daily dose of lead for the average coffee drinker would be 1.60 µg, which exceeds the current MADL by over three fold. Using the maximum measured lead concentration of 8.6 µg/L and the same average daily coffee intake of 0.78 L, the daily dose of lead from coffee consumption is estimated to be 6.71 µg/day, which is 13.4 times the current MADL. Even consuming one eight-ounce serving of coffee with this concentration of lead would exceed the MADL four-fold, resulting in a daily dose of lead of 2.03 µg.

Tea

Using the mean lead concentration of 0.2 µg/L from all cups except the Green Decorative mug and the average daily tea intake among tea drinkers of 0.36 L, the daily dose of lead from tea consumption would be 0.071 µg,

Table 2 Range of measured lead concentrations by cup

Cup Name	Lead Concentration (µg/L)			
	Min	Max	Mean	Median
Black Logo1	<0.2	0.2	0.18	0.2
Black Logo2	0.2	0.2	0.2	0.2
Green Decorative	0.8	8.6	2.05	1.45
Yellow Decorative	0.2	0.4	0.23	0.2
Red Decorative	<0.2	0.3	0.21	0.2
Glass	<0.2	0.3	0.2	0.2

Concentrations reported to be below the limit of detection (<0.2 µg/L) were assumed to have a value of the limit of detection divided by the square root of two (0.1 µg/L) in statistical analyses which is below both the current Proposition 65 NSRL of 15 μg/day and the MADL of 0.5 μg/day. Using the mean lead concentration of 2.05 μg/L from the Green Decorative mug, the daily dose of lead for the average tea drinker would be 0.728 μg, which exceeds the current MADL. Using the maximum lead concentration of 8.6 μg/L and the same average daily tea intake of 0.36 L, the daily dose of lead from tea consumption would be 3.05 μg/day, or approximately six times the current MADL.

BLLs resulting from exposure scenario

Table 3 represents the predicted BLLs for adults and fetuses based on the measured lead concentrations in this study. Two lead concentrations were used to estimate BLLs; one scenario used the maximum concentration of lead measured from all beverages (8.6 μg/L), and the other used the mean concentration of lead measured from all beverages, excluding results from the Green Decorative mug (0.2 μg/L). Based on the average consumption of coffee (0.78 L per day, or approximately 780 g assuming coffee has the same density as water), the predicted BLLs for adults ingesting coffee daily ranged from 0.0 to 1.3 μg/dL and from 0.0 to 1.2 μg/dL in fetuses, assuming baseline BLLs of 0 and 1 μg/dL, respectively. The contribution of lead, therefore, from coffee from the cups was estimated to increase BLLs above background by a maximum of 0.3 μg/dL in adults and by 0.2 μg/dL in fetuses. All of the estimated BLLs were below the BLL of concern of 5 μg/dL set by the CDC and 10 μg/dL set by EPA, and did not raise BLLs by more than +1.0 μg/dL, which is California's proposed

benchmark for risk assessment (Carlisle and Dowling 2007; CDC 2016a; EPA 2016).

Discussion

Lead was detected in over 90% of our beverage samples; most samples resulted in concentrations similar to the analytical limit of detection (0.2 μg/L). However, in one of the mugs tested (Green Decorative), the results ranged from 0.8 to 8.6 μg/L. This finding indicates that under the conditions of this study, lead may leach from mugs into hot beverages such as coffee and tea, or even hot water, and result in individual lead exposures well above maximum allowable dose levels set by the State of California. Although this finding only pertained to one of the five mugs tested, it was unexpected, given that these mugs were randomly selected from an office environment, and that all but one were purchased in the U.S. From the small sample size of this study no conclusion can be drawn about the prevalence of mugs with leachable lead in the U.S. market; however, the findings do indicate a need for further research with greater sample sizes (and thereby more robust statistical analyses) in this area. Regarding blood lead levels, we found that background lead exposure in the models primarily contributed to the BLLs for both adults and fetal exposures. As shown in Table 3, lead ingestion from coffee using the highest concentration measured in this study increased the BLL estimated in adults by 0.3 μg/dL. Given the limitations of this study, these results should be considered a screening-level assessment.

The U.S. FDA regulates lead content in ceramics used with foods, and in mugs specifically. In 1970, the FDA

Table 3 Estimated blood lead levels following exposure to lead from mugs

	Exposure Scenario			
Background (baseline BLL)	0.0 μg/dL[a]		1.0 μg/dL[b]	
Coffee Ingestion (consumption; concentration)	780.0 g/day; 0.00022 ppm[c]	780.0 g/day; 0.0086 ppm[d]	780.0 g/day; 0.00022 ppm[c]	780.0 g/day; 0.0086 ppm[d]
Maternal Blood Lead levels (μg/dL)	0.0	0.3	1	1.3
Fetal Blood Lead levels (μg/dL)	0	0.3	0.9	1.2

Key

"Low" Exposure	"High" Exposure

[a]Chosen value for background
[b]Default value from the ALM model
[c]Maximum Pb concentration in beverages from results
[d]Mean Pb concentration excluding green decorative mug

conducted a survey of imported pottery and found "high levels" of lead leaching from the products (USFDA 1979a, p. 51237). The FDA instituted a compliance program for domestic and international pottery in 1971, limiting the amount of lead that leached from pottery into a leaching solution to 7 µg/L (USFDA 1979a). In 1979, the FDA revised the guidelines for lead leached from ceramic foodware based on a recommended tolerable total lead intake value of 100 µg/day for infants up to 6 months of age and of 150 µg/day for children from 6 months to 2 years of age, based on the endpoint of altered heme synthesis (USFDA 1979b; USFDA 1989). In response to new data and updated international reference values, in 1989 the FDA adopted a range of 6 to 18 µg/day as the provisional tolerable lead intake from food for a 10 kg child, and proposed that the guidelines for ceramic foodware again be lowered (USFDA 1989). The agency noted at that time that it was not possible to establish a threshold for lead toxicity (USFDA 1989).

In 1992, the lead release guidelines were amended to include levels specific to cups and mugs "because these articles are frequently used under conditions that may enhance lead leaching" (USFDA 1992, p. 29734). In particular, the FDA noted that cups and mugs are "generally used to hold acidic beverages, such as...coffee or tea" (USFDA 1992, p. 29735). The FDA reported that the acidity in conjunction with the higher temperatures of these beverages enhances the lead leaching rate (USFDA 1992). According to the 1992 guidance, the FDA "may take enforcement action" when cups or mugs exceed a lead level of 0.5 µg/mL in a 4% acetic acid leaching solution in any of six mug or cup units examined (USFDA 1992, p. 29735). This limit currently is still in place for cups and mugs.

The FDA reported that based on its request for information and subsequent review, the amount of lead that will leach into a leaching solution is approximately 2.5 to 5 times the amount that will leach into "hot coffee during 15 to 30 min" (USFDA 1992, p. 29735). Applying this estimation to our results, we can estimate that 2 to 43 µg/L lead would leach from the Green Decorative mug into leaching solution, which is 0.002 to 0.043 µg/mL, and well within the FDA's limit of 0.5 µg/mL. Our results, then, suggest that mugs in compliance with federal regulatory limits for lead may still well-exceed California's Safe Harbor Levels.

Furthermore, the State of California recently proposed reducing the lead MADL to 0.2 µg/day based on modeling of exposure levels that would result in maximum blood lead levels of 15 µg/dL (Cal/EPA 2015a; Cal/EPA 2015b). In 2015, the State clarified that the MADL is intended to be a daily exposure dose, but that alternative exposure doses (increased doses) are plausible within the law, if exposures can be determined not to occur daily

(Cal/EPA 2015a). Nonetheless, given the proposed MADL, and assuming mugs are used daily, coffee consumption from the four mugs with lower associated lead levels in this study (mean of 0.2 µg/L, or daily intake of 0.156 µg) nearly results in exposures above the proposed MADL. The lower MADL would also further widen the discrepancy between California and federal compliance levels.

Given the existing low MADL, and the fact that the proposed MADL for lead is less than half of the current MADL, understanding the implications for product testing is important. Many analytical methods are not sensitive enough to detect the presence of lead at meaningful concentrations in terms of exposure levels in compliance with the Safe Harbor Level. Inherent variability in measurements at such low levels of analytical detection also exists, which must be characterized and understood properly in order to rely on them for regulatory compliance. For example, the two 10 min coffee samples in the Green Decorative mug were 1.7 and 8.6 µg/L, a seemingly wide range. Also, coffee that had been in the glass cup for 10 min had a detected lead concentration of 0.3 µg/L, but the same coffee after 60 min in the glass resulted in a detected lead concentration of only 0.2 µg/L. These ranges and "reductions" highlight the uncertainty and normal variations in the measurements.

In addition, definitively identifying and segregating the relative contributions of lead from different sources in this study is not possible. Lead is ubiquitous in our environment, and its presence in glass, equipment, tea, coffee, and drinking water cannot be ruled out. Several studies, for example, have reported concentrations of lead in solid coffee beans or solid residues of coffee infusions ranging from 0.053 to 1.239 µg/g (Nędzarek et al. 2013; Onianwa et al. 1999; Federal Republic of Germany and Federal Länder unknown; Santos et al. 2004; Othman 2010). One study reported lead concentrations in liquid coffee of 2.37 and 2.57 µg/L (Ong 2014), and another study reported lead concentrations below the limit of detection of 1.5 µg/L (Ashu and Chandravanshi 2011). Additionally, studies have reported lead in tea leaves or residues from tea infusions of 0.046 to 15.479 µg/g (Li et al. 2015; Shekoohiyan et al. 2012; Shokrzadeh et al. 2008; Onianwa et al. 1999; Othman 2010; Al-Othman et al. 2012; Zheng et al. 2014). Although we did not detect any lead in our direct tap water or boiled tap water control samples, the presence of lead in drinking water is a known concern, as shown by the recent state of emergency issued in Flint, Michigan, because of its drinking water lead content. The U.S. EPA reported that 1,831 (8%) of 22,808 residential water samples collected in Flint between September 2015 and June 2016 were above the action level of 15 ppb (State of Michigan 2016). Samples reported above the action level ranged from 16 to 22,905 ppb.

These other highly variable potential sources of lead exposure must be considered when assessing total lead intake; in this study, though, the beverage lead content appeared to be most strongly determined by the mug.

To our knowledge, ours is one of only a handful of studies in the peer-reviewed literature to evaluate lead leaching into coffee or tea from lead-containing ceramics. In 1985, Wallace et al. tested Italian-originating coffee mugs found in a U.S. household. The authors reported that approximately 4000 µg of lead were leached into a 250 mL serving of coffee (16,000 µg/L) over 15 min at a temperature of 65 °C and a pH of 5.1 (Wallace et al. 1985). Wallace et al. stated that the mugs were "badly degraded," and that a similar new cup released only 200 µg of lead (800 µg/L) in the same conditions (Wallace et al. 1985, p. 290). Ajmal et al. (1997) investigated lead leaching into tea from ceramic mugs in India. They reported that measured lead concentrations in the tea were below the limit of detection (Ajmal et al. 1997). To our knowledge, then, the current study is the first to evaluate lead leaching into coffee and tea from ceramic mugs purchased in the 21st century in the U.S. (with the exception of the Red Decorative mug, which was purchased in Europe).

A greater number of studies have investigated lead leaching from ceramics associated with various other foods and beverages (Sheets 1997; Mohamed et al. 1995; Levin et al. 2008; Markowitz 2000; Gonzalez de Mejia and Craigmill 1996; Feldman et al. 1999; Belgaied 2003; Hight 1996; Valadez-Vega et al. 2011). In studies that measured lead concentrations leached from a variety of ceramics using 4% acetic acid (the same leaching solution used by the FDA to evaluate ceramics), reported values reached up to 2004.7 ppm (2,004,700 µg/L), significantly higher than values we obtained in this study (Gonzalez de Mejia and Craigmill 1996). Many of these studies also evaluated the amount of lead that leached into various acidic and non acidic foods, such as salsa, beans, tamarind juice, pickle juice, wine, and milk products, with results reaching up to 244 ppm (244,000 µg/L). The highest value was associated with salsa, a highly acidic food (Gonzalez de Mejia and Craigmill 1996). The study, however, noted that there was a mean background level of lead of 0.93 ± 0.13 ppm in the salsa.

Overall, the potential for lead ingestion from contaminated ceramic mugs is minimal when compared to other sources, such as food. The Agency for Toxic Substances and Disease Registry (ATSDR) reported that the average daily intake of lead from food sources in the general population is approximately 56.5 µg/day (ATSDR 2007a). In comparison, the maximum daily lead intake from drinking 3.3 eight-ounce cups of coffee based on the data collected in our study resulted in a dose of 6.71 µg, over eight times less than the average daily lead intake from food. Nonetheless, exposure to lead should be minimized to the extent possible. The U.S. EPA does not publish a safe dose for lead because it felt it was "inappropriate to develop a reference dose (RfD) for inorganic lead (and lead compounds) because some of the health effects associated with exposure to lead occur at blood lead levels as low as to be essentially without a threshold" (ATSDR 2007a, p. 403). Similarly, WHO withdrew its provisional tolerable weekly intake (PTWI) for lead in 2010 because it did not believe establishing a value that would be "health protective" would be possible (WHO 2010).

This study is limited in that a small number of mugs were randomly selected from the authors' work environment, and were not purchased for the purpose of evaluating the full range of lead contamination in ceramic mugs. We also were not able to test multiple mugs from the same manufacturer or origin, with the exception of the black logo mugs. Overall, then, this study can be considered a pilot study, and the results should be considered as such until additional research can be conducted, and more samples collected. This preliminary investigation, however, provides data on potential lead exposures from daily beverage consumption among typical consumers, with particular implications for pregnant or breastfeeding women. This potential source of lead exposure is less well-characterized than are some other lead exposure sources (e.g., paint, gasoline), yet is relevant to a substantial portion of the U.S. population.

Acknowledgements
We would like thank Kevin Towle for his assistance with statistical analysis of our data. We would like to acknowledge the analytical work performed by Pat-Chem Laboratories.

Authors' contributions
All named authors contributed significantly to the development and writing of this manuscript. Generally, GA took the lead in designing the study; LG took the lead in estimating blood lead levels; MF took the lead in the statistical analysis; and SG assisted with the design of the study and took the lead in the comparison to regulatory limits. All authors were involved in executing the study. All authors read and approved the final manuscript.

Authors' information
G.A. has a bachelor of science in molecular toxicology. She performs human health risk assessments in her current position as a consultant. She has a particular interest in consumer product safety and compliance.
L.G. has a bachelor of science in environmental toxicology. She currently works as a consultant and performs exposure assessment and human health risk assessments, specifically regarding consumer products and occupational hazards.
M.F. has a masters of public health in environmental health science. Her areas of interest include exposure assessment, risk assessment, product sustainability, and environmental health.
S.G. has a Masters of Health Science in industrial hygiene and a PhD in Environmental Health Sciences. She also received a certificate in Risk Sciences and Public Policy. She is a Principal Health Scientist at Cardno ChemRisk, and is interested in characterizing exposure and potentially related health risks to consumers, workers and communities.

Competing interest
The authors declare that they have no competing interest, and that no external funding was provided for writing this manuscript.

References

Ajmal M, Khan A, Nomani AA, Ahmed S. Heavy metals: leaching from glazed surfaces of tea mugs. Sci Total Environ. 1997;207(1):49–54.

Al-Othman ZA, Yilmaz E, Sumayli HM, Soylak M. Evaluation of trace metals in tea samples from Jeddah and Jazan, Saudi Arabia by atomic absorption spectrometry. Bull Environ Contam Toxicol. 2012;89(6):1216–9. doi:10.1007/s00128-012-0842-1.

Ashu R, Chandravanshi BS. Concentration levels of metals in commercially available Ethiopian roasted coffee powders and their infusions. Bull Chem Soc Ethiop. 2011;25(1):11–24.

ATSDR. Toxicological profile for lead. Atlanta: U.S. Dept. of Health and Human Services, Public Health Service, Agency for Toxic Substances and Disease Registry; 2007a.

ATSDR. Lead - ToxFAQs. August 2007. Substances and disease registry. Atlanta: Division of Toxicology and Human Health Sciences; 2007b.

Belgaied JE. Release of heavy metals from Tunisian traditional earthenware. Food Chem Toxicol. 2003;41(1):95–8.

Brown MJ, Margolis S. Morbidity and Mortality Weekly Report. Volume 61 (Supplement). Lead in drinking water and human blood lead levels in the United States. Atlanta: U.S. Department of Health and Human Services, Centers for Disease Control and Prevention; 2012.

Cal/EPA. The center for environmental health petition requesting repeal or amendment of the safe harbor level for lead. October 14, 2015. Sacramento: California Environmental Protection Agency, Office of Environmental Health Hazard Assessment; 2015a.

Cal/EPA. Pre-regulatory proposal: possible amendments to section 25805. Specific regulatory levels: chemicals causing reproductive toxicity. October 14, 2015. Sacramento: California Environmental Protection Agency, Office of Environmental Health Hazard Assessment; 2015b.

Carlisle J, Dowling K. Development of health criteria for school site risk assessment pursuant to health and safety code section 901(g): child-specific benchmark change in blood lead concentration for school site risk assessment. California Environmental Protection Agency, Office of Environmental Health Hazard Assessment, Integrated Risk Assessment Branch; 2007.

CDC. Blood lead levels keep dropping; New guidelines recommended for those most vulnerable. 1997. http://www.cdc.gov/media/pressrel/lead.htm Accessed 18 May 2016.

CDC. National Center for Health Statistics (NCHS). National Health and Nutrition Examination Survey: NHANES 2011-2012 Dietary Data. U.S. Department of Health and Human Services, 2014. https://wwwn.cdc.gov/nchs/nhanes/search/DataPage.aspx?Component=Dietary&CycleBeginYear=2011. Accessed 18 May 2016.

CDC. Lead - New blood lead level information. What Do parents need to know to protect their children? 2016a. https://www.cdc.gov/nceh/lead/acclpp/blood_lead_levels.htm Accessed 23 June 2016.

CDC. Lead. 2016b. http://www.cdc.gov/nceh/lead/ Accessed 18 May 2016.

Consumer Reports. Results of Our apple juice and grape juice tests. 2012. https://www.consumerreports.org/content/dam/cro/magazine-articles/2012/January/Consumer%20Reports%20Arsenic%20Test%20Results%20January%202012.pdf. Accessed 18 May 2016.

EPA. 40 CFR parts 141 and 142. [FRL-3823-5]. RIN 2040-AB51. Drinking water regulations. Maximum contaminant level goals and national primary drinking water regulations for lead and copper. Final rule. Fed Reg. 1991;56(110):26460–564.

EPA. Lead at superfund sites. 2016. https://www.epa.gov/superfund/lead-superfund-sites Accessed 30 June 2016.

Federal Republic of Germany, Federal Länder (unknown) Results of the German Food Monitoring of the Years 1995-2002. Federal Office of Consumer Protection and Food Safety, Bundesamt fur Verbraucherschuta und Levensmittelsicherheit (BVL), Berlin

Feldman N, Lamp C, Craigmill A. Lead leaching in ceramics difficult to predict. Calif Agric. 1999;53(5):20–3.

Gonzalez de Mejia E, Craigmill AL. Transfer of lead from lead-glazed ceramics to food. Arch Environ Contam Toxicol. 1996;31(4):581–4.

Hight SC. Lead migration from lead crystal wine glasses. Food Addit Contam. 1996;13(7):747–65. doi:10.1080/02652039609374463.

Levin R, Brown MJ, Kashtock ME, Jacobs DE, Whelan EA, Rodman J, Schock MR, Padilla A, Sinks T. Lead exposures in U.S. Children, 2008: implications for prevention. Environ Health Perspect. 2008;116(10):1285–93. doi:10.1289/ehp.11241.

Li L, Fu Q-L, Achal V, Liu Y. A comparison of the potential health risk of aluminum and heavy metals in tea leaves and tea infusion of commercially available green tea in Jiangxi, China. Environ Monit Assess. 2015;187(5):1–12.

Markowitz M. Lead poisoning. Pediatr Rev. 2000;21(10):327–35.

Mason LH, Harp JP, Han DY. Pb neurotoxicity: neuropsychological effects of lead toxicity. Biomed Res Int. 2014;2014:840547. doi:10.1155/2014/840547.

Mohamed N, Chin YM, Pok FW. Leaching of lead from local ceramic tableware. Food Chem. 1995;54(3):245–9.

Monnot AD, Christian WV, Abramson MM, Follansbee MH. An exposure and health risk assessment of lead (Pb) in lipstick. Food Chem Toxicol. 2015;80:253–60.

Nędzarek A, Tórz A, Karakiewicz B, Clark JS, Laszczyńska M, Kaleta A, Adler G. Concentrations of heavy metals (Mn, Co, Ni, Cr, Ag, Pb) in coffee. Acta Biochim Pol. 2013;60(4):623–7.

OEHHA. 27 CCR § 25821. Level of exposure to chemicals causing reproductive toxicity. 2016a.

OEHHA. Proposition 65 No Significant Risk Levels (NSRLs) for Carcinogens and Maximum Allowable Dose Levels (MADLs) for Chemicals Causing Reproductive Toxicity. April 2016. Sacramento: California Environmental Protection Agency, Office of Environmental Health Hazard Assessment; 2016b.

Ong K. Determination of lead and cadmium in foods by graphite furnace atomic absorption spectroscopy. 2014. https://www.perkinelmer.com/lab-solutions/resources/docs/APP_PinAAcle-900H-Lead-Cadmium-in-Food-011965_01.pdf Accessed 14 Sept 2016.

Onianwa P, Adetola I, Iwegbue C, Ojo M, Tella O. Trace heavy metals composition of some Nigerian beverages and food drinks. Food Chem. 1999;66(3):275–9.

Othman ZAA. Lead contamination in selected foods from Riyadh city market and estimation of the daily intake. Molecules. 2010;15(10):7482–97.

Philip AT, Gerson B. Lead poisoning- Part I. Incidence, etiology, and toxicokinetics. Clin Lab Med. 1994;14(2):423–44.

Santos EE, Lauria DC, Porto da Silveira CL. Assessment of daily intake of trace elements due to consumption of foodstuffs by adult inhabitants of Rio de Janeiro city. Sci Total Environ. 2004;327(1):69–79.

Sheets RW. Extraction of lead, cadmium and zinc from overglaze decorations on ceramic dinnerware by acidic and basic food substances. Sci Total Environ. 1997;197(1–3):167–75.

Shekoohiyan S, Ghoochani M, Mohagheghian A, Mahvi AH, Yunesian M, Nazmara S. Determination of lead, cadmium and arsenic in infusion tea cultivated in north of Iran. Iranian J Environ Health Sci Eng. 2012;9(1):37.

Shokrzadeh M, Saberyan M, Saeedi Saravi S. Assessment of lead (Pb) and cadmium (Cd) in 10 samples of Iranian and foreign consumed tea leaves and dissolved beverages. Toxicol Environ Chem. 2008;90(5):879–83.

State of Michigan. Taking action on Flint water: Flint Resdiential Testing Report - results collected through June 27, 2016. 2016. http://www.michigan.gov/flintwater/0,6092,7-345-76292_76294_76297—,00.html.

The White House. President obama signs michigan emergency declaration. January 16, 2016. 2016. https://www.whitehouse.gov/the-press-office/2016/01/16/president-obama-signs-michigan-emergency-declaration. Accessed 30 June 2016.

USFDA. 21 CFR chapter I, subchapter B [Docket No. 79N-0200]. lead in food; advance notice of proposed rulemaking: request for data. Advance notice of proposed rulemaking. Fed Reg. 1979a;44(171):51233–42. August 31, 1979.

USFDA. Limits on leachable lead and cadmium from ceramic ware for food use: availability of administrative guidelines. Fed Reg. 1979b;44(156):47162. August 10, 1979.

USFDA. 21 CFR Part 109 [Docket No. 89N-0014] RIN: 0905-AC91. Lead from Ceramic Pitchers. Proposed Rule. Fed Reg. 1989;54(104):23485–9. June 1, 1989.

USFDA. Lead in ceramic foodware; revised compliance policy guide; availability. Notice. Fed Reg. 1992;57(129):29734–6. July 6, 1992.

USFDA. 21 CFR Parts 103, 129, 165, and 184. [Docket No. 88P-0030]. RIN 0910-AA11. Beverages: Bottled Water. Final Action. Fed Reg. 1995;60(218):57076–130. November 13, 1995.

USFDA. FDA Comments on Subcontract Number: 70000073494, Somogyi 2010, "Caffeine Intake by the U.S. Population". 2012. http://www.fda.gov/downloads/aboutfda/centersoffices/officeoffoods/cfsan/cfsanfoiaelectronicreadingroom/ucm333191.pdf. Accessed 23 June 2016.

USFDA. CPG Sec. 545.450 Pottery (Ceramics); Import and Domestic - Lead
    Contamination. 2015. http://www.fda.gov/ICECI/ComplianceManuals/
    CompliancePolicyGuidanceManual/ucm074516.htm. Accessed 23 June 2016.

Valadez-Vega C, Zúñiga-Pérez C, Quintanar-Gómez S, Morales-González JA,
    Madrigal-Santillán E, Villagómez-Ibarra JR, Sumaya-Martínez MT, García-
    Paredes JD. Lead, cadmium and cobalt (Pb, Cd, and Co) leaching of glass-
    clay containers by pH effect of food. Int J Mol Sci. 2011;12(12):2336–50. doi:
    10.3390/ijms12042336.

Wallace DM, Kalman DA, Bird TD. Hazardous lead release from glazed
    dinnerware: a cautionary note. Sci Total Environ. 1985;44(3):289–92.

WHO. Exposure to lead: a major public health concern. 2010. http://www.who.
    int/ipcs/features/lead..pdf?ua=1. Accessed 18 May 2016.

Zheng H, Li JL, Li HH, Hu GC, Li HS. Analysis of trace metals and perfluorinated
    compounds in 43 representative tea products from South China. J Food Sci.
    2014;79(6):C1123–9. doi:10.1111/1750-3841.12470.

18

Incidence of enteric pathogens in ugba, a traditional fermented food from African oil bean seeds (*Pentaclethra macrophylla*)

Princewill Chimezie Okorie[1,2]*, Nurudeen Ayoade Olasupo[2], Felicia Ngozi Anike[3], Gloria Nwakego Elemo[1] and Omoanghe Samuel Isikhuemhen[3]

Abstract

Background: The lack of good production practices and probable post-fermentation contaminations contribute to bacterial pathogen load in ugba, a fermented food made from African oil been seeds (*Pentaclethra macrophylla* Benth). Some of these bacteria are not easily detected using standard culturing techniques. The study used molecular-based approaches involving PCR, cloning and sequencing as well as culture-based methods to investigate the occurrence of pathogenic bacteria within the microbiome of ugba. Six samples (OK1-OK6) were purchased from different local markets within Lagos and Abia States in Nigeria and used in the study.

Results: A total of 14 pathogenic bacteria were identified among the bacteria diversity found in ugba. Two pathogens (11%) were uniquely identified by cultural and biochemical characteristics, 12 pathogens (67%) were uniquely identified by culture-independent (PCR-clone-based) method and 4 pathogens (22%) were found in both methods. Enteric pathogens were common and *Bacillus* sp. was conspicuously absent. *Escherichia coli* and *Staphylococcus aureus* detected by cultural method in this study and elsewhere were not picked up by culture-independent method. This is the first report of the presence of *Acinetobacter baumanii*, *Clostridium sartagofum*, *Enterococcus casseliflavus*, *Comamonas testosteronii*, and *Aeromonas* sp. in ugba as identified by PCR-clone-based techniques. Though the genera *Salmonella* and *Proteus* have been associated with ugba in previous studies, the species-level identities were determined in this research by the culture-independent method used.

Conclusions: This study highlights the importance of using appropriate technologies and correct species identification strategies in studying microbiological quality and food safety issues in fermented foods in a developing country like Nigeria.

Keywords: Food safety, Pathogenic bacteria, Fermented foods, African oil bean seeds, Ugba

Background

'Ugba' is fermented African oil bean seed (*Pentaclethra macrophylla* Benth.) that is consumed as a delicacy and used as food flavoring condiment. The seeds are fermented under alkaline conditions (Olasupo et al. 2016). It is rich in protein and other essential nutrients, serving mainly as a source of protein, with distinctive economic, social and cultural role among the consumers (Ogueke et al. 2010). Ugba is obtained through the processing of the large brown glossy seeds of the African oil bean seed involving- boiling the seeds in water for 4–12 h to soften the hard brown testa (shell), removing the shell, washing the kernel and slicing into long thin strips, and then mixing with salt (Odunfa and Oyeyiola 1985). The slices are packaged in small wraps with banana leaves (*Musa sapientum*) and left to ferment at room temperature for 2–5 days to yield the final product, ugba (Olasupo et al. 2016). There is no streamlined method, safety guideline or standards and so production practices and packaging

* Correspondence: 61okorie@gmail.com
[1]Department of Biotechnology, Federal Institute of Industrial Research Oshodi, Lagos, Nigeria
[2]Department of Microbiology, Faculty of Science, Lagos State University, Lagos, Nigeria
Full list of author information is available at the end of the article

is on individual/family basis. Although fermentation is a means of providing nutritious and palatable food, the safety of fermentation products especially in non-standardized production practices is a major concern that needs adequate research.

Most studies of African fermented foods have focused on isolation and identification of desirable microorganisms involved in the fermentation process. (Gadaga et al. 2004; Okorie and Olasupo 2013b). There is limited information on the occurrence and growth of pathogens in African fermented foods compared to developed countries in Europe and North America. The traditional processing method of fermenting the African oil bean seed into ugba is riddled with issues of product safety and quality inconsistency. The growth and occurrence of organisms of public health importance in ugba is of great concern. Natural fermentation process used routinely in the fermentation of ugba allows participation of diverse microorganisms which may include contaminants. Therefore, the participation of pathogenic and spoilage microorganisms during its production is not surprising especially during fermentation under very poor hygiene condition. In previous studies, a few pathogens have been detected through the cultural-based method. Most of them except *Eschericia coli* and *Staphylococcus aureus* were identified only to the genus level (Ogueke et al. 2010; Eze et al. 2014). Anyanwu et al. (2016) and Ogbulie et al. (2014) have both reported the presence of *E. coli*, *Klebsiella* and *Staphylococcus* species in samples of ugba. The presence of these organisms in the product constitute a source of concern to the consumers who are increasingly becoming safety conscious.

Current records of possible pathogens in African fermented foods have been based on isolation of pure cultures, followed by phenotypic identification of isolated microorganisms (Ejiofor et al. 1987; Enujiugha and Akanbi 2008; Nwagu et al. 2010; Ogueke and Aririatu 2004; Okorie and Olasupo 2013a). In the case of ugba, studies reported in the literature on microbial quality were based on traditional culture and identification by phenotypic and biochemical methods. These organisms were detected using the routine culture-based method with its inherent weaknesses in detecting and resolving the identity of some organisms, especially the viable but unculturable ones. However, more than 96% of the earth's microorganisms are believed to be unculturable in the laboratory (Davey and Kell 1996). It is, therefore, suspected that the scope of contamination of ugba could be wider than have been recorded by these studies. The deployment of DNA-based techniques in the study of microbiological quality of ugba is needed to reveal the extent of microbiological hazard associated with its consumption. This is the first study that employed DNA

based techniques in microbial evaluation of ugba. The main goal was to screen ugba samples using culture-based and culture independent (PCR-clone-based) methods and identify the pathogens that may be present. The diversity among total bacterial load was also evaluated.

Methods
Sample collection and preparation
Six samples (OK1 – Ok6) of fermented African oil bean seeds (Ugba) were purchased from different local markets in Nigeria. Five grams (5 g) of each sample were agitated vigorously with 10 ml of phosphate buffered saline for 5 min and centrifuged at 7500 rpm for 4 min. The supernatants were used for bacterial isolation and also dried down for DNA extraction. Samples were stored at -20 °C until used.

Culture-Dependent-Phenotypic screening
Bacteria isolation
The supernatants were serially diluted and plated onto an all-purpose medium - nutrient agar and four different selective media - MacConkey agar, violet red bile agar, *Salmonella/Shigella* agar and *Staphylococcus* agar using the spread plate method. All media were prepared according to manufacturer's (Oxoid) protocol. Colonies were purified on same medium used for isolation, characterized and identified.

Characterization and identification of isolates
Isolates from each sample were characterized based on cultural, morphological and biochemical tests including sugar fermentation. For fermentation tests 8 sugars were tested at 1% concentration. Five milliliters of each sugar solution (glucose, sucrose, maltose, mannitol, lactose, fructose, sorbitol, xylose and arabinose) were filter sterilized and inoculated with the isolates individually. Other biochemical tests performed were gram reaction, catalase, oxidase, indole production, methyl red, Voges-Proskauer, urea hydrolysis, starch hydrolysis, gelatin hydrolysis, casein hydrolysis, nitrate reduction, coagulase and citrate utilization (Collins and Lyne 1989). These tests were performed according to standard microbiology protocols. Members of *Enterobacteriaceae* were further characterized using the analytical profile index kit (API 20E kit, Biomereux, France), according to manufacturer's specifications.

Culture-Independent (PCR-clone based Screening)
DNA extraction
DNA was extracted from dried supernatants (pellets) of 6 ugba samples (OK1-OK6). Prior to use, pellets were separately suspended in 100 µl of nuclease free water. Genomic DNA was isolated from each sample using QIAmp DNA

Mini Kit (Qiagen, USA) according to the manufacturer's protocol. DNA concentration and purity were determined using Nanodrop (ND 1000) Spectrophotometer.

PCR Amplification of genomic DNA

PCR were carried out on genomic DNA from each sample to amplify genes coding for bacterial 16 s rRNA using 16 s primer sequence; Forward: 5´-CCTACGGG AGGCAGCAG-3´ and Reverse 5´-CCGTCAATTCCTT TRAGTTT-3´. Amplification was performed in a thermal cycler (Eppendorf, Germany) in 25 µl reaction volume consisting of 12.5 µL GoTaq Green Master Mix (Promega), 1.25 µL each of forward and reverse primers (Eurofins genomics), 9 µL water and 1 µL template DNA under the following conditions: denaturation 94 °C 35 s, annealing 49 °C 35 s, and extension 72 °C 1 min for 30 cycles.

Cloning and sequencing of amplified fragments

PCR products from each sample were cloned onto pGEM-T Easy Vector (Promega, USA), followed by transformation into JM109 competent *Escherichia coli* cells (Promega). A total of 384 positive clones were selected (64 clones × 6 samples). Colony PCR was carried out on each clone with same primer pairs used for initial PCR. Amplified fragments were purified with ExoSAP-IT (Affymetrix, USA). Sequencing of forward and reverse strands was performed by Sanger method at Eurofins Genomics (Kentucky, USA). Sequencing primers were same as used for PCR amplification of the DNA fragments.

Sequence analysis and bacteria identification

Sequencing data were assembled, trimmed, and quality-checked by DNA Sequence Assembler v4 (2013), Heracle BioSoft, (www.DnaBaser.com). The resulting sequences were subsequently compared with deposited sequence information using the BLAST algorithm (http://blast.ncbi.nlm.nih.gov/Blast.cgi/ and http://www.ezbiocloud.net/)). 16S rRNA identification at the genus level was done at ≥97% identity, and species level, ≥99% identity. Scores due to poor sequence quality were considered as non-interpretable results.

Results

Culture-dependent and PCR-clone-based sequence methods were used to evaluate the microbial community of fermented African oil bean seeds. Bacterial diversity detected in this study is shown in (Fig. 1). A wide range of bacteria including pathogens were identified. The cultural method revealed diverse but limited number of bacteria. Six groups of bacteria that are of potential public health importance were identified using culture-based method (Table 1). These organisms were detected across the samples screened with *Staphylococcus* species and *Escherichia coli* isolated in five out of the six samples examined, *Klebsiella* and *Proteus* in four samples, *Pseudomonas* in two samples while *Salmonella* was detected in only one sample.

Three hundred and sixty-four (364) clones were selected using blue/white cloning technique (Fig 2). White bacteria colonies indicated positive cloning of the target 16S rRNA fragment while blue colonies indicated negative cloning. More than 50% of cloned 16S fragment were positive. The sequencing of cloned fragments revealed wider bacteria diversity and more pathogens in fermented ugba.

The relative abundance of pathogenic bacteria identified by culture-independent method (clone based technique) and their relative abundance in each sample is shown in (Fig. 3). *Acinetobacter baumanii* was the most

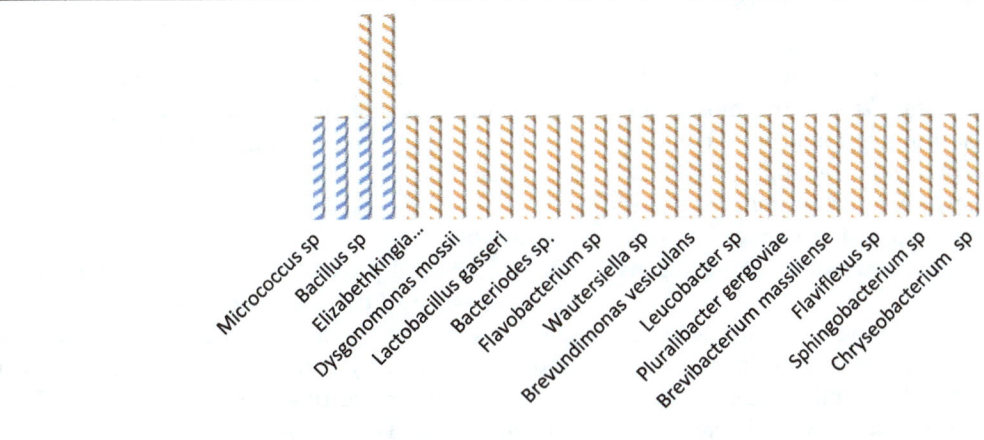

Fig. 1 Non-pathogenic bacteria uniquely detected by cultural and clone-based methods in ugba (fermented African oil bean seeds (*Pentaclethra macrophylla* Benth.). blue bars- bacteria detected by cultural method; orange bar- bacteria detected by clone-based method; blue and orange bar: bacterial detected by both methods

Table 1 Phenotypic and biochemical features of pathogenic bacteria isolated from ugba, an indigenous Nigerian food produced from African oil bean (*Pentaclethra macrophylla* Benth)

Characteristics	Bacteria isolates					
	1	2	3	4	5	6
Color/pigment	yellow	black	pink	pink	pink	green
Gram reaction	+	−	−	−	−	−
Cell shape	coccus	rod	rod	rod	rod	rod
Catalase test	+	+	+	+	+	+
Oxidase test	−	−	−	−	−	+
Indole production	−	+	−	−	+	−
Motility	−	+	−	+	+	+
Methyl red test	−	+	+	−	+	−
Voges-Proskauer	+	−	+	+	−	−
Hydrolysis of						
Urea	+	−	+	+	+	−
Starch	−	−	+	+	−	−
Gelatin	−	−	−	−	+	+
Casein	−	−	+	−	−	−
Nitrate reduction	+	+	+	−	−	+
Coagulase test	+	−	+	−	−	−
Citrate utilization	−	−	+	+	+	+
Sugar Fermentation						
Glucose	+	+	+	+	+	+
Sucrose	+	+	+	+	+	+
Maltose	+	+	+	+	+	+
Mannitol	+	+	+	+	−	+
Lactose	+	+	+	+	−	−
Fructose	+	−	+	+	−	+
Sorbitol	−	+	+	+	−	−
Xylose	−	+	+	+	+	−
Arabinose	−	+	+	+	−	+
Probable identity	*Staphylococcus aureus*	*Escherichia coli*	*Klebsiella* sp.	*Enterobacter* sp	*Proteus* sp.	*Pseudomonas* sp.

abundant bacteria. It was detected in 4 out of 6 samples in 42 clones. *Enterococcus feacalis* was the most common pathogen among the sample; found in 5 out of 6 samples. It was the second most abundant with 27 clones. *Clostridium sartagofum* and *Aeromonas* sp. were the least abundant pathogens; found in only one sample and one clone each. The abundance of other pathogens ranged between 1 and 42.

The presence of potential pathogens in the samples was detected by both cultural and culture-independent methods (Fig. 4). However, culture-independent method (PCR-clone-based method) revealed a wider bacterial diversity (> 2 fold) than the culture-based method. Sixty seven percent (67%) of pathogens were uniquely identified by clone-based method, 11% by culture-based method and 22% by both methods (Fig. 5). Culture-

independent method resolved the identity of detected pathogenic bacteria to species level (Fig. 4) while culture-based method identified most of them to genus level (Table 1).

Discussion

The lack of good production practices for fermented African oil bean seeds is a major setback in food safety standards of the product. African fermented oil bean seed is produced by cottage industries in ways defined by each producer, creating lapses that could lead to food contamination. These lapses in our vigilance for food safety have been responsible for outbreaks of serious food poisoning in different regions particularly in the continent of Africa (Campbell-Platt 1997; Olasupo et al. 2002). Our work seeks to draw attention to the issue of

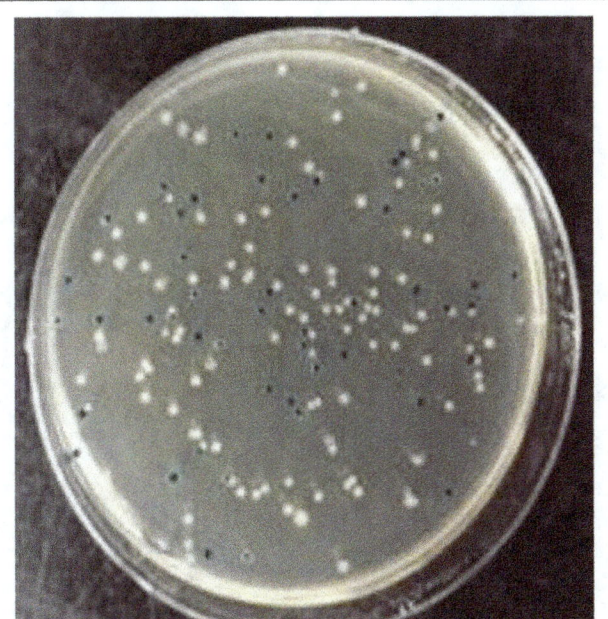

Fig. 2 Blue/white cloning of PCR products from samples of ugba (fermented African oil bean seeds *Pentaclethra macrophylla* Benth.). PCR fragments from DNA isolated from ugba samples were ligated unto pGEM-T plasmid and transformed into competent *E. coli* cells, and plated on Ampicillin-X-gal containing medium

safety and quality of traditionally fermented foods in Africa.

The consumption of fermented African oil bean seeds in Nigeria dates back to centuries ago. However, the issue of safety remains a major challenge, like some other traditional fermented foods produced in Africa. The method for its preparation and fermentation has remained unchanged and, still depends mainly on spontaneous fermentation process and use of local non-sterile utensils. Other potential sources of variability in bacterial type and load could be the food handler's hygienic status and lack of standardized manufacturing practice, leading to possible introduction of pathogenic organisms. The presence of some pathogens have been reported in fermented African oil bean seeds and this has included *Escherichia coli* and species of *Staphylococcus, Klebsiella,* and *Proteus* (Anyanwu et al. 2016; Eze et al. 2014; Isu and Njoku 1997; Ogbulie et al. 2014). Other pathogens that have been reported in other African fermented foods include *Bacillus cereus, Salmonella* sp. *Vibrio cholera, Aeromonas* sp., *Campylobacter* and *Shigella* (Gadaga et al. 2004). In a similar study on safety of traditional Nigerian fermented foods, Olasupo et al. (2002) reveals that microorganisms of public health concern that have been associated with Nigerian fermented foods include *Staphylococcus aureus, Klebsiella* sp., *E. coli, Salmonella* sp., *Bacillus subtilis,* and *Enterococcus faecalis.*

To ensure food safety, it is critically important that quality control standards including accurate methods for identifying pathogens be adopted and followed, in order to understand the extent of danger posed by the consumption of such food. Previous efforts in identifying possible pathogens in ugba have been predominantly culture-based, involving the isolation and identification methods using phenotypic tools. This identification scheme has the inherent weakness of not being able to detect and identify viable but unculturable organisms in a food matrix. The results obtained in this study using

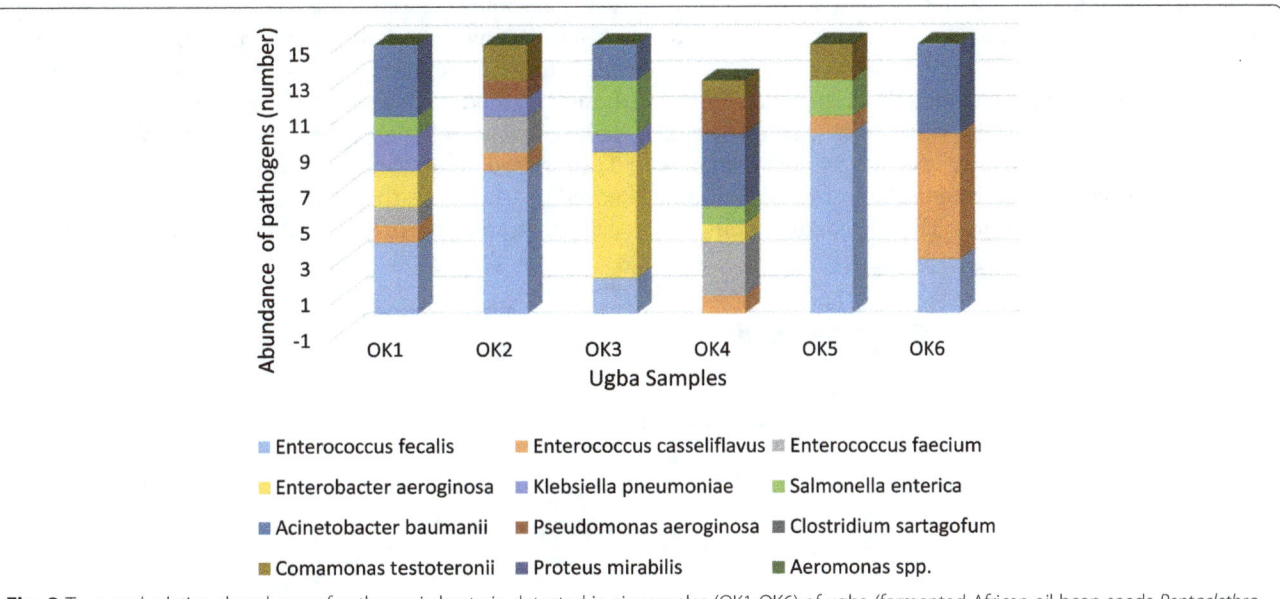

Fig. 3 Type and relative abundance of pathogenic bacteria detected in six samples (OK1-OK6) of ugba (fermented African oil bean seeds *Pentaclethra macrophylla* Benth.) by clone-based method

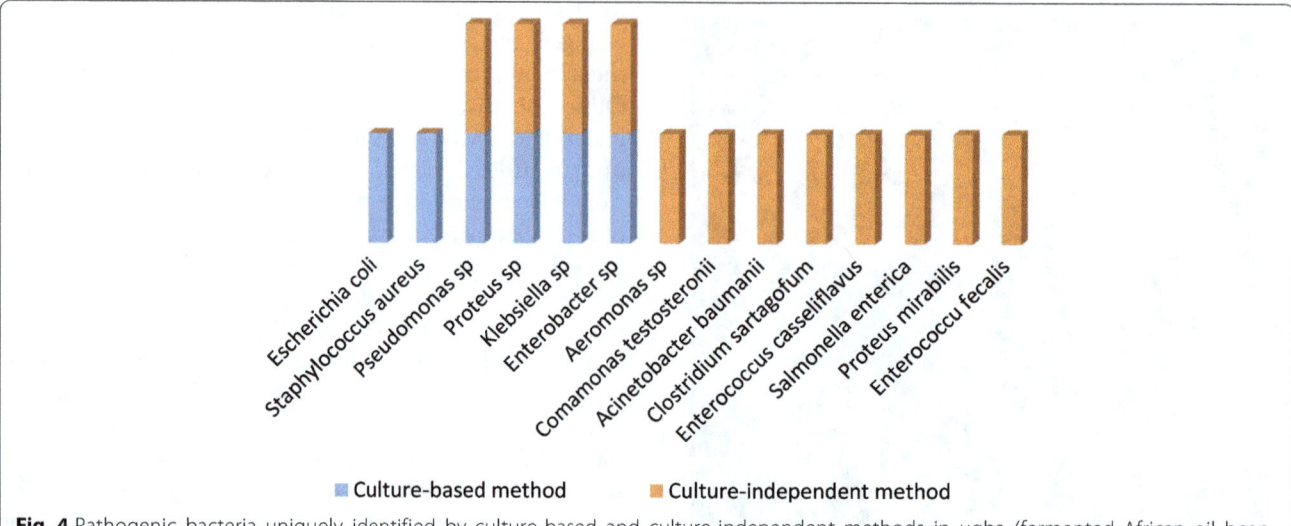

Fig. 4 Pathogenic bacteria uniquely identified by culture-based and culture-independent methods in ugba (fermented African oil bean seeds- *Pentaclethra macrophylla* Benth.). blue bars- pathogenic bacteria detected by culture based method; orange bar- pathogenic bacteria detected by clone-based method; blue and orange bar: pathogenic bacterial detected by both methods

the phenotypic method of screening of cultured bacteria are consistent with what has been reported in the literature. However, as expected, DNA sequence based identification scheme employed in this study has shown that the range of organisms that could be of public health importance associated with ugba, could be much wider than what has been reported in literature and in this study, using culture-based methods. A variety of pathogens were reported, for the first time, in this study as being found in fermented African oil bean seed (ugba) using the PCR-clone library technique. Finally, while bacteria belonging to the genera *Salmonella* and *Proteus* have been associated with ugba in the past, their species-level identities were determined in the present study as *Salmonella enterica* and *Proteus mirabilis* respectively.

It is believed that molecular methods have a superior ability to detect and identify organisms that are viable but may not be culturable or are in very low numbers and could not be detected by pure culture isolation and phenotypic characterization. This accounted for the

wider bacteria diversity and potential pathogens detected in this study. This belief is shared by Gao and Moore (Gao and Moore 1996); Schloss and Handelsman (Schloss and Handelsman 2005) who have demonstrated that molecular techniques, such as representational difference analysis, consensus sequence–based PCR, and complementary DNA library screening, have led to the identification of several previously unculturable infectious agents. Rhoads et al. (2012), in a study evaluating culturing versus 16S rRNA sequencing as tools for identifying bacterial species in human chronic wound infection, identified 145 unique genera using molecular methods, whereas only 14 unique genera were identified using aerobic culture methods.

Davey (Davey and Kell 1996) and Gunasekera et al. 2003 noted that a major disadvantage of using culture-based methods in the analysis of food samples is their failure to detect viable but non-culturable organisms. They raised doubts about the effectiveness of culture-based methods in the recovery of sub-lethally injured cells that may occur in heat treated products such as

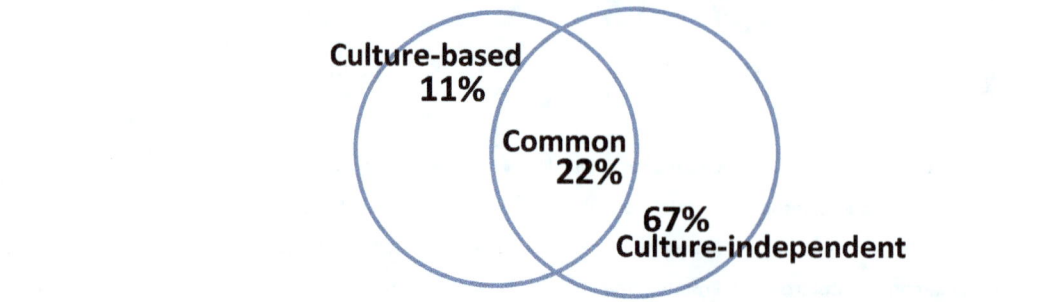

Fig. 5 Differences and Commonality (%) of pathogenic bacteria detected by culture-based and culture-independent methods in ugba (fermented African oil bean seeds- *Pentaclethra macrophylla* Benth.)

pasteurized milk. Since the fermentation process of ugba is usually terminated by boiling the final product in water, it is possible that the boiling process could have injured but not eliminated some of the bacteria and rendered them unculturable. However, because the molecular methods are culture independent, such cells are detected and identified. This and post fermentation handling and processing could therefore account for the wider range of bacteria diversity and potentially pathogenic bacterial genera/species detected by molecular techniques, against what was detected by cultural methods in ugba.

In addition to some other bacteria identified in literature, this study has reported for the first time various types of possible pathogenic bacteria in *ugba* including *Enterococcus faecalis, Enterococcus casseliflavus, Enterobacter aeroginosa, Klebsilla pneumoniae, Salmonella enterica, Proteus mirabilis, Aeromonas* sp. *Enterobacter* sp., *Enterococcus faecium, Comamonas testosteronii* and *Clostridium sartagofum*. Many of these organisms are part of the gastrointestinal microbiota of the human population, and their presence in the product is suggestive of fecal contamination. This is an indication of the poor hygiene in the processing environment and lack of good manufacturing practice. Fecal-oral route of transmission for many bacterial food-borne diseases is very significant; therefore, basic hygienic measures are an essential step for improving food safety. This study therefore, underscores the need to develop a standardized protocol for the production of this product, to ensure its quality and safety.

It remains possible that the presence of these bacteria of possible public health importance identified by this study is due to post-fermentation contamination. It is also possible however that their presence in ugba could be from the raw materials especially water, used in the production process. Since fermentation of African oil bean seeds has been shown to be an alkaline process, up to pH 8.2 (Ogueke et al. 2015; Olasupo et al. 2016), the antimicrobial effect often associated with most fermented food products due to their low pH is probably lacking in ugba. It is therefore possible that these organisms survived the fermentation process.

Whether the bacteria identified by this study were post-fermentation contaminants or they survived the fermentation process, their presence in ugba pose health risks to the consuming public. This is especially so as ugba purchased in the market place is sometimes eaten without pre-heating or cooking. Although, the enterotoxigenic potential of the detected organisms are yet to be determined, their presence in foods could serve as indication for the need to promote awareness about the possible health hazards due to handling and processing of traditional fermented foods.

Conclusion

This study clearly demonstrates the presence of pathogenic bacteria in ugba. This constitutes health and food safety concerns for consumers. The molecular tools employed have expanded the scope of possible pathogens associated with ugba. This is the first time that *Enterococcus casseliflavus, Acinetobacter baumanii, Clostridium sartagofum. Comamonas testosteronii,* and *Aeromonas* sp. are reported to be found on ugba samples. However, it is necessary to study larger sample size to determine the occurrence and abundance of these organisms in ugba across Nigeria. Nevertheless, it is pertinent to develop a standardized protocol for the production of ugba, in order to improve its quality and safety. It also underscores the need to maintain improved hygienic standards and good manufacturing practice (GMP) in the preparation of fermented condiments generally and ugba in particular.

Acknowledgements
The authors are grateful to Miss Janelle Robnson, of Agricultural and Technical University, North Carolina, USA and Mr. Olatope S. O. of Federal Institute of Industrial Research, Lagos Nigeria for their assistance in the execution of this work. None of the authors have any competing interest in the manuscript.

Funding
Self funded.

Authors' contributions
PCO, NAO, GNE, OSI designed the study. PCO and FNA conducted the experiment under the supervision of OSI and NAO. PCO, OSI and FNA performed the processing of data. All authors read and approved the final manuscript.

Competing interest
The authors declare that they have no competing interests.

Author details
[1]Department of Biotechnology, Federal Institute of Industrial Research Oshodi, Lagos, Nigeria. [2]Department of Microbiology, Faculty of Science, Lagos State University, Lagos, Nigeria. [3]Department of Natural Resources and Environmental Design, North Carolina Agricultural and Technical State University, Greensboro, North Carolina, USA.

References
Anyanwu NCJ, Okonkwo OL, Iheanacho CN, Ajide B. Microbiological and nutritional qualities of fermented ugba (Pentaclethra macrophylla Bentham) sold in Mbaise, Imo state, Nigeria. Ann Res Rev Biol. 2016;9(4):1–8.

Campbell-Platt G. Editorial: Food control and its impact on food safety. Food Contr. 1997;1:1–3.

Collins CH, Lyne PM. Microbiological methods. 6th ed. London: Butterworths; 1989. p. 129–30.

Davey HM, Kell DB. Flow cytometry and cell sorting of heterogeneous microbial populations: the importance of single cell analyses. Microbiol Rev. 1996;60: 641–96.

Ejiofor MAN, Oti E, Okafor JC. Studies on the fermentation of seeds of the African oil bean tree (Pentaclethra macrophylla Benth). Int Tree Crops J. 1987;4:135–44.

Enujiugha VA

Enujiugha VN, Akanbi CT. Quality evaluation of canned fermented African oil bean seed slices during ambient storage. Afr J Food Sci. 2008;2:54–9.

Eze VC, Onwuakor CE, Ukeka E. Proximate composition, biochemical and microbiological changes associated with fermenting African oil bean (Pentaclethra macrophylla Benth.) seeds. Amer J Microbiol. 2014;2(5):674–81.

Gadaga TH, Nyanga LK, Mutukumira AN. The occurrence, growth and control of pathogens in African fermented foods. Afr J Food Agric Nutr Dev. 2004;4(1): 406–13.

Gao SJ, Moore PS. Molecular approaches to the identification of unculturable infectious agents. Emerg Infect Dis. 1996;2(3):159–67.

Gunasekera TS, Dorasch MR, Slado MB, Veal DA. Specific detection of *Pseudomonas* spp in milk by fluorescent in situ hybridization using ribosomal RNA directed probes. J Appl Microbiol. 2003;94:936–45.

Isu NR, Njoku HO. An evaluation of the microflora associated with fermented African oil bean (*Pentaclethra macrophylla* Bentham) seeds during ugba production. Plt Foods Hum Nutr. 1997;51:145–57.

Nwagu TN, Amadi C, Alaekwe O. Role of bacteria isolates in the spoilage of fermented African Oil Bean Seed Ugba. Pakistan J. Biol. Sci. 2010;13:497–503.

Odunfa SA, Oyeyiola GF. Microbiological study of the fermentation of ugba: A Nigerian indigenous fermented food flavor. J Plt Foods. 1985;6:155–63.

Ogbulie TE, Nsofor CA, Nze FC. Bacteria species associated with ugba (*Pentaclethra macrophylla* Bentham) produced traditionally and in the laboratory and the effect of fermentation on product of oligosaccharide hydrolysis. Nig. Food J. 2014;32(2):73–80.

Ogueke CC, Anosike F, Owuamanam CI. Prediction of amino nitrogen during ugba (*Pentaclethra macrophylla*) production under different fermentation variables: A response surface approach. Nig. Food J. 2015;33:61–6.

Ogueke CC, Aririatu LE. Microbial and organoleptic changes associated with ugba stored at ambient temperature. Nig Food J. 2004;22:133–40.

Ogueke CC, Nwosu JN, Owuamanam CI, Iwouno JN. Ugba, the fermented African oilbean seeds; its production, chemical composition, preservation, and health benefits. Pakistan J Biol Sci. 2010;13:489–96.

Olasupo NA, Okorie PC, Oguntoyinbo FA. The biotechnology of ugba, a Nigerian traditional fermented food condiment. Front Microbiol. 2016;7:1153–61.

Olasupo NA, Smith SI, Akinsinde KA. Examination of microbial status of selected indigenous fermented foods in Nigeria. J Food Safety. 2002;22:85–93.

Okorie PC, Olasupo NA. Growth and extracellular enzyme production by microorganisms isolated from ugba- an indigenous Nigerian fermented food. Afr J Biotechnol. 2013a;12(26):4158–67.

Okorie PC, Olasupo NA. Controlled fermentation and preservation of ugba – an indigenous Nigerian fermented food. Spring. 2013b;2:470–8.

Rhoads DD, Cox SB, Rees EJ, Sun Y, Wolcott RD. Clinical identification of bacteria in human chronic wound infection: culturing vs 16s rDNA sequencing. BMC Infect Dis. 2012;12:321–8.

Schloss PD, Handelsman J. Metagenomics for studying unculturable microorganisms: cutting the Gordian knot. Gen Biol. 2005;2005(6):229–35.

19

Heavy metal accumulation and health risk assessment in wastewater-irrigated urban vegetable farming sites of Addis Ababa, Ethiopia

Desta Woldetsadik [1*], Pay Drechsel[2], Bernard Keraita[3], Fisseha Itanna[4] and Heluf Gebrekidan[1]

Abstract

Background: Wastewater irrigation for vegetable production is a highly prevalent practice in Addis Ababa and a number of articles have been published on wastewater-irrigated soils and vegetables contaminated with heavy metals. However, to the best of our knowledge, an insight into assessment of human health risks associated with the consumption of vegetable crops grown on wastewater-irrigated soils is non-existent in the city. Long-term effect of wastewater irrigation on the build-up of heavy metals in soils and selected vegetable crops in Addis Ababa urban vegetable farming sites (10) was evaluated. By calculating estimated daily intakes (EDIs) and target hazard quotients (THQs) of metals, health risk associated with the consumption of the analyzed vegetables was also evaluated.

Results: The heavy metal concentrations in irrigation water and soils did not exceed the recommended maximum limits (RMLs). Moreover, Cd, Co, Cr, Cu, Ni and Zn concentrations in all analyzed vegetables were lower than the RML standards. In contrast, Pb concentrations were 1.4–3.9 times higher. Results of two way ANOVA test showed that variation in metals concentrations were significant ($p < 0.001$) across farming site, vegetable type and site x vegetable interaction. The EDI and THQ values showed that there would be no potential health risk to local inhabitants due to intake of individual metal if one or more of the analyzed vegetables are consumed. Furthermore, total target hazard quotients (TTHQs) for the combined metals due to all analyzed vegetables were lower than 1, suggesting no potential health risk even to highly exposed local inhabitants.

Conclusions: There is a great respite that toxic metals like Pb and Cd have not posed potential health risk even after long term (more than 50 years) use of this water for irrigation. However, intermittent monitoring of the metals from irrigation water, in soil and crops may be required to follow/prevent their build-up in the food chain.

Keywords: Vegetable farming, Wastewater irrigation, Heavy metal, Health risk, Target hazard quotient, Addis Ababa

Background

Wastewater (untreated, partially treated or diluted) has been widely used for agriculture in most urban and peri-urban cities of developing countries (Scott et al. 2004). Market proximity, high opportunities for income generation, reliable and free irrigation water supply, and minimum artificial fertilizer requirement are the often cited benefits of irrigation within cities (Drechsel et al.

2006; Qadir et al. 2010). However, long-term application of partially treated or untreated wastewater could result in accumulation of heavy metals in the soil (Elgallal et al. 2016). Effluents from household and industries, drainage water, atmospheric deposition, and traffic-related emissions transported with storm water into the sewage and/or irrigation system carry number of pollutants and enrich the urban waste water with heavy metals (Saha et al. 2015; Zia et al. 2016; Woldetsadik et al. 2017). The consumption of food crops grown in wastewater-irrigated areas is one of the principal factor contributing human exposure to

---
* Correspondence: destowol@yahoo.com
[1]School of Natural Resources Management and Environmental Sciences, Haramaya University, PObox: 138, Dire Dawa, Ethiopia
Full list of author information is available at the end of the article

pathogens. In addition, the cultivation of crops for human consumption on wastewater-irrigated soil can potentially lead to the uptake and accumulation of trace metals in the edible plant parts resulting potential risk to human (Rattan et al. 2005; Xue et al. 2012; Ahmad et al. 2016; Zia et al. 2016). Heavy metals are very harmful because of their non-biodggrdable nature, long half-lives and their high bioaccumulation potential (Duruibe et al. 2007; Shah et al. 2012). Several researchers reported that serious health problems may develop as a result of excessive accumulation of heavy metals and even essential trace elements such as Cu and Zn in human body (Oliver 1997; Jarup 2003; Kabata-Pendias and Mukherjee 2007; Luo et al. 2011; Khan et al. 2015).

The increase of 'wastewater irrigation' is however in most cases not farmers' choice (Raschid-Sally and Jayakody 2008). In Africa, the number of people without access to adequate water and sanitation facilities has risen swiftly in recent decades as the continent's rapid urbanization outpaced its capacity to provide the essential water and sanitation services. In Addis Ababa, large volumes of untreated water are released to water bodies which farmers use for irrigation (Weldesilassie et al. 2011a, 2011b). According to Nuttal N. Fast pace of African urbanization affecting water supplies and sanitation. United Nations Environment Program and March 21 (2011)), not only liquid waste provides a challenge, but also solid waste dumped along Addis Ababa main river, near bridges and shores of small tributaries where it is washed into the river. Despite all potential risks, irrigated farming of high value crops is livelihood to many urban residents since it provides employment and income (Weldesilassie et al. 2009). About, 60% of the city's vegetable consumption, particularly leafy vegetables, is supplied by urban farmers who irrigate their crops using polluted river water (Nuttal N. Fast pace of African urbanization affecting water supplies and sanitation. United Nations Environment Program and March 21 2011).

Wastewater irrigation for vegetable production is a highly prevalent practice in the city and a number of articles have been published on wastewater-irrigated soils and vegetables contaminated with heavy metals starting from the 90′s (Itanna 1998, 2002; Alemayehu 2006; Weldegebriel et al. 2012; Aschale et al. 2015; Mekonnen et al. 2015). However, to the best of our knowledge, an insight into assessment of human health risks associated with the consumption of vegetable crops grown on wastewater-irrigated soils is non-existent. It has, for instance, been concluded from the data of heavy metal concentrations in vegetable crops on human health risk without analyzing the pattern for dietary intakes of these metals (Weldegebriel et al. 2012; Aschale et al. 2015). However, information about dietary intake of metals is equally important for assessing their

potential risk to human health. Within this context our study tried to quantify the concentrations of heavy metals in irrigation water, soils and selected vegetables on a representative range of Addis Ababa's urban vegetable farming sites and estimate daily intake and target hazard quotient (THQ) of heavy metals through consumption of these vegetables.

Methods
Study area
This study was conducted in Addis Ababa, Ethiopia, where urban farmers have been practicing vegetable production at various urban farming sites along the Akaki River ('Tinishu' and 'Teleku' Akaki Rivers). The practice has been started in late 1940s. There are two form vegetable production: producers' cooperatives and individual bases. Currently, more than 800 ha of land are irrigated for vegetable production using water from the Akaki River (Weldesilassie and Nigussie 2011). The areas covered are ten prominent vegetable farming sites, locally known as Sore Amba, Lekunda, Peacock-Urael, Peacock-Bole, Kera, Mekanissa, Lafto, Hana-Mariam, Akaki 08, Akaki (Fig. 1) located at five sub-city administrative areas: Kolfe Keraniyo, Chirkos, Bole, Nefas Silk Lafto and Akaki Kaliti, which lies in 038° 41′ E to 038° 47′ E and 08° 52′ N to 9° 02′ N (Woldetsadik et al. 2017). The streams in consideration are highlighted with blue color.

With the exception of Akaki 08 and Akaki farming sites, at all other sites the manual construction of traditional weirs using sand bags and coarse stones is the most common method to block the water flow till it can enter a system of irrigation channels which follow gravity to support farms further downstream. In these farming sites, vegetable crops, mainly leafy vegetable such as lettuce, Ethiopian Kale and swiss chard, are grown using furrow irrigation method, by manually opening and closing furrows constructed within the farms. In addition to furrow irrigation technique, flood irrigation, by which fields are flooded in a controlled manner by manually opening and closing of a bund, is also used at Sore Amba, Lekunda, Peacock- Urael, and Peacock-Bole farming sites. At Akaki 08 and Akaki farming sites, the vast majorities of farmers use diesel motor pumps to extract water directly from the river and transport to farm using connected plastic pipes. Some farmers at Lafto farming sites also follow similar water extraction methods (Woldetsadik et al. 2017). Lettuce, swiss chard and Ethiopian kale were selected for this study since they are the major vegetable crops grown in the study sites.

Sample collection and preparation
At all farming sites, irrigation water was collected at a point where farmers fetch/collect, or where it enters the

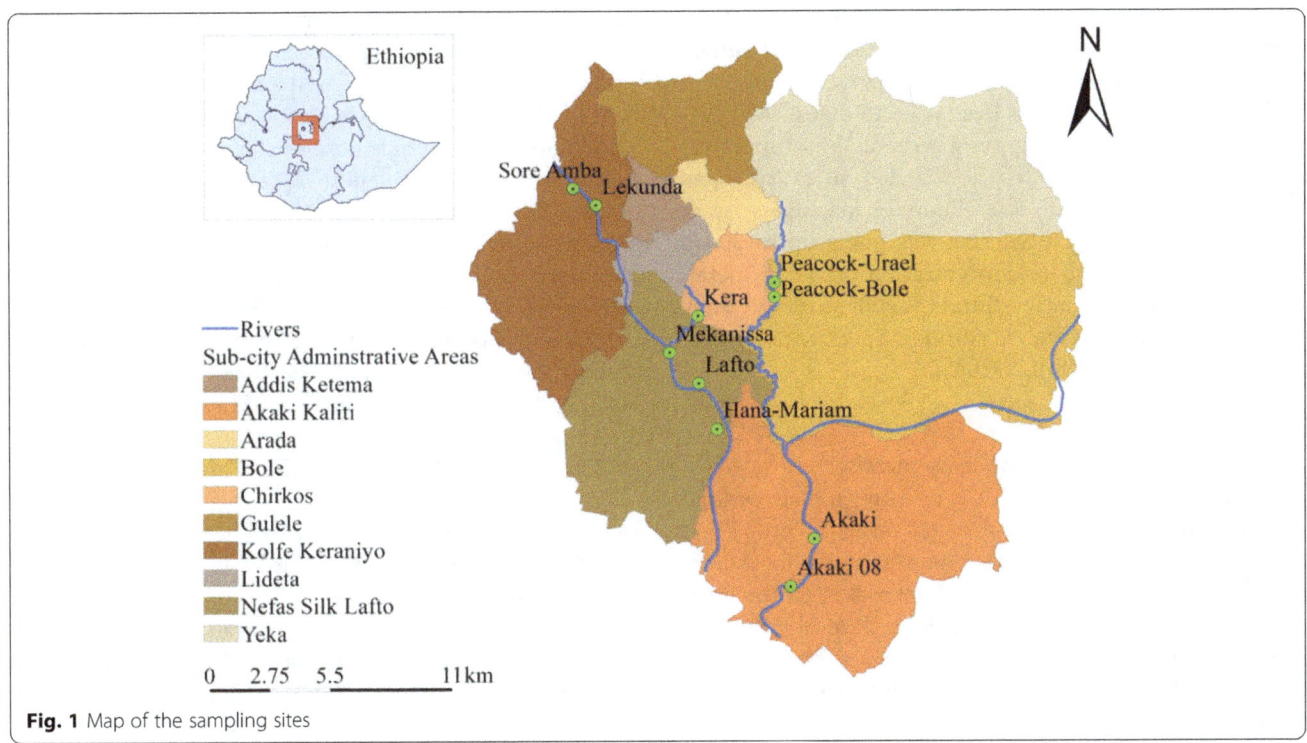

Fig. 1 Map of the sampling sites

farm via canals. From each site, quadruplicate composite samples from 4 different fetching points/inlets to farm were collected in 500 ml plastic bottles, pre-treated with 5 ml of concentrated HNO_3 to prevent microbial degradation of heavy metals, and transported in an icebox to laboratory where they were stored at 4 °C until analysis. So a total of 40 irrigation water samples were collected.

At each farming site, 4 different farmers vegetable farms were selected based on the type of vegetables they grow. From each vegetable farm, 15 surface subsamples (0–20 cm) (3 plots per farm * 5 subsamples from each plot) were collected and made into a single composite sample. So a total of 4 composite soil samples were made per farming site and packed into polyethylene bags and then transported to the laboratory for preparation. The samples were air-dried, passed through a 2 mm sieve and then put into zipped lock polyethylene bags and stored at ambient temperature before further analysis.

Standing vegetable samples (*Lactuca sativa var. crispa, Brassica carinata A. Br.* and *Beta Vulgaris var. cicla*) were also collected from the same vegetable farms where soils were collected. The same sampling technique was followed except only 6 subsamples (for each vegetable type) were used to prepare the composite samples in case of vegetables. A total 120 composite samples (10 farming sites * 3 vegetables * 4 composite samples) were collected, packed into polyethylene bags and transported to the laboratory. Vegetable samples were properly washed with deionized water to remove all visible soil particles. After removing the extra water from the surface of vegetables, the samples were then cut into pieces with a knife. All the samples were then oven-dried at 80 °C for 48 hours. Dried samples were powdered using a pestle and mortar.

Analyses

Fifty ml of water sample was digested with HNO_3 (10 ml) (APHA 2005). After cooling, the digested sample was filtered and the digest was maintained to 50 ml with distilled water. The digest was analyzed for heavy metals with Graphite Furnace Atomic Absorption Spectrophotometer (GFAAS, Thermo Scientific, USA).

Soil particle size distribution was determined by hydrometer method (Gee and Bauder 1986). Soil pH (McLean 1982) was determined from a suspension of 1: 2.5 of soil: water ratio. The cation exchange capacity was determined by leaching method with ammonium acetate solution (1 M NH_4OAc). The organic carbon was determined by dichromate oxidation method and subsequent titration with ferrous ammonium sulphate (Walkley and Black 1934). Soil organic matter (OM) was calculated by multiplying soil organic carbon by 1.724 assuming that average C concentration of organic matter is 58%. For heavy metal analysis of soil, 0.25 g of samples were placed into 50 ml vessels, followed by addition of 10 ml concentrated HNO_3. The mixtures were left to cold digest in a fume cupboard over night and then heated in 1.6 KW microwave oven for 30 min. After cooling to room temperature, 10 ml of double distilled water was added into the vessel and filtered via a 0.45 μm cellulose nitrate filter paper. Finally, the filtrate was subjected to the total element analysis using ICP-OES (Ciros CCD, SPEC-TRO Analytical Instruments GmbH, Kleve, Germany).

Nitric acid and H_2O_2 has been used to digest the vegetable samples, 1 g of ground vegetable sample was digested with 5 ml of nitric acid and 3 ml of hydrogen peroxide. The extract was filtered, insolubles were removed and finally the volume was made up to 50 ml with distilled water. The concentration of heavy metals (Pb, Cd, Cu and Co) in the filtrate was determined using Graphite Furnace Atomic Absorption Spectrophotometer (GFAAS, Thermo Scientific, USA).

Data analysis
Estimated daily intake (EDI) of heavy metals
The estimated daily intake (EDI) of heavy metals was determined based on both the metal levels in crops and the amount of consumption of the respective food crop. The EDI of metals was evaluated according to the average concentrations of each metal in each food crops and the respective daily consumption rate (Zhuang et al. 2009). The EDI of the metals for adults was determined by the following equation:

$$EDI = C_{metal} \times W_{food}/B_w$$

Where C_{metal} is the concentration of heavy metals in vegetable crops; W_{food} represents the daily average consumption of crops in the 5 sub-city administrative areas and B_w is the body weight. A short survey was undertaken to assess vegetable intake patterns of adults and understand how green salads are commonly washed at home. This short survey was carried out in 5 sub-city administrative areas (Kolfe Keraniyo, Chirkos, Bole, Nefas Silk Lafto and Akaki Kaliti). Questionnaire interview were administered to gather information on daily intake pattern of selected leafy vegetables and common washing methods used before serving salad (Woldetsadik et al. 2017). A total of 200 adults were involved in the survey. The minimum and maximum age and body weight record in the questionnaire survey for adults were 18–73 years and 42–84 kg, respectively. Based on the results, the average daily vegetables intakes for adults ranged from 11.9 to 16.3, 27.4 to 36.7 and 22.3 to 37.2 g day^{-1} for *Lactuca sativa var. crispa*, *Brassica carinata A. Br.*and *Beta Vulgaris var. cicla*, respectively. The conversion factor to convert fresh green vegetable weights to dry weights was 0.085 (Zhuang et al. 2009). The metal intakes were compared with the tolerable daily intakes of metals recommended by WHO (1993).

Target hazard quotient (THQ)
The health risks to local inhabitants associated with the intake of Cd, Cu, Ni, Co, Pb, Zn and Cr through the consumptions of wastewater-irrigated vegetables (*Lactuca sativa var. crispa*, *Brassica carinata A. Br.* and *Beta Vulgaris var. cicla*) were based on Target Hazard Quotients (THQs). The THQ is a ratio of determined dose of a pollutant to a reference dose level. If the ratio is less than 1, the exposed population is unlikely to experience obvious adverse effects. Non-carcinogenic health risks for humans associated with the consumption of these vegetables were assessed by calculating THQ. The method to estimate THQ was provided in USEPA Region III Risk--Based Concentration Table (USEPA. Integrated Risk Information System-database. Philadelphia PA, Washington 2007) and in Chien et al. (2002) and Zhuang et al. (2009):

$$THQ = C_n \times I \times 10^{-3} \times Efr \times ED/RfD \times B_w \times AT$$

where C_n represents the mean metal concentration in a specific vegetable on fresh weight basis (mg kg^{-1}); I is ingestion rate (g person^{-1} d^{-1}); EFr is exposure frequency (365 days year^{-1}); ED is exposure duration (70 years); RfD is the oral reference dose (mg kg^{-1} d^{-1});BW is the average body mass, adult (65 kg); AT is averaging time for noncarcinogens (365 days year$^{-1} \times$ number of exposure years).

Statistical analysis
Data of heavy metal concentrations in vegetables were checked for homogeneity of variance and normality. The data of heavy metal concentrations in all analyzed vegetables across the various farming sites were subjected to two-way analysis of variance (ANOVA) to assess the significance of differences in heavy metal concentrations by site, vegetable type and their interaction. Pearson correlation analyses were also carried out to assess the relationships of soil and vegetable metal concentrations. All statistical analyses were computed with SPSS software version 16.

Results and discussion
Heavy metals in irrigation water
Mean concentrations of selected heavy metals in irrigation water samples collected from various urban farming sites of Addis Ababa are given in Table 1. Across the ten sampling sites, the metals concentrations were far below the recommended maximum limit for irrigation water set by FAO (Ayers and Westcot 1985). The mean concentrations of Cd, Co, Cr, Cu, Ni, Pb and Zn were 3.54–58.8, 2.11–13.6, 2.26–6.74, 2.78–29.3, 3.71–33.5, 105–938 and 17.8–48.8 times below the recommended maximum limit. As compared to the concentrations reported in the present study, Aschale et al. (2015)) reported lower mean ranges of Cd (0.04–0.06 µg L^{-1}), Co (2.1–2.7 µg L^{-1}), Cu (3.3–6.6), Ni (3.9–6.5 µg L^{-1}), Pb (1.4–5.1 µg L^{-1}) and Zn (10.9–22.5 µg L^{-1}) but higher mean range of Cr (2.4–255 µg L^{-1}) in irrigation water samples of the same vegetable farming sites. Similarly,

Table 1 Mean heavy metal concentration of irrigation water of Addis Ababa vegetable farming sites

Site	Total concentration ($\mu g\ L^{-1}$)						
	Cd	Co	Cr	Cu	Ni	Pb	Zn
Sora Amba	0.43(0.05)	3.68(0.72)	17.7(2.27)	6.83(1.08)	11.5(2.08)	5.33(0.96)	40.9(6.35)
Lekunda	0.80(0.07)	14.6(1.55)	29.3(3.41)	38.6(3.62)	16.2(2.25)	21.8(2.28)	58.5(5.09)
Peacock-Urael	0.37(0.06)	7.28(0.92)	8.53(1.08)	30.7(3.42)	24.4(2.56)	6.82(0.58)	49.7(4.79)
Peacock-Bole	0.17(0.05)	5.48(0.94)	3.09(0.51)	8.04(1.31)	5.97(0.86)	13.8(3.06)	54.0(5.78)
Kera	2.82(0.62)	16.3(2.37)	44.2(6.38)	71.5(6.73)	29.6(3.55)	47.7(4.98)	88.4(12.9)
Mekanissa	0.81(0.05)	8.73(1.35)	19.8(2.75)	27.7(3.39)	53.9(5.50)	9.48(1.93)	112(16.4)
Lafto	1.48(0.26)	21.6(3.64)	14.2(1.15)	78.3(8.32)	36.5(3.68)	36.9(2.75)	56.9(5.12)
Hana-Mariam	1.05(0.25)	23.7(3.00)	35.1(4.92)	17.7(2.73)	9.48(1.90)	19.4(2.02)	69.6(2.45)
Akaki08	0.57(0.04)	3.38(0.64)	24.4(3.09)	49.2(6.22)	16.2(2.67)	18.6(2.32)	62.8(6.13)
Akaki	0.33(0.04)	5.57(0.47)	14.8(2.69)	36.1(4.10)	8.44(1.76)	16.8(2.00)	44.8(4.03)
RML ($\mu g\ L^{-1}$)	10	50	100	200	200	5000	2000

Figures in parentheses represent standard deviation

RML Recommended Maximum Limit for irrigation water by FAO (Ayers and Westcot 1985)

Alemayehu (2006) reported lower levels of Cd, Co, Cr, Cu, Ni and Zn in Akaki river/irrigation water. As a consequence of very few localized industrial activities and the dilution of wastewater with stream water, low levels of metals in irrigation water samples were recorded. Furthermore, wastewater discharged into the river and irrigation canals are more of domestic origins. Related study in Accra has shown similar phenomena. Unlike the usual trends of observing low levels of metals in irrigation water of Addis Ababa's urban vegetable farming sites (Itanna 1998; Alemayehu 2006; Aschale et al. 2015), Weldegebriel et al. (2012) have reported Cd, Co, Cu, Ni and Zn concentrations as high as 33 $\mu g\ L^{-1}$, 626 $\mu g\ L^{-1}$, 370 $\mu g\ L^{-1}$, 216 $\mu g\ L^{-1}$ and 618 $\mu g\ L^{-1}$, respectively. Despite the low levels of metals in diluted wastewater, continuous use of this water for irrigation could contribute the accumulation of metals into the soil.

Heavy metals in soils

The physico-chemical parameters determined for wastewater irrigated soils in urban vegetable farming sites of Addis Ababa are listed in Table 2. Across the vegetable farming sites, the mean values of soil pH varied from 5.99 to 7.16. The mean organic matter content was highest at Sore Amba (4.6%) followed by Lekunda (4.1%), Mekanissa (3.8%) and lowest in soils from Akaki (2.6%). The CEC value was highest in soils of Akaki 08 (54.5). As compared to other vegetable farming sites, the lowest CEC value (34.9) was exhibited from soils of Peacock-Urael. The CEC results concurred the findings of Weldegebriel et al. (2012). The mean clay content ranged between 34.8 and 60.2%, with the highest and lowest at Sore Amba and Peacock-Urael sites, respectively. The mean Cd, Co, Cr, Cu, Ni, Pb and Zn concentrations in soils from the sampling areas ranged 0.95–3.61, 28.6–

58.6, 55.9–140, 24.2–51.6, 31.5–61.7, 22.1–107 and 119–203, respectively. With the exception of Mean Cr(140 mg kg^{-1}) at Lekunda, Ni(61.7 mg kg^{-1}) at Kera, Cd (3.61 mg kg^{-1}), Pb (107 mg kg^{-1}) and Zn(203 mg kg^{-1}) at Lafto, the mean concentrations of the metals in soils of the studied sites were below the threshold levels for agricultural soils. For sewage-irrigated site (Lafto), greater levels of metals were observed than those sites having no specific application of sewage. But even the upper limits of the metal concentrations (Co, Cr, Cu and Ni) were below the maximum threshold levels. The mean levels of Cd, Co, Cu, Ni, Pb and Zn recorded during the present study were comparable or slightly higher than those reported in previous studies (Itanna 1998, 2002; Weldegebriel et al. 2012; Aschale et al. 2015). In the studied sites, the soils had been irrigated by wastewater for more than 60 years, which showed higher or comparable levels of metals compared to wastewater-irrigated agricultural soils in other African (Mapanda et al. 2005; Lente et al. 2012) and Asian cities (Ahmad et al. 2016; Xue et al. 2012). Conversely, others reported high levels of heavy metals in soils under wastewater cropping system, e.g. in Kolkata city, India (Saha et al. 2015) and Beijing city, China (Liu et al. 2005). High metal levels were also obtained in soils irrigated with wastewater in Harare, Zimbabwe (Muchuweti et al. 2006). However, periodic monitoring of mobile fractions of metals, together with physico-chemical properties of soils and agricultural practices, is required to prevent excessive uptake by vegetable crops.

Heavy metals in vegetables

Concentrations of heavy metals in edible parts of the analyzed vegetables are summarized in Table 3. Mean Cd concentrations were highest in vegetables harvested from Kera and Lafto farming sites, with levels ranging

Table 2 Physicochemical characteristics of soils irrigated with wastewater in urban vegetable farming sites

| | Total concentration (mg kg⁻¹) | | | | | | | | | | Particle size | | |
Site	Cd	Co	Cr	Cu	Ni	Pb	Zn	pH (H₂O)	OM(%)	CEC(cmol$_{(+)}$kg⁻¹)	Sand	Silt	Clay
Sora Amba	0.95(0.17)	28.6(2.14)	94.0(14.5)	24.2(1.57)	31.5(2.62)	22.1(1.87)	119(14.2)	6.51	4.6	42.9	11.8	28	60.2
Lekunda	2.72(0.20)	43.7(2.37)	140(10.3)	30.7(4.91)	60.3(4.10)	36.7(4.40)	150(12.1)	6.59	4.14	39.7	17.2	34.3	48.5
Peacock-Urael	2.58(0.21)	38.8(2.43)	67.4(3.42)	27.4(2.89)	46.2(2.50)	27.8(2.87)	157(6.41)	6.8	3.3	34.9	15	50.2	34.8
Peacock-Bole	1.55(0.10)	37.1(1.83)	55.9(6.52)	28.5(2.35)	46.0(3.99)	25.8(3.60)	120(9.69)	7.16	2.92	39.6	23.8	27	49.2
Kera	2.95(0.42)	58.6(3.74)	76.3(6.74)	49.9(6.20)	61.7(9.15)	81.1(10.9)	160(8.35)	6.11	3.22	43.9	16.7	43.3	40
Mekanissa	2.27(0.31)	38.6(3.58)	61.6(7.32)	43.3(4.41)	48.7(4.75)	29.6(3.59)	145(26.4)	6.57	3.79	39.5	25	31.2	43.8
Lafto	3.61(0.38)	44.9(3.15)	78.0(10.4)	51.6(8.26)	49.1(6.21)	107(10.7)	203(19.5)	5.99	3.62	44.7	16.8	42.5	40.7
Hana-Mariam	1.37(0.20)	28.8(3.27)	56.3(2.52)	38.3(4.92)	39.9(4.85)	33.1(1.88)	130(16.6)	6.63	3.06	40.3	23.2	25	51.8
Akaki08	1.99(0.24)	40.5(3.75)	66.2(5.26)	32.3(4.53)	43.8(5.97)	42.1(1.67)	156(10.2)	7.1	2.88	54.5	15.5	27.3	57.2
Akaki	1.19(0.27)	43.4(2.38)	69.1(8.51)	27.9(1.59)	46.6(3.27)	35.9(5.22)	154(28.1)	6.93	2.6	49.1	17.7	28.5	53.8
RML ᵃ(mg kg⁻¹)	3	50	100	100	50	100	300						

Figures in parentheses represent standard deviation
[a]Source: Ewers 1991

Table 3 Metal concentrations in vegetables grown in wastewater-irrigated urban farming sites

Site	Vegetable	Mean concentration (mg kg^{-1})						
		Cd	Co	Cr	Cu	Ni	Pb	Zn
Sora Amba	*Lactuca sativa var. crispa*	0.54(0.06)	0.42(0.08)	5.28(0.50)	23.9(0.69)	3.49(0.40)	10.5(0.88)	57.7(8.91)
	Brassica carinata A. Br.	0.34(0.03)	0.32(0.03)	4.47(0.71)	12.0(1.26)	3.13(0.50)	7.16(1.03)	80.2(16.6)
	Beta Vulgaris var. cicla	0.39(0.05)	0.54(0.08)	2.85(0.33)	17.0(1.16)	5.94(0.36)	8.63(0.74)	84.8(9.21)
Lekunda	*Lactuca sativa var. crispa*	0.73(0.11)	0.70(0.07)	6.29(0.63)	34.3(1.54)	5.44(0.59)	10.7(0.82)	87.1(10.7)
	Brassica carinata A. Br.	0.45(0.11)	0.51(0.10)	3.11(0.54)	16.0(1.82)	2.87(0.49)	6.52(0.68)	76.8(5.32)
	Beta Vulgaris var. cicla	0.62(0.08)	0.91(0.06)	3.81(0.30)	38.9(1.42)	4.92(0.37)	12.6(1.50)	105(8.26)
Peacock-Urael	*Lactuca sativa var. crispa*	0.56(0.11)	0.48(0.12)	2.38(0.18)	13.7(1.02)	3.28(0.46)	8.88(0.95)	63.4(5.22)
	Brassica carinata A. Br.	0.53(0.20)	0.65(0.12)	1.56(0.14)	14.5(1.24)	2.78(0.42)	6.12(0.83)	89.6(10.2)
	Beta Vulgaris var. cicla	0.76(0.11)	0.76(0.04)	2.36(0.22)	24.3(1.06)	5.24(0.52)	9.79(0.83)	87.0(8.70)
Peacock-Bole	*Lactuca sativa var. crispa*	0.40(0.10)	0.53(0.07)	3.08(0.10)	23.2(1.03)	5.21(0.55)	12.9(0.61)	56.9(3.90)
	Brassica carinata A. Br.	0.35(0.04)	0.52(0.12)	1.17(0.20)	21.4(1.24)	3.01(0.53)	7.90(0.88)	66.3(6.12)
	Beta Vulgaris var. cicla	0.31(0.01)	0.59(0.05)	3.43(0.36)	23.6(1.74)	4.54(0.39)	13.2(0.96)	82.5(7.70)
Kera	*Lactuca sativa var. crispa*	1.59(0.13)	0.81(0.06)	8.01(0.60)	36.2(1.67)	2.78(0.31)	12.7(0.87)	94.4(11.6)
	Brassica carinata A. Br.	0.87(0.12)	0.78(0.07)	4.06(0.68)	21.5(1.54)	2.34(0.42)	8.57(0.63)	105(9.80)
	Beta Vulgaris var. cicla	1.09(0.11)	1.23(0.16)	5.53(0.79)	25.1(1.83)	4.12(0.29)	15.9(0.90)	129(10.2)
Mekanissa	*Lactuca sativa var. crispa*	0.78(0.08)	1.45(0.08)	5.07(0.67)	31.0(8.32)	7.86(0.58)	9.22(1.57)	67.7(4.93)
	Brassica carinata A. Br.	0.71(0.13)	0.63(0.08)	6.32(0.68)	15.5(1.44)	4.00(0.34)	6.74(1.20)	91.3(9.22)
	Beta Vulgaris var. cicla	0.86(0.07)	1.86(0.17)	6.21(0.55)	31.3(3.73)	6.67(0.55)	8.79(1.31)	78.9(11.5)
Lafto	*Lactuca sativa var. crispa*	1.79(0.12)	1.30(0.15)	6.95(0.32)	35.0(1.30)	4.30(0.56)	8.46(1.47)	82.5(10.9)
	Brassica carinata A. Br.	1.17(0.06)	0.71(0.10)	6.57(0.42)	27.8(2.58)	5.19(0.71)	9.50(1.57)	109(11.9)
	Beta Vulgaris var. cicla	1.65(0.09)	1.61(0.07)	7.62(0.48)	37.1(4.08)	7.99(0.84)	13.8(1.37)	117(9.42)
Hana-Mariam	*Lactuca sativa var. crispa*	0.49(0.15)	0.91(0.05)	5.40(0.80)	20.9(1.18)	7.08(0.41)	9.14(1.56)	72.2(7.84)
	Brassica carinata A. Br.	0.44(0.09)	0.78(0.07)	2.09(0.38)	17.4(1.09)	3.78(0.32)	4.14(0.50)	85.5(9.95)
	Beta Vulgaris var. cicla	0.68(0.12)	1.83(0.16)	4.58(0.35)	30.3(1.14)	10.3(0.66)	7.19(1.66)	77.7(11.7)
Akaki08	*Lactuca sativa var. crispa*	0.80(0.17)	0.78(0.08)	2.88(0.28)	24.7(1.36)	5.30(0.39)	9.23(0.65)	67.6(5.01)
	Brassica carinata A. Br.	0.39(0.08)	0.70(0.10)	5.13(0.80)	22.9(1.50)	3.14(0.21)	4.95(0.78)	84.1(6.82)
	Beta Vulgaris var. cicla	0.58(0.09)	1.47(0.10)	3.93(0.39)	44.3(4.44)	6.40(0.47)	8.61(1.80)	98.9(5.81)
Akaki	*Lactuca sativa var. crispa*	1.05(0.14)	0.94(0.12)	5.39(0.62)	19.6(1.83)	4.42(0.92)	6.92(1.18)	79.8(8.87)
	Brassica carinata A. Br.	0.72(0.09)	0.73(0.11)	3.92(0.58)	13.3(1.00)	3.29(0.33)	4.84(0.53)	92.6(9.34)
	Beta Vulgaris var. cicla	0.71(0.11)	1.18(0.14)	3.89(0.15)	34.2(4.23)	5.98(0.68)	11.8(1.04)	87.5(9.46)
RML (mg kg^{-1} dry weight)		2.35[a]	50[b]	27.1[c]	235[d]	800[d]	3.53[a]	588[d]

RML; 0.085 was taken as conversion factor, to convert fresh green vegetable weight to dry weight (Qureshi et al. 2016)
Sources: [a](FAO/WHO-codex alimentarius commission 2001; EC 2006)
[b](Pendias and Pendias 1992)
[c](Weigert 1991)
[d](Mapanda et al. (2007) based on UK and FAO/WHO standards)
Figures in parentheses represent standard deviation

from 0.87 mg kg^{-1} (*Brassica carinata A. Br.*) to 1.79 mg kg^{-1} (*Lactuca sativa var. crispa*) dry weights. But even the highest concentrations did not exceed the RML standard. The high accumulation of Cd in vegetables at the two sites may be attributed to the acidic nature of the soils (Table 2), resulting in greater Cd availability (Kachenko and Singh 2006). This was further supported by the significant correlations ($p = 0.696$–

0.748) of vegetable Cd concentrations with soil Cd. Higher Cd levels which surpassed the recommended maximum limit were reported by Weldegebriel et al. (2012) in vegetables harvested from Kera and Goffa urban vegetable farming sites. Conversely, lower levels of Cd in vegetables at various vegetable farming sites of Addis Ababa were reported by Aschale et al. (2015) and Mekonnen et al. (2015). Cadmium levels exceeding the

RMLs were reported by Mapanda et al. (2007) and Gupta et al. (2010). Similar high level was also found in Radish (Bigdeli and Seilsepour 2008). Results of two way ANOVA test showed that variation in Cd concentrations were significant across farming site, vegetable type and site x vegetable interaction (Table 4). Among the analyzed vegetables, Cd accumulation was significantly high ($p < 0.05$) in *Lactuca sativa var. crispa*. As a result, we emphasize the differences in physiology of metal uptake, exclusion, accumulation, as well as foliage deposition and retention (Zurera et al. 1987; Cui et al. 2004; Zia et al. 2016).

The results obtained in the present study showed that the concentrations of Co in the vegetables were between 0.32 and 1.86 mg kg^{-1} DW (Table 3), the lowest concentration was found in *Brassica carinata A. Br.* and highest in *Beta Vulgaris var. cicla*. The concentrations were substantially lower than its RML. Yet, there were significant differences in Co concentrations in the analyzed vegetable ($p < 0.05$). Among the metals under the consideration of the present study, Co showed minimum in all vegetables next to Cd. Weldegebriel et al. (2012), Aschale et al. (2015) and Mekonnen et al. (2015) have also found lowest concentrations of Co as compared to Cr, Cu, Mn, Ni, Pb and Zn. In the analyzed vegetables, concentrations of Cr were substantially lower than the RML standard. Mean Cr concentrations among vegetable crops was in the order of *Lactuca sativa var. crispa* > *Beta Vulgaris var. cicla* > *Brassica carinata A. Br.*. The observed mean (overall) Cr concentrations were 5.1, 3.84 and 4.42 mg kg^{-1} in *Lactuca sativa var. crispa*, *Brassica carinata A. Br.*, *Beta Vulgaris var. cicla*, respectively. Similar low levels were also obtained in previous studies (Itanna 1998; Liu et al. 2005; Aschale et al. 2015; Mekonnen et al. 2015; Zia et al. 2016;). The two way ANOVA test revealed significant differences by farming site, vegetable type and their interaction (Table 4).

Mean concentrations of Cu in vegetables across the 10 vegetable farming sites were varied and all below the RML standard (Table 3). At Lekunda, the concentrations

ranged from 16.0 (*Brassica carinata A. Br.*) to 38.9 (*Beta Vulgaris var. cicla*) mg kg^{-1}, in Kera farming site they ranged from 21.5 (*Brassica carinata A. Br.*) to 36.2 (*Lactuca sativa var. crispa*) mg kg^{-1} and in Akaki 08 they ranged from 22.9 (*Brassica carinata A. Br.*) to 44.3 (*Beta Vulgaris var. cicla*) mg kg^{-1} dry weights. An earlier study by Weldegebriel et al. (2012) in vegetable farming site around Goffa showed concentration of Cu in *Lactuca sativa* that were twice that of those sampled in the present study. Lower concentrations of Cu in various vegetables were also reported in selected vegetable farming sites in Addis Ababa and its outskirts (Aschale et al. 2015; Mekonnen et al. 2015). Studies in other African cities have shown varied results but all substantially below the RML standard. In Harare, Muchuweti et al. (2006) showed elevated levels of Cu in various crops while Lente et al. (2012) in Accra and Mapanda et al. (2007) in Harare showed lower concentrations. Similarly, concentrations of Ni were substantially lower than the RML standard. When the concentrations of Ni in *Lactuca sativa* grown at wastewater -irrigated sites of Addis Ababa, Ethiopia (Aschale et al. 2015) were compared with the values recorded in the present study, the values of the previous study were 2–4 fold lower. Similar lower results were obtained in Accra (Lente et al. 2012) and Varanasi, India (Ghosh et al. 2012).

Despite the relatively low analyzed Pb concentrations in water and soil, marked differences were observed for Pb accumulation in the leaves of the analyzed vegetables, which exceeded 1.4–3.9 times the RML standard of the respective crops. This confirms the Pb levels of vegetables analyzed previously in Addis Ababa's urban vegetable farming sites (Weldegebriel et al. 2012) and with results reported from other African cities (Muchuweti et al. 2006; Odai et al. 2008; Lente et al. 2012). According to Hamilton et al. (2005), plant roots can adsorb Pb but may not translocate it to shoots, a possibility is that like in Ghana, high Pb levels on wastewater as well as control sites (groundwater irrigated urban farms) are attributable more to vehicular exhaust fumes (Affum et al. 2008) than to irrigation water. In fact, high soil pH, clay and organic matter content are not supporting Pb uptake via roots. On the other hand, Kylander et al. (2003) analyzed in Accra a Pb distribution following traffic density as also shown in other studies (Baye and Hymete 2010; Osma et al. 2013; Teju et al. 2012). Since lead is not biodegradable, and highly immobile, once soil has become contaminated, it remains a long-term source of dust exposure, although lead-free gasoline dominates today's market. This finding showed that the washing of leaves before analysis requires more attention, and has to go beyond the removal of visible dust particles.

Zinc accumulation varied in the vegetables across the 10 farming sites, from 57.7 for *Lactuca sativa var. crispa*

Table 4 Results of two way ANOVA test for heavy metal levels in vegetables harvested from wastewater-irrigated urban vegetable farming sites

Heavy metals	Farming site	Vegetables	Site x vegetable
Cd	138.56[***]	67.17[***]	7.75[***]
Co	107.86[***]	330.23[***]	21.32[***]
Cr	100.90[***]	56.60[***]	19.73[***]
Cu	54.79[***]	250.25[***]	20.44[***]
Ni	66.08[***]	313.59[***]	17.60[***]
Pb	32.05[***]	168.79[***]	7.67[***]
Zn	22.61[***]	59.37[***]	3.36[***]

[***]Level of significance: $p < 0.001$

and 129 mg kg^{-1} for *Beta Vulgaris var. cicla*. The values were 4.55–10.2 times lower than the RML standard. Zinc concentrations in Kera farming site were consistently higher than vegetables sampled over all other farming sites, ranging from 94.4 (*Lactuca sativa var. crispa*) and 129 (*Beta Vulgaris var. cicla*) mg kg^{-1} dry weights. In general, the level of metal accumulation in *Beta Vulgaris var. cicla* was higher than the other vegetables. Briefly, *Beta Vulgaris var. cicla* accumulated significantly ($p < 0.05$) higher levels of Co, Cu, Ni, Pb and Zn, while *Lactuca sativa var. crispa* exhibited significantly ($p < 0.05$) higher levels of Cd and Cr.

Daily intake of metals and target hazard quotient

The estimated daily intakes (EDIs) of metals for adults in wastewater -irrigated vegetable farming sites at 5 sub-city administrative areas via the consumption of leafy vegetable are presented in Table 5. The provisional tolerable daily intakes (PTDIs) for Cd, Cr, Cu, Ni, Pb and Zn are 0.06 mg, 0.2 mg, 3 mg, 0.3 mg, 0.214 mg and 60 mg, respectively (National Research Council 1989). For each individual metal measured in the present study, none of the EDIs exceeded its corresponding PTDIs, nor did approach the doses. The highest EDI of Cd (0.122 µg d^{-1}) through the consumption of the analyzed vegetables was from Nefas Silk Lafto sub-city administrative area. The EDIs of Cd for Selected sub-city administrative areas in Addis Ababa (0.046–0.122 µg d^{-1}) were substantially lower than the values reported for other countries: Tanzania (21.6 µg d^{-1}) (Bahemuka and Mubofu 1999), China (59 µg d^{-1}) (Zhuang et al. 2009), Pakistan (5.29 µg d^{-1}) (Mahmood and Malik 2014) and India (32 µg d^{-1}) (Chopra and Pathak 2015). This discrepancy could be partly attributed to others (Bahemuka and Mubofu 1999; Zhuang et al. 2009; Mahmood and Malik 2014; Chopra and Pathak 2015) analyzing more vegetable types than we did in the present study. The present study showed that the contribution of these vegetables to the daily intake of Cd was less than 0.3% of its corresponding PTDI. It is, however, worth considering the contribution other food groups to Cd or other metals dietary intakes.

The total EDIs of Co ranged from 0.052 to 0.116 µg d^{-1}, much lower than the values estimated in other countries: Ghana 5.3 µg d^{-1} (Lente et al. 2012) and Pakistan 541 µg d^{-1} (Mahmood and Malik 2014). In the present study, the vegetable that contributed the greatest quantity of Co to the intake was *Beta Vulgaris var. cicla*. In Accra, cabbage grown on wastewater-irrigated site contributed 3.14 µg to the daily intake (Lente et al. 2012).

Table 5 Estimated Dietary Intake (EDI) for individual heavy metals caused by the consumption of different vegetables grown on wastewater-irrigated soils at 5 sub-city administrative areas

Administrative areas	Vegetable	Cd	Co	Cr	Cu	Ni	Pb	Zn
Kolfe Keraniyo	*Lactuca sativa var. crispa*	1.44E-05	1.27E-05	1.31E-04	6.59E-04	1.01E-04	2.40E-04	1.64E-03
	Brassica carinata A. Br.	1.56E-05	1.65E-05	1.49E-04	5.50E-04	1.18E-04	2.69E-04	3.08E-03
	Beta Vulgaris var. cicla	1.59E-05	2.28E-05	1.05E-04	8.79E-04	1.71E-04	3.34E-04	2.98E-03
	Total	4.59E-05	5.20E-05	3.85E-04	2.09E-03	3.89E-04	8.43E-04	7.71E-03
Chirkos	*Lactuca sativa var. crispa*	2.84E-05	1.44E-05	1.43E-04	6.46E-04	4.96E-05	2.27E-04	1.68E-03
	Brassica carinata A. Br.	4.26E-05	3.85E-05	2.00E-04	1.06E-03	1.15E-04	4.22E-04	5.19E-03
	Beta Vulgaris var. cicla	4.57E-05	5.15E-05	2.32E-04	1.05E-03	1.72E-04	6.70E-04	5.40E-03
	Total	1.17E-04	1.04E-04	5.74E-04	2.76E-03	3.37E-04	1.32E-03	1.23E-02
Bole	*Lactuca sativa var. crispa*	1.14E-05	1.20E-05	6.48E-05	4.38E-04	1.01E-04	2.58E-04	1.43E-03
	Brassica carinata A. Br.	1.98E-05	2.62E-05	6.13E-05	8.05E-04	1.30E-04	3.15E-04	3.50E-03
	Beta Vulgaris var. cicla	2.34E-05	2.96E-05	1.26E-04	1.04E-03	2.14E-04	5.01E-04	3.70E-03
	Total	5.47E-05	6.77E-05	2.52E-04	2.29E-03	4.44E-04	1.07E-03	8.63E-03
Nefas Silk Lafto	*Lactuca sativa var. crispa*	2.12E-05	2.54E-05	1.21E-04	6.02E-04	1.33E-04	1.86E-04	1.54E-03
	Brassica carinata A. Br.	4.24E-05	3.88E-05	2.75E-04	1.11E-03	2.38E-04	3.73E-04	5.24E-03
	Beta Vulgaris var. cicla	5.81E-05	9.65E-05	3.35E-04	1.80E-03	4.55E-04	5.42E-04	4.99E-03
	Total	1.22E-04	1.61E-04	7.31E-04	3.51E-03	8.26E-04	1.10E-03	1.18E-02
Akaki Kaliti	*Lactuca sativa var. crispa*	1.94E-05	1.80E-05	8.68E-05	4.65E-04	1.02E-04	1.69E-04	1.55E-03
	Brassica carinata A. Br.	2.96E-05	3.82E-05	2.40E-04	9.62E-04	1.71E-04	2.60E-04	4.69E-03
	Beta Vulgaris var. cicla	2.93E-05	5.96E-05	1.76E-04	1.77E-03	2.79E-04	4.59E-04	4.20E-03
	Total	7.82E-05	1.16E-04	5.03E-04	3.20E-03	5.52E-04	8.88E-04	1.04E-02

The total EDIs of Cr, Cu and Ni ranged from 0.252–0.731, 2.09–3.51, and 0.337–0.826 µg d^{-1}, respectively. Similarly, the findings of the present study concerning EDIs of these metals show that the values are substantially lower than their corresponding PTDIs and are free of potential risk. Conversely, other estimates made from other countries have shown that the daily intakes for Cr, Cu and Ni were higher than their corresponding PTDIs (Maleki and Zarasvand 2008; Gupta et al. 2012). Although the concentrations of Pb in all analyzed vegetables were far above the RML standard, the total EDIs (0.094–1.32 µg d^{-1}) were substantially lower than the PTDI standard. The highest total EDI of Pb (1.32 µg d^{-1} at Chirkos) was found to contribute 0.6% to the PTDI. Higher dietary exposure estimate for Pb through the consumption of vegetables grown on wastewater-irrigated fields were reported by Singh et al. (2010) and Mahmood and Malik (2014). Thus, in the context of the present study, intake of these contaminated (Pb) vegetables is unlikely to induce health risks arising from Pb.

Based on the consumption of selected vegetables grown on polluted river water-irrigated vegetable farming sites, the total EDIs of Zn (7.71–11.8 µg d^{-1}) were relatively high as compared to the other metals. But these EDI values contributed less than 0.02% to the corresponding PTDI standard. Yet, it can be clearly observed that our estimates for Zn are far lower than those reported from other countries (Khan et al. 2008; Lente et al. 2012; Mahmood and Malik 2014). Overall, a large daily consumption of these vegetables is unlikely to pose detrimental health risks to the consumer associated with individual metal intake. However, it is worth considering other food crops which may contribute to metal exposure and further studies are required to completely understand the risk involved.

The health risk associated with the consumption of selected leafy vegetables grown on wastewater -irrigated vegetable farming sites was evaluated using Target Hazard Quotient (THQ). The THQ has been recognized as a useful parameter for the evaluation of risk associated with the consumption of contaminated (metals) food crops (Zheng et al. 2007; Zhuang et al. 2009). Target Hazard Quotient value of less than 1 indicates a relative absence of health risk associated with the consumption of metal contaminated food crops (USEPA 2007). The THQ values ranged from 0.042–0.108, 0.005–0.014, 0.0002–0.0004, 0.048–0.078, 0.015–0.037, 0.194–0.298 and 0.026–0.037 for Cd, Co, Cr, Cu, Ni, Pb and Zn, respectively (Table 6). From the above data, it is apparent that the consumption of the examined vegetables do not expose local inhabitants to a potential health risk from dietary Cd, Co, Cr, Cu, Ni, Pb and Zn. The results obtained in the present study did not concur with THQ

values recorded by Zheng et al. (2007), Zhuang et al. (2009) and Hu et al. (2014). Among the metals THQ values, the greatest values were obtained for Pb for the consumption of wastewater-irrigated vegetables at 5 sub-city administrative areas and were in the order: Chirkos (0.298) > Nefas Silk Lafto (0.244) > Bole (0.243) > Akaki Kaliti (0.197) > Kolfe Keraniyo (0.194).

The total metal THQ (sum of individual metal THQ for the analyzed vegetables) is shown in Fig. 2. The TTHQs of the metals ranged from 0.33 to 0.53. Comparing sub-city administrative areas, the TTHQs of the metals decrease in the order of Chirkos > Nefas Silk Lafto > Akaki Kaliti > Bole > Kolfe Keraniyo. The present result indicate that Pb and Cd were the major component contributing to the TTHQs, in agreement with separate assessments for areas near Huludao Zinc plant in Huludao, China (Zheng et al. 2007) and in the vicinity of Dabaoshan mine in Shaoguan city, China (Zhuang et al. 2009). In the studied sites, the consumption of all analyzed vegetables resulted in TTHQ values of less than 1, compared to the high TTHQ values obtained from emerging economies (Abbasi et al. 2013; Qureshi et al. 2016). Compared with our previous study (Woldetsadik et al. 2017) which focused on microbial hazards of wastewater irrigation, it is clear that heavy metals pose relatively no risk to local inhabitants through the consumption of leafy vegetables grown on wastewater- irrigated vegetable farming sites. However, it is worth considering the effects that may result from the interaction of the metals.

Conclusions

From this study, it was evident that the concentrations of metals in irrigation water and soil were lower than the RML standards. Wastewater dilution may be the important reason for lower levels of metals in irrigation water. Based on 1:100 dilution, the process is predicted to bring 3 mg kg^{-1} metal level down to just 0.03 mg kg^{-1}, as compared to only 2 log units reduction for pathogens, and still above thresholds, indicating metals discharged to streams will dissipate by dilution and incorporate into sediments. Hence, a more differentiated view is required as the readers might associate the term wastewater with raw effluent. Significant variations in metal concentrations between the analyzed vegetables reflect the difference in their uptake capabilities. With the exception of Pb, the concentrations of the other metals in all analyzed vegetables were far below the various international RML standards. From the health point of view, the EDI and THQ values showed that there would be no potential health risk to local inhabitants due to intake of individual metal if one or more of the analyzed vegetables are consumed. Furthermore, hazard quotients for the combined metals (TTHQ) due to all analyzed vegetables

Table 6 THQ for individual heavy metals through the consumption of different vegetables grown on wastewater-irrigated soils at 5 sub-city administrative areas

Administrative areas	Vegetable	Cd	Co	Cr	Cu	Ni	Pb	Zn
Kolfe Keraniyo	*Lactuca sativa var. crispa*	1.33E-02	1.17E-03	8.07E-05	1.52E-02	4.67E-03	5.54E-02	5.05E-03
	Brassica carinata A. Br.	1.43E-02	1.50E-03	9.05E-05	1.25E-02	5.37E-03	6.13E-02	9.38E-03
	Beta Vulgaris var. cicla	1.48E-02	2.12E-03	6.49E-05	2.04E-02	7.93E-03	7.76E-02	9.25E-03
	Total	4.23E-02	4.80E-03	2.36E-04	4.82E-02	1.80E-02	1.94E-01	2.37E-02
Chirkos	*Lactuca sativa var. crispa*	2.48E-02	1.26E-03	8.33E-05	1.41E-02	2.17E-03	4.95E-02	4.91E-03
	Brassica carinata A. Br.	3.86E-02	3.48E-03	1.21E-04	2.39E-02	5.23E-03	9.55E-02	1.57E-02
	Beta Vulgaris var. cicla	4.18E-02	4.72E-03	1.41E-04	2.41E-02	7.89E-03	1.53E-01	1.65E-02
	Total	1.05E-01	9.46E-03	3.45E-04	6.21E-02	1.53E-02	2.98E-01	3.71E-02
Bole	*Lactuca sativa var. crispa*	1.03E-02	1.08E-03	3.89E-05	9.86E-03	4.54E-03	5.82E-02	4.29E-03
	Brassica carinata A. Br.	1.82E-02	2.40E-03	3.75E-05	1.85E-02	5.96E-03	7.23E-02	1.07E-02
	Beta Vulgaris var. cicla	2.11E-02	2.67E-03	7.59E-05	2.35E-02	9.63E-03	1.13E-01	1.11E-02
	Total	4.96E-02	6.15E-03	1.52E-04	5.19E-02	2.01E-02	2.43E-01	2.61E-02
Nefas Silk Lafto	*Lactuca sativa var. crispa*	1.87E-02	2.23E-03	7.08E-05	1.32E-02	5.86E-03	4.09E-02	4.52E-03
	Brassica carinata A. Br.	3.72E-02	3.41E-03	1.61E-04	2.44E-02	1.04E-02	8.18E-02	1.53E-02
	Beta Vulgaris var. cicla	5.18E-02	8.60E-03	1.99E-04	4.00E-02	2.03E-02	1.21E-01	1.48E-02
	Total	1.08E-01	1.42E-02	4.31E-04	7.77E-02	3.65E-02	2.44E-01	3.47E-02
Akaki Kaliti	*Lactuca sativa var. crispa*	1.67E-02	1.55E-03	4.98E-05	1.00E-02	4.39E-03	3.65E-02	4.44E-03
	Brassica carinata A. Br.	2.63E-02	3.40E-03	1.43E-04	2.14E-02	7.60E-03	5.78E-02	1.39E-02
	Beta Vulgaris var. cicla	2.61E-02	5.32E-03	1.05E-04	3.95E-02	1.25E-02	1.02E-01	1.25E-02
	Total	6.91E-02	1.03E-02	2.97E-04	7.09E-02	2.44E-02	1.97E-01	3.09E-02

were lower than 1, which signifies no potential health risk even to highly exposed inhabitants. These results emphasize the need for further investigations of other crops from the study sites. Still, health risk exposure of children through the consumption of local vegetables should also be investigated due to their high sensitivity to metal exposure. There is a great respite that toxic metals like Pb and Cd have not pose potential health risk even after long term (more than 50 years) use of this water for irrigation. Our previous study indicated that faecal contamination level of lettuce irrigated with wastewater is above the threshold of safe consumption. Hence, it is imperative to focus on and off farm mitigation measures including proper vegetable washing that helps reduce potential pathogenic risks. However, intermittent monitoring of the metals from irrigation water, in soil and crops may be required to follow/prevent their build-up in the food chain.

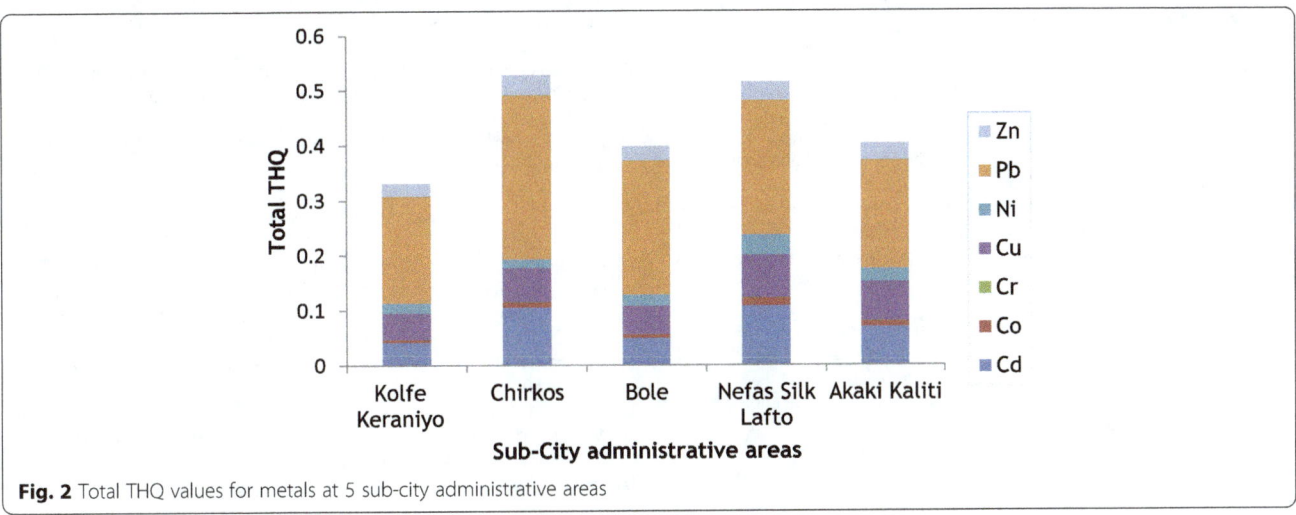

Fig. 2 Total THQ values for metals at 5 sub-city administrative areas

Abbreviations

CEC: Cation exchange capacity; EDI: Estimated daily intake; RML: Recommended maximum limit; OM: Organic matter; THQ: Target hazard quotient; TTHQ: Total target hazard quotient

Acknowledgements

This work was supported by the International Water Management Institute (IWMI-CGIAR), Blacksmith Institute (Pure Earth) and the Ministry of Education of Ethiopia. We are grateful to the staff of soil chemistry laboratory at Debre Zeit Agricultural Research Center and to the staff of soil laboratory in Bochum for the laboratory assistance. We also wish to acknowledge the field work support of development agents at various sub-city administrative areas of Addis Ababa.

Funding

This work was funded by the International Water Management Institute (IWMI-CGIAR), Blacksmith Institute (Pure Earth) and the Ministry of Education of Ethiopia. The International Water Management Institute (IWMI-CGIAR), Blacksmith Institute (Pure Earth) and the Ministry of Education of Ethiopia had no role in the design, data collection, analysis or publication of the manuscript.

Authors' contributions

DW, PD, BK, FI and HG conceived and designed the study. DW conducted the study. DW, PD, BK and FI contributed to the analysis and interpretation of data. DW drafted the manuscript. DW, PD, BK, FI and HG revised the draft manuscript. All authors read and approved the final manuscript.

Competing interests

The authors declare that they have no competing interests.

Author details

[1]School of Natural Resources Management and Environmental Sciences, Haramaya University, PObox: 138, Dire Dawa, Ethiopia. [2]International Water Management Institute, Colombo, Sri Lanka. [3]Department of Global Health, University of Copenhagen, Copenhagen, Denmark. [4]Department of Crop Science, University of Namibia, Windhoek, Namibia.

References

Abbasi MA, Iqbal J, Khan MA, Shah MH. Health risk assessment and multivariate apportionment of trace metals in wild leafy vegetables from lesser Himalayas. Pakistan Ecotoxicol Environ Saf. 2013;92:237–44.

Affum HA, Oduro-Afriyie K, Nartey VK, Adomako D, Nyarko BJB. Biomonitoring of airborne heavy metals along a major road in Accra. Ghana Environ Monit Assess. 2008;137:15–24.

Ahmad K, Ashfaq A, Khan ZI, Ashraf M, Akram NA, Yasmin S, Batool AI, Sher M, Shad HA, Khan A, Rehman SU, Ullah MF, Noorka IR. Health risk assessment of heavy metals and metalloids via dietary intake of a potential vegetable (Coriandrum sativum L.) grown in contaminated water irrigated agricultural sites of Sargodha, Pakistan. Human Ecol Risk Assess. 2016;22:597–610.

Alemayehu T. Heavy metal concentration in the urban Environment of Addis Ababa, Ethiopia. Soil Sediment Contam. 2006;15:1–12.

APHA. Standard Methods for the Examination of Water and Wastewater, 21st Ed. Washington, DC: 2005.

Aschale M, Sileshi Y, Kelly-Quinn M, Hailu D. Assessment of Potentially Toxic Elements in Vegetables Grown along Akaki River in Addis Ababa and Potential Health Implications. Food Sci Q Manage. 2015;40:42–52.

Ayers RS, Westcot DW. Water quality for agriculture, FAO Irrigation and Drainage Paper 29. Rome: Food and Agriculture and Organization; 1985.

Bahemuka TE, Mubofu EB. Heavy metals in edible green vegetables grown along the sites of the Sinza and Msimbazi Rivers in Darselam. Tanzania Food Chem. 1999;66:63–6.

Baye H, Hymete A. Lead and Cadmium accumulation in medicinal plants collected from environmentally different sites. Bull Environ Contam Toxicol. 2010;84:197–201.

Bigdeli M, Seilsepour M. Investigation of metals accumulation in some vegetables irrigated with waste water in Shahre Rey-Iran and toxicological implications. Am-Euras J Agric Environ Sci. 2008;4:86–92.

Chien LC, Hung TC, Choang KY, Choang KY, Yeh CY, Meng PJ, Shieh MJ, Han BC. Daily intake of TBT, Cu, Zn, Cd and As for fishermen in Taiwan. Sci Total Environ. 2002;285:177–85.

Chopra AK, Pathak C. Accumulation of heavy metals in the vegetables grown in wastewater irrigated areas of Dehradun, India with reference to human health risk. Environ Monit Assess. 2015;187(7):1–8.

Codex Alimentarius Commission. In report of the 33rd session of the Codex Committee on Food Additives and Contaminants, Jt. FAO/WHO Food Stand. Program. ALINORM 01/12A, 2001;1–289.

Cui YJ, Zhu YG, Zhai RH, Chen DY, Huang YZ, Qui Y, Liang JZ. Transfer of metals from near a smelter in Nanning. China Environ Int. 2004;30:785–91.

Drechsel P, Graefe S, Sonou M, Cofie O. Informal irrigation in West Africa: An overview. IWMI Research Report No. 102. International Water Management Institute: Colombo, Sri Lanka; 2006.

Duruibe JO, Ogwuegbu MOC, Egwurugwu JN. Heavy metal pollution and human biotoxic effects. Int J Phys Sci. 2007;2(5):112–18.

Elgallal M, Fletcher L, Evans B. Assessment of potential risks associated with chemicals in wastewater used for irrigation in arid and semiarid zones: A review. Agr Water Manage. 2016;177:419–31.

European Commission. Standards for lead, cadmium and mercury. Commission Regulation (EC) No 1881/2006. Accessed on 23 July 2015: 2006 http://eurlex.europa.eu/LexUriServ/LexUriServ.do?uri = CONSLEG: 2006R1881:20100701:EN:PDF

Ewers U. Standards, guidelines and legislative regulations concerning metals and their compounds. In: Merian E, editor. Metals and their compounds in the environment, relevance. Weinheim: VCH; 1991. p. 458–68.

Gee GW, Bauder JW. Particle-size analysis. In: Klute A, editor. Methods of Soil Analysis. Part 1: Physical and Mineralogical Methods. Monograph no 9. Madison: American Society of Agronomy Inc., and Soil Science Society of America Inc; 1986. p. 383–411.

Ghosh AK, Bhatt M, Agrawal H. Effect of longterm application of treated sewage water on heavy metal accumulation in vegetables grown in Northern India. Environ Monit Assess. 2012;184:1025–36.

Gupta S, Satpati S, Nayek S, Garai D. Effect of wastewater irrigation on vegetables in relation to bioaccumulation of heavy metals and biochemical changes. Environ Monit Assess. 2010;165:169–77.

Gupta N, Khan DK, Santra SC. Heavy metal accumulation in vegetables grown in a long-term wastewater-irrigated agricultural land of tropical India. Environ Monit Assess. 2012;184:6673–82.

Hamilton AJ, Boland AM, Stevens D, Kelly J, Radcliffe J, Ziehrl A, Dillon PJ, Paulin R. Position of the Australian horticultural industry with respect to the use of reclaimed water. Agr Water Manag. 2005;71:181–209.

Hu WY, Chen Y, Huang B, Niedermann S. Health risk assessment of heavy metals in soils and vegetables from a typical greenhouse vegetable production system in China. Hum Ecol Risk Assess. 2014;20:1264–80.

Itanna F. Metal concentrations of some vegetables irrigated with industrial liquid waste at Akaki. Ethiopia SINET: Ethiop J Sci. 1998;21:133–44.

Itanna F. Metals in leafy vegetables grown in Addis Ababa and toxicological implications. Ethiop J Health Dev. 2002;16:295–302.

Jarup L. Hazards of heavy metal contamination. Br Med Bull. 2003;68:167–82.

Kabata-Pendias A, Mukherjee AB. Trace elements from soil to human. NewYork: Springer; 2007.

Kachenko AG, Singh B. Heavy metals contamination in vegetables grown in urban and metal smelter contaminated sites in Australia. Water, Air, and Soil Pollut. 2006;169:101–23.

Khan S, Cao Q, Zheng YM, Huang YZ, Zhu YG. Health risks of heavy metals in contaminated soils and food crops irrigated with wastewater in Beijing. China Environ Pollut. 2008;152:686–92.

Khan MU, Muhammad S, Malik RN, Khan SA, Tariq M. Heavy metals potential health risk assessment through consumption of wastewater irrigated wild plants: a case study. Hum Ecol Risk Assess. 2015;22:141–52.

Kylander ME, Rauch S, Morrison GM, Andam K. Impact of automobile emissions on the levels of platinum and lead in Accra. Ghana J Environ Monit. 2003;5:91–5.

Lente I, Keraita B, Drechsel P, Ofosu-Anim J, Brimah AK. Risk assessment of heavy-metal contamination on vegetables grown in long-term wastewater irrigated urban farming sites in Accra. Ghana Water Qual Expos Health. 2012;4:179–86.

Liu WH, Zhao JZ, Ouyang ZY, Söderlund L, Liu GH. Impacts of sewage irrigation on heavy metal distribution and contamination in Beijing. China Environ Int. 2005;31:805–12.

Luo CL, Liu CP, Wang Y, Liu X, Li FB, Zhang G, Li XD. Heavy metal contamination in soils and vegetables near an e-waste processing site, south China. J Hazard Mater. 2011;186:481–90.

Mahmood A, Malik RN. Human health risk assessment of heavy metals via consumption of contaminated vegetables collected from different irrigation sources in Lahore. Pakistan Arabian J Chem. 2014;7:91–9.

Maleki A, Zarasvand MA. Heavy metals in selected edible vegetables and estimation of their daily intake in Sanandaj, Iran. Southeast Asian J Trop Med Public Health. 2008;39(2):335–40.

Mapanda F, Mangwayana EN, Nyamangara J, Giller KE. The effect of long term irrigation using wastewater on heavy metal contents of soils under vegetables in Harare. Zimbabwe Agric Ecosyst Environ. 2005;107:151–65.

Mapanda F, Mangwayana EN, Nyamangara J, Giller KE. Uptake of heavy metals by vegetables irrigated using wastewater and the subsequent risks in Harare. Zimbabwe Phys Chem Earth Parts A/B/C. 2007;32(15–18):1399–405.

Mclean EO. Soil pH and lime requirement. In: Page AL, Miller RH, Keeney DR, editors. Methods of soil analysis, Part 2. Chemical and microbiological properties. Madison: American Society of Agronomy; 1982. p. 199–223.

Mekonnen KN, Ambushe AA, Chandravanshi BS, Abshiro MR, McCrindle RI. Assessment of potentially toxic elements in Swiss chard and sediments of Akaki River. Ethiopia Toxico Environ Chem. 2015;96(10):1501–15.

Muchuweti M, Birkett JW, Chinyanga E, Zvauya R, Scrimshaw MD, Lester JN. Heavy metal content of vegetables irrigated with mixtures of wastewater and sewage sludge in Zimbabwe: implications for human health. Agric Ecosyst Environ. 2006;112:41–8.

National Research Council. RDA. Recommended dietary allowances. 10th ed. Washington: National Academy Press; 1989.

Nuttal N. Fast pace of African urbanization affecting water supplies and sanitation. United Nations Environment Program, March 21, 2011. Accessed 10 Oct 2015. http://www.unep.org/Documents.Multilingual/Default.Print.asp?DocumentID = 664&ArticleID = 8666&l = en.

Odai SN, Mensah E, Sipitey D, Ryo S, Awuah E. Heavy metals uptake by vegetables cultivated on urban waste dumpsites: case study of Kumasi. Ghana Res J Environ Toxicol. 2008;2:92–9.

Oliver MA. Soil and human health: A review. Eur J Soil Sci. 1997;48:573–92.

Osma E, Serin M, Leblebici Z, Aksoy A. Assessment of heavy metal accumulations (Cd, Cr, Cu, Ni, Pb and Zn) in some vegetables and soils in Istanbul Turkey. Pol J Environ Stud. 2013;22:1449–55.

Pendias AK, Pendias H. Elements of Group VIII. In: Trace Elements in Soils and Plants. CRC Press: Boca Raton; 1992. p. 271–76.

Qadir M, Wichelns D, Raschid-Sally L, McCornick PG, Drechsel P, Bahri A, Minhas PS. The challenges of wastewater irrigation in developing countries. Agr Water Manag. 2010;97(4):561–68.

Qureshi AS, Hussain MI, Ismail S, Khan QM. Evaluating heavy metal accumulation and potential health risks in vegetables irrigated with treated wastewater. Chemosphere. 2016;163:54–61.

Raschid-Sally L, Jayakody P. Drivers and Characteristics of Wastewater Agricutlure in Developing Countries: Results from a Global Assessment. IWMI Research Report No. 127. Colombo: International Water Management Institute; 2008.

Rattan RK, Datta SP, Chhonkar PK, Suribabu K, Singh AK. Long-term impact of irrigation with sewage effluents on heavy metal content in soils, crops and groundwater-a case study. Agric Ecosyst Environ. 2005;109:310–22.

Saha S, Hazra GC, Saha B, Mandal B. Assessment of heavy metals contamination in different crops grown in longterm sewageirrigated areas of Kolkata, West Bengal. India Environ Monit Assess. 2015;187:4087–99.

Scott CA, Faruqui NI, Raschid-Sally L. Wastewater use in irrigated agriculture: Management challenges in developing countries. In: Scott CA, Faruqui NI, Raschid- Sally L, editors. Wastewater Use in Irrigated Agriculture: Confronting the Livelihood and Environmental Realities. Wallingford: CABI Publishing; 2004. p. 1–10.

Shah MT, Ara J, Muhammad S, Khan S, Tariq S. Health risk assessment via surface water and sub-surface water consumption in the mafic and ultramafic terrain, Mohmand agency, northern Pakistan. J Geochem Explo. 2012;118:60–7.

Singh A, Sharma RK, Agrawal M, Marshall FM. Health risk assessment of heavy metals via dietary intake of foodstuffs from the wastewater irrigated site of a dry tropical area of India. Food Chem Toxicol. 2010;48:611–19.

Teju E, Megerssa N, Chandravanshi BS, Zewge F. Determination of the level of lead in the roadside soils of Addis Ababa. Ethiopia SINET Ethiop J Sci. 2012;35(2):81–94.

USEPA. Integrated Risk Information System-database. Philadelphia PA; Washington: 2007.

Walkley A, Black IA. An examination of the Degtjareff method for determining soil organic matter, and a proposed modification of the chromic acid titration method. Soil Sci. 1934;37:29–38.

Weigert P. Marian E. In: Metals and their compounds in the environment, occurrence, analysis and biological relevance. Weinheim: VCH; 1991. p. 458–68.

Weldegebriel Y, Chandravanshi BS, Wondimu T. Concentration levels of metals in vegetables grown in soils irrigated with river water in Addis Ababa. Ethiopia Ecotoxicol Environ Saf. 2012;77:57–63.

Weldesilassie A, Nigussie M. Vegetable Production, Marketing and Quality of Irrigation Water and Vegetable Produced within and around Addis Ababa city, Ethiopia, Unpublished report. 2011.

Weldesilassie A, Frör O, Boelee E, Dabbert S. The economic value of improved wastewater irrigation: a contingent valuation study in Addis Ababa. Ethiopia J Agric Res Econ. 2009;34(3):428–49.

Weldesilassie A, Amerasinghe P, Danso G. Assessing the Empirical Challenges of Evaluating the Benefits and Risks of Irrigating with Wastewater. Water Int. 2011a;36(4):441–54.

Weldesilassie A, Boelee E, Drechsel P, Dabbert S. Wastewater use in crop production in periurban areas of Addis Ababa: impacts on health in farm households. Environ Dev Econ. 2011b;16(1):25–49.

WHO. Evaluation of certain food additives and contaminants, 41st Report of the Joint FAO/WHO Expert Committee on Food Additives. Geneva: World Health Organization; 1993. Technical Report Series.

Woldetsadik D, Drechsel P, Keraita B, Itanna F, Erko B, Gebrekidan H. Microbiological quality of lettuce (Lactuca sativa) irrigated with wastewater in Addis Ababa, Ethiopia and effect of green salads washing methods. Int J Food Contam. 2017;4:3 doi:10.1186/s40550-017-0048-8.

Xue ZJ, Liu SQ, Liu YL, Yan YL. Health risk assessment of heavy metals for edible parts of vegetables grown in sewage-irrigated soils in suburbs of Baoding City. China Environ Monit Assess. 2012;184:3503–13.

Zheng N, Wang Q, Zhang X, Zheng D, Zhang Z, Znang S. Population health risk due to dietary intake of heavy metals in the industrial area of Huludao City, China. Sci Total Environ. 2007;387:96–104.

Zhuang P, Zou B, Li NY, Li ZA. Heavy metal contamination in soils and food crops around Dabaoshan mine in Guangdong, China: implication for human health. Environ Geochem Health. 2009;31:707–15.

Zia MH, Watts MJ, Niaz A, Middleton DR, Kim AW. Health risk assessment of potentially harmful elements and dietary minerals from vegetables irrigated with untreated wastewater. Pakistan Environ Geochem Health. 2016;38:1–22.

Zurera G, Estrada B, Rincon F. Pozo R (1987) Lead and cadmium contamination levels in edible vegetables. B Environ Contam Toxicol. 1987;38(5):805–12.

20

Microbial profile of common spices and spice blends used in Tamale, Ghana

Noel Bakobie[1], Amponsah Samuel Addae[1], Abudu Ballu Duwiejuah[1,2*], Samuel Jerry Cobbina[1] and Solomon Miniyila[3]

Abstract

Background: The main purpose of using spice to grill meat is to add aroma, colour, flavour, taste and pungency. However, the purpose is sometime befitted when spice is contaminated with pathogenic bacteria that result in foodborne illnesses and toxicological effect.

Results: The study was necessitated by paucity information on handling practices and microbial load common spices used for grilling meat, Ghana. A total of twenty spice samples were collected from five popular and widely patronised joints in the Tamale in Ghana. Detection and identification of potential pathogens was carried out following standard procedures. *E. coli* count ranged from 0 to 3.14 $\log_{10}$ cfu/ ml with a mean of 1.17 ± 1.07 $\log_{10}$ cfu/ ml. Contamination level for *Salmonella* spp ranged between 0 and 0.9 $\log_{10}$ cfu/ ml with a mean of 0.38 ± 0.31 $\log_{10}$ cfu/ ml. Coliform bacteria were present in almost all the spices sampled. Faecal coliform and *E. coli* presence was an indication of contamination by fresh faecal matter. The possible sources of spice contamination include storage equipment, handling, unhygienic surroundings, vehicular transmission, atmospheric particles and air-microbes.

Conclusions: There is a possible risk to public health associated with consumption of spicy meat from the selected joints. There is the need to maintain good sanitary practice and hygienic quality during production stages of spice in order to avoid or reduce prevalence of food borne illnesses in Tamale and Ghana as a whole.

Keywords: Contamination, foodborne illness, pathogenic bacteria, spice, Ghana

Background

Street food is sold in every nook and cranny of major towns and cities in both developed and developing countries. Street foods are mostly prepared in open-air public spaces either stationary or itinerant, either on foot or from mobile outlets, removable outlets, and fixed outlets without enclosed space by food vendors to accommodate consumers (Food and Agriculture Organization of the United Nations 2016). Street food vending and consumption have recently flourished for some decades in Africa largely due to increase in urban population and wide spread of urban boundaries and urban sprawl (Food and Agriculture Organization of the United Nations 2016).

Food safety is a concern for the general public as almost everybody for the past year has ever suffered from foodborne illnesses at least once (WHO 2015). Diarrhoea has been one of the major causes of hospital attendance in Ghana, and 16% deaths of children younger than five years in African (Bruce et al. 2005). Food-borne disease outbreaks due to importation were reported to have increased between 2009 and 2010 in which fish and spice were the most common sources (Centre for Disease Control CDC 2012).

Most of these street foods are sold to consumers with spices that are traditionally or locally prepared. Spices are group of food additives utilised to enhance sensory quality of some foods by making those foods more palatable (Debs-Louka et al. 2013). Sodium and fat content is very low in spice causing the increased in their consumption (Srinivasan 2005). However, conditions under which they are produced and basic operations such as drying, harvesting, threshing, transport and storage are potential sources of contamination with bacteria, fungi and insects (Alam-Khan and Abrahem 2010; Sádecká 2007). This can lead to an increase in certain

* Correspondence: abalu096@gmail.com
[1]Department of Ecotourism and Environmental Management, Faculty of Natural Resources and Environment, University for Development Studies, Tamale, Ghana
[2]Department of Biotechnology, Faculty of Agriculture, University for Development Studies, Tamale, Ghana
Full list of author information is available at the end of the article

foodborne illnesses and intoxications (Buckenhuskes and Rendlen 2004) and food spoilage (Ahene et al. 2011).

Spice has low moisture content making it safe, but when in contact with some food products that has high water, microbial populations could be stimulated because of increase in water activity (Menlove and Sainsbury 2002). The interventions in production process are mostly based on reduction or elimination of some contamination by means of heat treatment (Zwietering et al. 2016).

Generally, in Ghana, mix-powdered spice is usually used in grilling meat such as chevon, chicken, guinea fowl, suya (dried smoke meat), pork and many more which are processed manually and sold to general public at various lorry terminals, by roadside or itinerant vendors (Mensah et al. 2002). A study by Addo (2005) reported fungal count of $3.08 \log_{10}$ cfu/g to $2.40 \log_{10}$ cfu/g in the ginger samples. Also, the total aerobic bacteria count in the spice ranged from $3.6 \log_{10}$ cfu/g in ginger sample to $3.7 \log_{10}$ cfu/g sample (in mixture of ginger and garlic) bacteria species which collectively contaminated the spice were *Enterobacter* spp, *Aeromonas salmonicida* and *Salmonella* spp. Sagoo et al. (2009) and Shamsuddeen (2009) also reported potential public health risk of spices and herbs after isolating high counts of *Bacillus cereus*, *Clostridium perfrigens* and *Escherichia coli*. Spices are usually sold as loose or packed in the local markets and consumed in cities, towns and countryside of Ghana. The study was necessitated by paucity information on handling practices and microbial load on common spices used for grilling meat, Ghana.

Methods

Study area

The research was carried out in the central part of Tamale. Tamale is the capital city of the Northern Region of Ghana and lies in latitude $9° 15'$ and $9° 05'$ N and in longitude $0° 45'$ and $1° 0'$ W at an altitude of 185 m above sea level. It is third most populous settlement in Ghana with 537,986 inhabitants according to 2010 census (Ghana Statistical Service 2012). As a fast growing city, a lot of people migrate to Tamale to do businesses (Abankwa et al. 2009). It is a home to both locals and internationals.

Sample collection and preparation

A total of twenty spice samples were collected from five popular and widely patronised selected sites in Tamale. Four (4) samples were collected from each vendor on weekly basis for a period of one month (February to March, 2014). A small plastic container cleaned with methylated spirit was used to collect samples from the vendors. The samples were stored in a sterile box containing ice cubes and taken to the CSIR- Water Research Institute laboratory (Tamale) for analysis. Strict standard practices and aseptic conditions were observed to avoid possible contamination during sample collection and transportation, prior to analysis. Vendors of

spicy meat were interviewed to ascertain possible sources of bacteria contaminant.

Using a sterilise spatula and weighing boat, five-grams of the spice samples were transferred into a sterile falcon tube with 45 ml of sterile Phosphate Buffer Saline 10× water and vortexes for about 2 min. It was then carefully swirled few times to ensure a homogenous mixture.

Microbial of determination

Microbiological analysis was done in accordance with American Public Health Association (APHA) (2008) standard procedures for analysis. A 3-fold serial dilution was performed for the enumeration of *Salmonella* spp, *E. coli*, total coliform, faecal coliform and *Clostridium perfringens* bacteria from spice samples. Hi-crome, MFC agar, M-Endo agar and Salmonella-Shigalla (SS) agar were used for detection of *Escherichia coli*, faecal coliform, total coliform and *Salmonella* spp, respectively whereas nutrient agar was used for *Clostridium perfringens* detection using pour plate method.

Faecal coliform was determined by filtering 60 ml of sample through a 0.45 um filter onto a Petri dish containing MFC and incubated at 44 ± 2 °C for 18–24 h and colonies were then counted. For detection for total coliform and *Escherichia coli*, sample was plated in M-Endo and Hi-chrome and incubated at 37 ± 2 °C for 18–24 h, respectively. Colonies of total coliform and *E. coli* were counted.

For *Salmonella* spp detection, 10 ml of the supernatant was filtered after priming membrane filter with 10 ml spice dissolved in water. Another 5 ml of spice dissolved in water was further used to rinse inner sides of funnel. Membrane filter was then transferred onto Salmonella-Shigalla agar for detection of the target microbes with the aid of a sterilised forceps. In *Clostridium perfringens* detection, 1 ml of the supernatant derived from the wash of spice samples were aliquot into a test tube after being heated in a water bath for about 15 min at about 50 ± 2 °C. Nutrient agars of about 20 ml were poured into the Petri dish, and then swirled both clockwise and anticlockwise. Colonies were counted with a colony counter.

Statistical analysis

Analysis of variance (ANOVA) was deployed to determine significant difference between microbial loads from five popular meat grilling joints. Pearson's correlation matrix was also conducted before changing the microbial count values to a logarithmic scale.

Results and Discussion

To prevent occurrence of food borne illnesses, it is important to ensure foods safety and hygiene at every stage of food processing and selling. The present study recorded faecal coliform count that ranged between 0 and $3.4 \log_{10}$ cfu/ ml with a mean of $1.17 \pm 1.07 \log_{10}$ cfu/ ml (Table 1).

Table 1 Microbial load of spices used for grilling meat in the Tamale

Sample Sites	TC	FC	E. coli	SS	C. perfringens
A1	0	2.7	1.81	0.03	0
B1	0.9	1.63	0	0.44	0
C1	0	0	2	0.34	0
D1	0.31	0	1.76	0	0
E1	0.9	1.95	0.68	0.2	0
A2	0	2.13	1.43	0	0
B2	0	1.84	0	0	0
C2	1.66	0.8	2.32	0.46	0
D2	0.22	1.31	0	0.9	0
E2	0.9	0	0	0.47	0
A3	0.23	0.77	3.14	0.3	0
B3	1.7	2.4	0.54	0.3	0
C3	0	1.7	0	0.64	0
D3	0.69	0	1.9	0.5	0
E3	0.4	0.62	2.2	0	0
A4	1.04	0.32	2.6	0.9	0
B4	0.36	0	0.66	0.72	0
C4	0	0	1.51	0.4	0
D4	0.3	1.2	0	0.8	0
E4	0.5	0	0	0	0
Min	0	0	0	0	0
Max	1.7	2.7	3.14	0.9	0
Mean	0.54	0.96	1.17	0.38	0
SD	0.55	0.94	1.07	0.31	0
GSB (2006)	0	0	0	0	0
WHO Limits (2011)	0	0	0	0	0

Note: All values are units of $\log_{10}$ cfu/ml: A-D: Samples sites and 1–4 means first to fourth sample

TC Total coliform, FC Faecal coliform, E. coli: Escherichia coli, SS Salmonella spp., C. perfringens Clostridium perfringens

Total coliform count ranged from 0 to 1.7 $\log_{10}$ cfu/ ml with a mean of 0.54 ± 0.55 $\log_{10}$ cfu/ ml (Table 1). *E. coli* count ranged between 0 and 3.14 $\log_{10}$ cfu/ ml with a mean of 1.17 ± 1.07 $\log_{10}$ cfu/ ml (Table 1).

Coliform bacteria were present in almost all the spice sampled. The faecal coliform and *E. coli* detected in the samples was an indication of contamination by fresh faecal matter. This might be due to inadequate hand washing by vendors and absence of good personal hygiene. This according to FDA (2011) can cause vibrio cholera, bloody diarrhoea, and kidney failure in children or people with weak immune system.

Most food poisoning is commonly cause by *Salmonellae* that are pathogens. Salmonellosis is principally a food borne disease (European Union EU 2004). Contamination level for *Salmonella* spp ranged from 0 to 0.9 $\log_{10}$ cfu/ ml with a mean of 0.38 ± 0.31 $\log_{10}$ cfu/ ml (Table 1).

Salmonella spp was detected in few samples. The contamination might be due to poor hygienic conditions, presence in food can cause abdominal clamps, fever and vomiting (Mankee et al. 2003). Salmonella presence is of concern in spices as the commodity is added in sufficient quantities to grilled meat for public consumption.

Clostridium perfringens can be found in dust, soils, vegetation among others environment media. *Clostridium perfringens* were not detected in any of the 20 samples. This might be attributable to poor growth parameter like favourable temperature. Since, its ability to grow in food is largely dependent on storage temperature (European Food Safety Authority 2004). This finding collaborate with that of Gardenguides (2002) that reported limited or no *Clostridium perfringens* count in spice in Australia. This seems to pose no risk to public health, however at slightly conducive temperature spore germination and cell multiplication can occur.

Similar studies conducted on spice used for selling local meat (kilichi) in Northern Nigeria revealed that the spices were greatly contaminated with *E. coli*, *Salmonella* spp and *Clostridium* spp (Shamsuddeen 2009). Sagoo et al. (2009) also isolated high counts of *Clostridium perfringens* and *E. coli* from spice and herb in United Kingdom which have potential public health risk. Spices have been associated in large scale outbreaks of food borne illnesses (Gustavsen and Breen 1984), the impact of pathogenic contaminated spice on incidence of food borne illnesses in Ghana cannot be override. The contamination of spices can be attributed to many factors. Since, Banerjee et al. (2003) suggested that spices such as curry powder, thyme, white pepper, paprika are subjected to microbial contamination at various stages of preparation. Equipment and food handlers have also been associated with contamination of food with various types of etiologic agent (Moro et al. 2001).

Possible sources of contaminants in spices

The study revealed that most vendors buy spices such as pepper, ginger, curry powder, paprika, nutmeg, chilli pepper (all in powdered form) from market and are then mixed in a plastic container. It was observed that vendors mixed spices with their bare hands without wearing hand groves and the prepared spices are left uncovered in the plastic container at the site. This exposes spice to contaminant such as dust, atmospheric particles and airborne microbes.

The study revealed that only one vendor out of five vendors had obtained permits from the Ghana Standard Board and Ghana Tourism Authority. Some reasons for not having permits were due to lack of funds and some vendors not aware that they needed to work with permits. Four out of five vendors had no formal education this might have influenced vendors' consciousness on personal hygiene, food handling and processing techniques. This according to

Barro et al. (2006), Mensah et al. (2002) and Bryan et al. (1997) are possible means of transmitting pathogens into foods. Pesewu et al. (2014) and Amponsah-Doku et al. (2010) attested that substantial persistence and proliferation of bacteria from food production/processing to consumption is a reflection of poor sanitary and food handling practices, and conditions at various stages of the chain.

Spice handlers should adopt and practice good handling processes to ensure food safety for consumers. The levels of some pathogenic bacteria are not acceptable as the pose serious health implication. This is similar to earlier report by Addo (2005) that there is serious health implication in the consumption of spices on the market.

In Tamale, it is evident that most food vendors are challenged with exogenous factors such as lack of public infrastructures for food vendors to comply with standard hygienic practices (such as clean water sources, public toilets), inadequate waste disposal service, poor sanitary conditions at vending sites (for example dust from dirt roads, open-air sewages, traffic fumes), poor storage conditions, contaminated inputs from farmers and market sellers, and unclean transportation conditions. These findings are echoed by FAO (2016). Tomlins and Johnson (2004) reported that street food pollution can partly originate from raw materials from rural small holders. Also, poor management of markets environment in Ghana has been linked as major cause of food contamination by Soriyi et al. (2008). Most fresh foods especially fermented food like beef and fish are highly vulnerable to pathogenic bacteria invasion and food poisoning.

Statistically, pathogenic bacteria count from the various selected sites is not significant difference (Table 2). The correlation matrix of bacteria count in spices demonstrates no interrelationships between pathogenic bacteria study. Hence, source of contamination of the pathogens is different, affirming the multiple sources of contamination of the spices observed.

Conclusion

The study profiled some pathogenic bacteria on common spices used for grilling meat in Tamale. People have the right to consume safe and suitable food. The study showed that mixtures of spices were contaminated with some pathogenic bacteria. *Salmonella* spp and *Escherichia coli* are potential enteric pathogens. There is a possible risk to public health associated with consumption of spicy meat from the selected joints. The possible sources of spice contamination include storage equipment, handling, unhygienic surroundings, vehicular transmission, atmospheric particles and air-microbes. The consumption of meat spiced with these contaminated spices can lead to health related problems such as cholera, diarrhoea, stomach cramps, typhoid fever and dysentery. It is recommended that periodic sanitary inspection and certification of vendors by regulatory body is absolutely necessary to ensure consistency.

Acknowledgements
The authors are grateful to all staff of CSIR- Water Research Institute, Tamale for their timely analysis of the samples.

Funding
The authors self-funded study from collection, analysis and interpretation of data and to the final writing of the manuscript.

Authors' contributions
NB, ASA, ABD have contributed in the study conception and design, data acquisition, and analysis and interpretation of data. SJC and SM participated in intellectual helping in different stages of the study. NB, ABD, SJC and SM participated in drafting of manuscript and preparation of final version. All Authors have read the manuscript and have agreed to submit it in its current form for consideration for publication. All read and approved the final manuscript.

Competing interests
The authors of this research paper have no competing interests.

Author details
[1] Department of Ecotourism and Environmental Management, Faculty of Natural Resources and Environment, University for Development Studies, Tamale, Ghana. [2]Department of Biotechnology, Faculty of Agriculture, University for Development Studies, Tamale, Ghana. [3]Ghana Integrated Water Sanitation and Hygiene World Vision, Tamale, Ghana.

References
Abankwa V, Grimard A, Somer K, Kuria F. United Nations Human Settlements Programme (UNHABITAT). 2009. Available: www.unhabitat.org/pmss/getElectronicVersion.aspx?nr=2929. Accessed on 9 September 2016.
Addo AA. Premiliminary studies on the microbiological and nutrient quality of three local spices on the Ghanaian market and the control of resident microflora by gamma irradiation: BSc. Hons Dissertation, Department of Botany, University of Ghana, Legon; 2005. p. 68
Ahene RE, Odamtten GT, Owusu E. Fungal and bacterial contaminants of six spices and spice products in Ghana. Afr J Environ Sci Technol. 2011;5(9):633–40.
Alam-Khan K, Abrahem M. Effect of irradiation on quality of spices. Int Food Res J. 2010;17:825–36.
American Public Health Association. Standard Methods for the Examination of Water and wastewater. 20th Edition, American Public Health Association,

Table 2 Comparative analysis of bacteria pathogens count

ANOVA		Sum of Squares	df	Mean Square	F	Sig.
TC	Between Groups	467.3	4	116.83	0.52	0.72
	Within Groups	3369.5	15	224.63		
FC	Between Groups	68853.3	4	17213.33	1.23	0.34
	Within Groups	210109.5	15	14007.3		
EC	Between Groups	607037.8	4	151759.5	1.83	0.18
	Within Groups	1246760	15	83117.33		
SS	Between Groups	18.22	4	4.56	0.67	0.63
	Within Groups	102.64	15	6.84		

American Water Works Association, Water Environment Federation. USA: United Book Press, Inc; 2008.

Amponsah-Doku F, Obiri-Danso K, Abaidoo RC, Drechsel P, Kondrasen F. Bacterial contamination of lettuce and associated risk factors at production sites, markets and street food restaurants in urban and peri-urban Kumasi, Ghana. Sci Res Essays. 2010;5(2):217–23.

Banerjee M, Sarkar PK. Microbiological quality of some retail spices in India. Food Res Int. 2003;36:469–74.

Barro N, Bello AR, Savadogo A, Ouattara CAT, Ilboudo AJ, Traore AS. Hygienic status assessment of dish washing waters, utensils, hands and pieces of money from street food processing sites in Ouagadougou (Burkina Faso). Afr J Biotechnol. 2006;5(11):1107–12.

Bruce J, Boschi-Pinto C, Shibuya K, Black R. WHO's estimates of the causes of death in children. Lancet. 2005;365:1147–52.

Bryan F, Jermini M, Schmitt R, Chilufya E, Mwanza M, Matoba A, Mfume E, Chibiya H. Hazards associated with holding and reheating foods at vending sites in a small town in Zambia. J Food Prot. 1997;60:391–8.

Buckenhuskes HJ, Rendlen M. Hygienic problems of phytogenic raw materials for food production with special emphasis to herbs and spices. Food Sci Biotechnol. 2004;13:262–8.

Centre for Disease Control (CDC). Diseases from imported food on the rise. 2012.

Debs-Louka E, El Zouki J, Dabboussi F. Assessment of the microbiological quality and safety of common spices and herbs sold in Lebanon. Journal Food Nutrition and Disorder. 2013;2(4):1–6.

European Food Safety Authority. Opinion of the scientific panel on biological hazards on a request from the commission related to *Clostridium* spp. in foodstuffs. EFSA J. 2004;199:1–65.

European Union (EU). Bacteriological safety and toxicological safety of dried herbs and spices. EU coordinated programme for the official control of foodstuffs 2004. 2004. p. 1–38. 3rd Trimester National microbiological Survey 2004 (04NS3).

FDA (Food and Drug Administration). 2010 Retail meat report national antimicrobial resistance monitoring system. Rockville: Food and Drug Administration; 2011.

Food and Agriculture Organization of the United Nations. Street food in urban Ghana; a desktop review and analysis of findings and recommendations from existing literature. In: Stefano Marras Mohamed A, editor. Food and Agriculture Organization of the United Nations Accra. 2016.

Ghana Statistical Service. Population census 2010. 2012. Available at www.ghanadistrict.com. Accessed on 6 June 2016.

Gustavsen S, Breen O. Investigation of an outbreak of *Salmonella oranienburg* infections in Norway, caused by contaminated black pepper. Am J Epidemiol. 1984;119(5):806–12.

Mankee A, Ali S, Chin A, Indalsingh R, Khan R, Mohammad F, Reheman R, Sooknanan S, Tota-Maharaj R, Simeon D, Adesiyun A. Bacteriological quality of doubles sold by street vendors in Trinidad and the attitudes, knowledge and perceptions of the public about its consumption and health risk. Food Microbiol. 2003;20:631–9.

Menlove A, Sainsbury J. The microbiological safety and quality of ready meals: ready meal technology. UK: Leather head Publishing; 2002.

Mensah P, Yeboah-Manu D, Owusu-Darko K, Ablordey A. Street foods in Accra Ghana: how safe are they? Bull World Health Organ. 2002;80:546–54.

Moro D, Oludmo AO, Salu OB, Famurewa O. Carriage of enteric pathogens among students of a tertiary institution in Lagos, Nigeria. JR Rev Se. 2001;2:73–6.

Pesewu GA, Agyei JN, Gyimah KI, Olu-Taiwo MA, Osei-Djarbeng S, Codjoe FS, Ayeh-Kumi PF. Bacteriological assessment of the quality of raw-mixed vegetable salads prepared and sold by street food vendors in Korle-Gonno, Accra Metropolis, Ghana. J Health Sci. 2014;2:560–6.

Sádecká J. Irradiation of spices - a review. Czech J Food Sci. 2007;25:231–42.

Sagoo SK, Little CL, Greenwood M, Mithani V, Grant KA, McLauchlin J, de Pinna E, Threlfall EJ. Assessment of the microbiological safety of dried spices and herbs from production and retail premises in the United Kingdom. Food Microbiol. 2009;26(1):39–43.

Shamsuddeen U. Microbiological quality of spices used in the production of kilishi, a traditionally dried and grilled meat product. Bayero Pure Applied Science. 2009;2:66–9.

Soriyi I, Agbogli HK, Dongdem JT. A pilot microbial assessment of beef sold in the Ashaiman market, a suburb of Accra, Ghana. Afr J Food Agric Nutr Dev. 2008;8(1):91–103.

Srinivasan K. Role of spices beyond food flavoring: nutraceuticals with multiple health effects. Food Rev Intl. 2005;21:167–88.

Tomlins KI, Johnson PNT. Developing food safety strategies and procedures through reduction of food hazards in street-vended foods to improve food security for consumers, street food vendors and input suppliers. Crop Post harvest Programme (CPHP) Project R8270. Funded by DFID. Ghana: Natural Resources Institute, UK & Food Research Institute; 2004.

WHO. Food safety: what you should know. World Health Day: 7 April 2015 SEA-NUT-196 Distribution: General World Health Organization. 2015.

Zwietering MH, Jacxsens L, Membre JM, Nauta M, Peterz M. Relevance of microbial finished product testing in food safety management. Food Control. 2016;60:31–43.

21

Assessment of pesticide residue levels among locally produced fruits and vegetables in Monze district, Zambia

Mildred Mwanja[1,2*], Choolwe Jacobs[3], Allan Rabson Mbewe[2] and Nosiku Sipilanyambe Munyinda[2]

Abstract

Background: The use of pesticides in fruits and vegetable production is beneficial for preventing, destroying or repelling pests that may damage these crops. The use of these chemicals however, often leads to the presence of residues in the fruits and vegetables after harvest. This study investigated farmers' compliance to applicable national standards by assessing pesticide residues in selected locally produced fruits and vegetables in two study sites in Monze, Zambia. The study used mixed methods (convergent parallel) design. We procured rape, cabbages, tomato and orange samples from conveniently sampled fruit and vegetable farmers around Hachaanga and St. Mary's areas in Monze, Zambia. Samples were analyzed for residues of dichlorvos using gas chromatography-mass spectrometry (GC-MS). Estimated average daily intakes (EADI) were calculated using standard formula. We also explored farmers' practices in dealing with regulatory issues in pesticide use and handling. A total of 14 key informant interviews with farmers, agriculture and public health officers and one policy maker were undertaken using a semi structured interview guide, were voice recorded, later transcribed and analyzed using Nvivo 10 software.

Results: Results revealed detectable residues in 63.3% of 30 tested samples out of which three samples (one each of cabbage, tomato and orange samples) exceeded the codex Alimentarius maximum residual limit (0.1 mg/kg). However, all samples had residues below the Zambia Food and Drugs standard (0.5 ppm). The EADIs were also below WHO/FAO allowable daily intake recommended in all fruit and vegetable samples; however hazard indices for cabbage and oranges were close to the value one. In regard to farmers' practices, results showed great variation in pesticide use and handling, limited knowledge, observation of reduced waiting periods and limited monitoring and regulation of pesticide use among farmers.

Conclusion: Our investigation found that all our samples had residues within the locally applicable regulation limits. All our EADIs were below the FAO/WHO limits. However, farmers' practices in pesticide use and handling were not conformity to guidelines. Therefore, there is need for educating food producers on handling and hazards of pesticides in Zambia.

Keywords: Pesticide residues, Maximum Residue Limit, Pesticide use and handling, Farmers' practices

Background

The use of pesticides in agricultural sector in Zambia has increasingly become an important aspect of agricultural technology and innovation, critical for agriculture development, economic growth and poverty reduction (Bwalya 2010). Globally, the use of pesticides in food production is common with many famers using commercial pesticides for pest control to increase yield and improve quality. The World Health Organization (WHO) reports that 20% of pesticide use in the world is concentrated in developing countries (PAN G 2012). Over the past couple of decades, a rapid increase in the quantity and use of pesticides in the agriculture sector has been observed. This trend is expected to continue for the coming decade due to social, economic and technological developments (Greish et al. 2011). Pests and diseases attack fruits and vegetables from cultivation through to storage. Pesticides are used to preserve their

---

* Correspondence: mwanjamildred@yahoo.com
[1]Environmental Health Department, Monze District Medical Office, P.O. Box 66144, Monze, Zambia
[2]School of Medicine Department of Public Health, Environmental Health Unit, University of Zambia, P.O Box 50110, Ridgeway, Lusaka, Zambia
Full list of author information is available at the end of the article

quality and prevent diseases. While the use of these chemical pesticides enhances the farmer's productivity, inappropriate pesticide use poses health hazards to consumers, other organisms and the environment (Wang et al. 2010).

Since the banning of organochlorine pesticides, organophosphates including dichlorvos became the most commonly used pesticides despite their World Health Organization classification as "hazardous" (PAN G 2012). Dichlorvos is known to be among pesticides frequently associated with documented cases of poisoning resulting in acute or chronic adverse health effects (Mowry et al. 2015). These pesticides are often applied indiscriminately and inappropriately (Ntow et al. 2006), resulting in adverse environmental and health effects. Several studies that have evaluated pesticide residues in fruits and vegetables revealed that fruits and vegetables may contain remnants of insecticides above the Maximum Residue Limits (MRL) set by the United Nations through the Codex Alimentarius Commission (Farag et al. 2011; Boland et al. 2004).

Good agriculture practices (GAP) in pesticide application can reduce the risk of pesticide contamination to users (Boland et al. 2004). However, farmers require knowledge on the right dosage, right ways of application and the suitable interval between harvesting and pesticide treatment. In most developing countries, farmers have low to moderate levels of knowledge about pesticides (Szpyrka et al. 2015; Armah 2011) and pesticide safety labels (Boobis et al. 2008). Farmers also misuse pesticide and poorly dispose of empty pesticide containers. Poor knowledge in most cases is attributed to lack of training and monitoring in pesticide use and handling (Damalas and Eleftherohorinos 2011). Pesticide users' inability to adhere to stipulated instructions leads to pesticide exposure not only to themselves but to the general public and the environment.

The Zambia Environmental Management Agency (ZEMA) provides for control of pesticide use through the licensing system and ensures that only acceptable pesticides are allowed in the country and found on the Zambian market. However, ZEMA's enforcement capacity is inadequate and cases of non-compliance are rampant (COMACO 2014) resulting in misuse of hazardous pesticides including dichlorvos, while the registration process rarely include testing and verification of the efficacy and hazard characteristics of the pesticide under consideration (COMACO 2014). The purpose of the monitoring program is to ensure that pesticide residues in fruits and vegetables do not exceed maximum residue levels allowed by government, that there is no misuse of pesticides that could result in unexpected residues in food. We set out to investigate dichlorvos residues in rape, cabbage, tomatoes and orange

samples in Monze Zambia and verify compliance with standards. We also explored farmers' practices in dealing with monitoring and regulation issues in pesticide use and handling.

Methods
Study setting
The research was conducted in Monze district in the southern province of Zambia. Monze town is situated 200 km south of Lusaka (Zambia's capital city) and 300 km north of Livingstone. Monze district covers an area of 6687 km^2. It lies at 1128.72 m above sea level, 16 degrees 18 min south and 26 degrees 28 min east. Monze district has a projected 2014 population of 206,693 (CSO 2010).

Economically, the district depends on farming, mainly subsistence crop farming (including fruits and vegetable growing) and animal husbandry. Pesticides are largely used to control pests and diseases to improve their market value. Vegetable growers are concentrated along the Magoye River that passes through the district on the east side of Monze town; mainly around Hachaanga, St Mary's and Chikuni areas.

Study design
Mixed method (convergent parallel) design was used in this study. The quantitative approach used the cross section analytical study design while the narrative study design was used for the qualitative approach. Samples were procured from conveniently sampled fruit and vegetable farmers around Hachaanga and St. Mary's areas of Monze. Prior to sample collection, farmers with fruits and vegetables at maturity stage and had sprayed them with dichlorvos were identified. Key Informant Interviews (KIIs) were conducted with farmers and public authorities to understand pesticide use in Hachaanga and St. Mary's areas of Monze.

Sampling
A total of 30 samples of cabbage, rape, tomatoes and oranges were procured from nine farmers as per FAO guidelines (Alba 2004); 2 kg of each vegetable and 1 kg of oranges per sample were collected 1 day after the application of the pesticide (observed withdraw period). Samples were prepared for residual analysis according to Codex Alimentarius Commission guidelines (FAO/WHO 2014). All the 30 samples were analyzed for residues of dichlorvos using gas chromatography-mass spectrometry (GC-MS).

Sample preparation and analysis
About 1 kg of each vegetable and orange sample was thoroughly washed and chopped using a laboratory knife. This was blended using a waring laboratory blender

with a variable speed to form a composite sample. A total of 50 g was then taken from the composite sample. After milling, a 20 g sample was placed into a 250 ml beaker, 5.0 g sodium chloride was added followed by adding dichloromethane, extracted by ultrasonic for 30 min. After that 10 g anhydrous sodium sulfate was added and stayed for 2 min. The extraction was transferred to a column packed with 4.0 g anhydrous sodium sulfate and rinsed twice (total of 5 ml) with dichloromethane, all the eluents were collected and evaporated to near dryness under nitrogen stream at 45 °C, the residue was re-dissolved with about 1 ml Hexane.

A florisil SPE column was conditioned with 6.0 ml mixture of hexane-acetone (4:1) and 5.0 ml hexane, the concentrated extraction was loaded on the top of the cartridge, and followed by eluting with 6.0 ml mixture of hexane-acetone (4:1), the eluent was collected, and evaporated to near dryness under nitrogen stream at 45 °C. The dried extract was then re-dissolved in acetonitrile solvent and final extract of 1 ml injected in the gas-mass spectrometry chromatography. The gas chromatography-mass spectrometry from Shimadzu was used for measurement of dichlorvos residues. External standard calibration was used to prepare the calibration curve, using concentrations of 5, 6, 7, 8, 9 and 10 µl/ml. The curve type was linear, original not forced; weighted method: None equation was:

$$Y = 2146.629x + 1151.629$$

$$R^2 = 0.9844324$$

$$R = 0.9921856$$

The amount of the pesticide in each sample was calculated based on the slope of the standard curve.

Data analysis and management
Pesticide residue values were reported as measured in mg/kg and compared to the codex Alimentarius MRL and the Zambia Food & Drug Act standards. Those that were found to be above MRLs were regarded to have violated the standards while those at or below the MRLs did not.

Stata version 13IC (StataCorp, College Station, Texas, USA) was used to derive median and interquartile ranges of pesticide residue levels in fruits and vegetable types.

Estimated average daily intake calculation
The estimated average daily intakes (EADI) were found by multiplying the median residual pesticide concentration (mg/kg) by the consumption rate (kg/day). To evaluate the safety of consumers, exposure estimates were assessed and compared to the acceptable daily intake (ADI) established by FAO/WHO (FAO/WHO

2004). Health risk indices were computed using data obtained from Zambia Food consumption and Micronutrient status survey report (Halimatou et al. 2014) as 21.9 kg/year for vegetables and 54.75 kg/year for fruits. This was based on assumptions that; everyone consumes vegetables on daily basis as Zambian diet is mainly vegetarian. One hundred percent of the vegetables are consumed after 24 h (observed withdrawal period) of pesticide application. The estimated hazard indices were calculated by dividing the EADI (mg/kg/day) by the corresponding value of WHO/FAO acceptable daily intake (Table 2).

Qualitative study
Population and data collection
Key informants were purposively sampled based on their knowledge on pesticide use, handling and regulation. Identification of farmers was done through agriculture extension officers manning the two study areas. Farmers with fruits and vegetables at maturity stage and had sprayed them with dichlorvos were eligible for KIIs and their produce for residue testing. Other key informants included agriculture extension officers, block extension, senior crop officer, public health department officer (from MOH) and ZEMA officer. Apart from one policy maker who was based in Lusaka, all the key informants were Monze residents. All KII were purposively selected as they were viewed to have information on the study subject. Only key informants who consented were interviewed.

Ten farmers and six officers were identified bringing the sample size to 16. All 16 eligible respondents were approached to participate in the study, but two (one farmer and one health officer from council) refused to consent, and this reduced the sample size to 14 as the two who declined to participate could not be replaced. Key informant interviews (KIIs) were conducted using semi-structured interviews using two different structured guides, one for farmers and another for policy makers or regulators. KIIs were audio recorded with the permission of the participants. Field notes were also taken during KIIs which lasted between 20 and 40 min. Secondary data reviewed included the Zambia Daily mail article by ZEMA which was posted on 15th December, 2015.

Key informant interviews with individual farmers were conducted in the local language (Tonga) and translated to English by the principle investigator who is a native Tonga speaker. Interviews with agriculture, public health and ZEMA officers were done in English. All KIIs were then transcribed and formatted in Microsoft word before importation into Nvivo10 software for the analysis process. Guiding questions focused on farmers' practices on pesticide use and handling, knowledge on pesticide use and handling, knowledge on health effects of pesticide residues and issues of training, pesticide regulation and monitoring.

Key Informant Interviews were analyzed using thematic analysis according to the six-stage process of thematic analysis starting with familiarization, generation of initial codes, searching for themes, reviewing themes, defining and naming themes and finishing with the final report (Lawson 2014). Data was transcribed as part of the familiarization stage. Transcribed data was transferred to Nvivo version 10 for arrangement, coding and merging into themes.

Transcripts were carefully and thoroughly read and re-read after which initial coding and categorization of themes was done by the principle investigator. Responses that were related through content and context were categorized as themes until no new themes emerged. Codes were categorized according to similar contents and then developed into broader themes.

Results
Compliance with maximum residue limits (MRL) in sampled fruits and vegetables
Out of 30 samples analyzed, 19 (63.3%) had pesticide residues while 11 (36.7%) had none detected (Table 1). Out of the samples that had residues, three had residues above codex Alimentarius Maximum Residue Limits (MRLs) but were within the standards in the Zambia Food and Drug Act. The results also showed that tomato samples were more contaminated (7/9) followed by cabbage samples (6/9). Of the analyzed rape and orange samples half were contaminated with dichlorvos residues (Fig. 1).

When compared to standard recommended daily intake, we found that estimated average daily intake of pesticides were all below the WHO/FAO allowable daily intake (0. 004) and poses no health hazard (Hazard Index <1) as shown in Table 2.

Findings from qualitative approach
Demographic characteristics of key informants
A total of 14 key informants were interviewed including nine farmers, four officers from ministry of agriculture and health and one policy maker form ZEMA, Lusaka.

All farmers did not have any alternative employment and two were females.

The major themes that emerged from KII included pesticide use and handling, knowledge of health effects of pesticides, farmers' training on pesticide use and handling and pesticide regulation and monitoring. The major drivers of pesticide residues in fruits and vegetables were lack of training of farmers in pesticide use and handling and lack of pesticide regulation and monitoring at grass root level. These lead to malpractices by the majority farmers as shown in Fig. 2. The details are as presented in subsequent sections.

Pesticide use and handling
Use of pesticides by fruit and vegetable growers around Hachaanga and St Mary's is common with almost every farmer depending on pesticides for pest control. All respondents confirmed using a number of chemicals in their fruits and vegetables. Doom and Phoskil (trade names for Dichlorvos and monocrotophose respectively) were reported among the frequently used chemicals as stated by Respondent 3 (male) from St Mary's:

"It is true, we use pesticides normally phoskil and doom, vilatel and so on. They are too numerous to mention. But normally we use phoskil and doom" **(Male respondent 3-St Mary's).**

The frequency of spraying reported ranged from every seven (7) days to fortnightly depending on the pesticide type and only one farmer reported spraying her oranges every three (3) months.

"...... So I make sure that I spray every week if I am on time. But when I see that the tomatoes are almost ready that is when I start spraying after every two weeks" **(Male Respondent 2-Hachaanga).**

The waiting period observed for Doom (dichlorvos) differed widely amongst the vegetable growers. The majority of the farmers were observing just 1 day, while a few observes 3 or 7 days.

Table 1 Pesticide residue levels in sampled fruits and vegetables collected from Hachaanga and St Mary's

Vegetable type	Number of samples without residues	Number of samples with residues		Median/IQR in mg/kg body weight/day
		Less 0.1 mg/kg (.006–.060)	Above 0.1 mg/kg	
Cabbage	3	5	1	0.04 (0–0.04)
Tomato	2	6	1	0.006 (0–0.04)
Rape	3	3	0	0.02 (0–0.04)
Oranges	3	2	1	0.02 (0–0.07)
Totals	11	16	3	

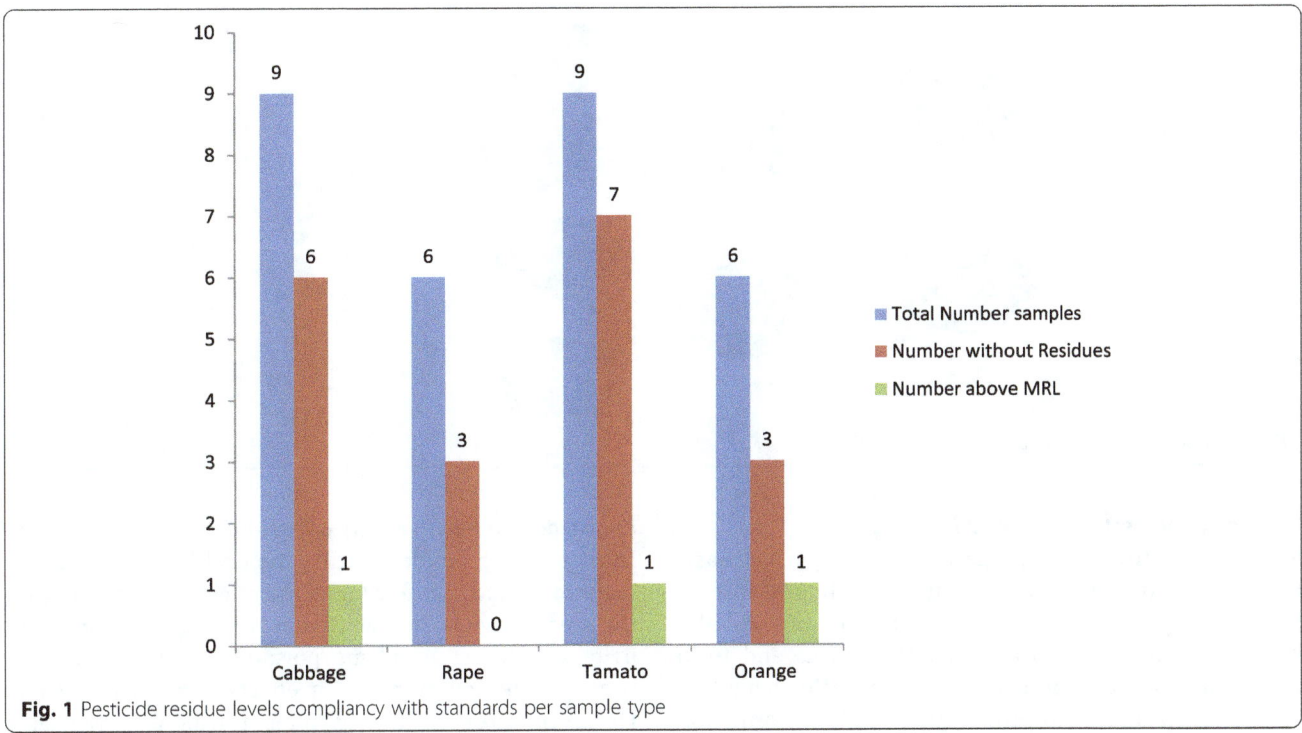

Fig. 1 Pesticide residue levels compliancy with standards per sample type

"According to the knowledge I got from GTZ, in the rainy season you can wait for 3 days only, because vegetables are watered by the rains in addition, there is plenty of water so the chemical becomes inactive quite fast. But in the dry season water is scarce and so vegetables are not adequately watered, therefore, you should wait for full 7 days" **(Male Respondent 2 St Mary's).**

Most of the respondents reported following label instructions on the dosage and were aware of the pros and cons of not following label instructions:

"These instructions are followed ... If as a farmer you do not follow the instruction then the health of people is put at stake. So, we as farmers, we should follow these instructions" **(Male Respondent 2-Hachaanga).**

Pesticide containers were mainly in form of plastic containers or plastic packs. Disposal methods that came out clearly were burying, disposal in pit latrine and burning. However, field observations showed that most of the farmers left pesticide empty containers in their fields buried with a thin layer of earth or glass (Fig. 3), contrary to ZEMA recommendations that pesticide containers should be buried in deep pits or burned.

One respondent from Hachaanga cited such careless behavior as a cause of poisoning and danger to children who follow their parents to the vegetable fields:

"... Yes because others just throw the empty tins anyhow. They do not bury them like I indicated, they just throw them away. Children then pick them and get poisoned. You see this child madam, she almost died, that one (pointing at her granddaughter who was watering vegetables) ... She picked a tin, started kujikilila (role-playing cooking) and eat the vegetables they had cooked in the tin and got poisoned". **(Female Respondent 1–Hachaanga)**

Table 2 Health risk assessment based on average daily intake of pesticide residues in fruits and vegetables

Vegetable type	Median value (mg/kg)	WHO/FAO ADI (mg/kg body weight)	EADI (mg/kg body weight)	Hazard index EADI/ADI
Cabbage	0.040	0.004	0.0024	0.60
Rape	0.020	0.004	0.0012	0.30
Tomatoes	0.006	0.004	0.00036	0.09
Oranges	0.020	0.004	0.0030	0.75

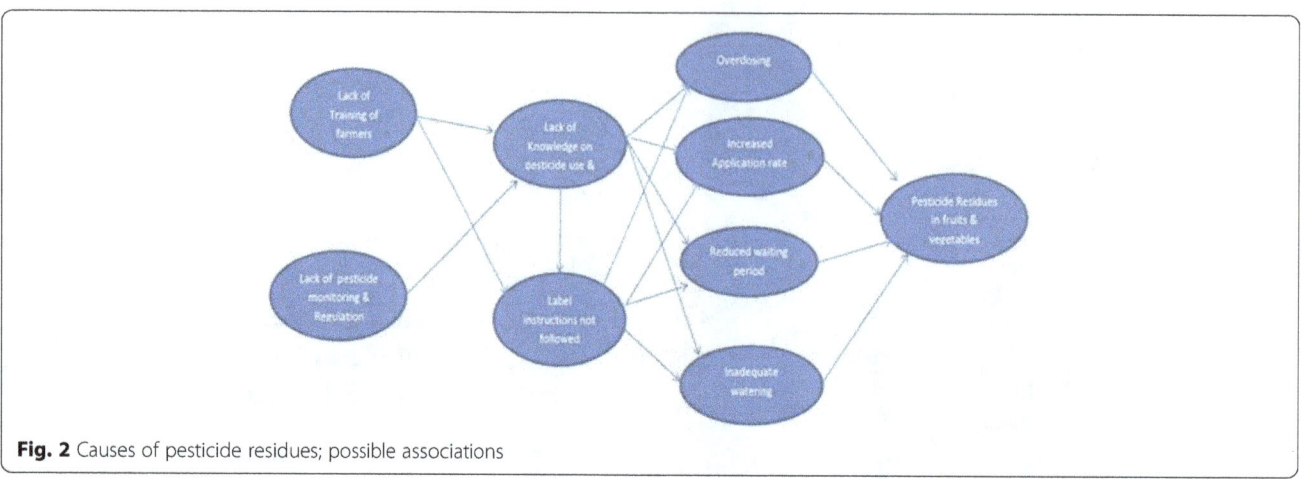

Fig. 2 Causes of pesticide residues; possible associations

Knowledge on pesticide use and handling

Respondents from Hachaanga and St Mary's had basic information on pesticide use and handling. Most of them acquired this knowledge from reading pesticide labels, from consultations amongst themselves and from agro dealers rather than from being trained or taught by relevant personnel. However much of the information was still limited.

During the time of the research, ZEMA (which is the pesticide regulating agency in the country) was equally concerned about lack of knowledge (which is as a result of lack of formal training) amongst farmers in the country and this is according to the Zambia Daily mail article of 15th December 2015 that we reviewed.

As ZEMA, we are concerned with the lack of knowledge on how to handle, use and dispose of agro-chemicals in the country. We are now working with various institutions to enhance knowledge and understanding on how to use agro-chemicals (**Zambia Daily mail article December 15, 2015**).

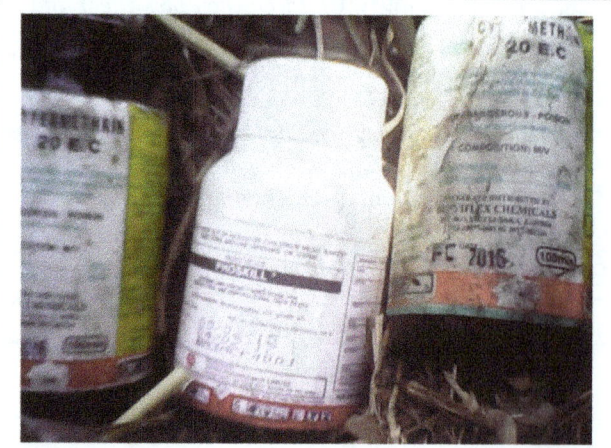

Fig. 3 Pesticide containers in vegetable field

Knowledge on health effects of pesticides

Most of the respondents knew the health effects of pesticides to humans. A few knew the effects of pesticides on the environment. Stomach pains, diarrhea, eye and skin irritations, chest problems poisoning and cancer were some of the common health effects of pesticides cited. However, most of the respondent fruit and vegetable farmers showed massive ignorance on causes of pesticide residues in fruits and vegetables.

"Yes, immediately after spraying you feel your chest congested, you don't feel good and even itching in the eyes because when you spray, when it is windy, when the wind is moving fast, you can feel it in your eyes as well. Yes, so you feel congestion on your throat, chest and eye irritation as well."
(**Female Respondent 4–Hachaanga**)

Training of farmers in pesticide use and handling

Despite agriculture officers confirming training of farmers in the use of pesticides, the majority of the farmers who are the end users of pesticides said they were not trained in pesticide use and handling.

"I have not undergone any major training apart from having meetings with fellow farmers where we encourage each other and learn from one another. Otherwise I have not gone for any training where I learnt about the use of these chemicals. Otherwise I just depend on my own knowledge and agro-dealers." (**Male Respondent 5–St Mary's**)

Pesticide regulation and monitoring

The majority interviewed reported that there was no monitoring amongst farmers on how they use and handle pesticides in their vegetable growing by relevant authorities:

"As farmers we just go buy these chemicals and come and spray. Like being checked on how we use these chemicals, they don't check on us." **(Male Respondent 1St Mary's)**

Key informant interviews also revealed inadequacies in regulation and monitoring of pesticide use among farmers. There was also on clear policy as to who is responsible for pesticide residue monitoring in fruits and vegetables. This is what key informant policy makers had to say.

"....that we regulate when something is manufactured, after that, if someone is going to transport, distribute or sale but we do not regulate the point of application. We do not go that far. But we end at distribution, if someone is distributing, then we end there". **(Female respondent policy maker, ZEMA)**

Discussion

Assessment of dichlorvos residue levels in locally produced fruits and vegetables in Monze-Zambia revealed that although more than half of the collected samples were contaminated, levels were all below the food and drug act (Zambia) standards. However 30% of the contaminated samples violated the codex Alimentarius standards. The estimated daily intakes were also found to be below what is allowable in all fruit and vegetable samples. However, hazard indices for cabbage and oranges were close to the value one. Root cause analysis for the qualitative revealed lack of training of farmers in pesticide use and handling and lack of pesticide monitoring and regulation among farmers as major drivers of pesticide residues in fruits and vegetables. Farmers reported great variation in pesticide use and handling, observation of reduced waiting periods, knowledge and training on pesticide use. Also, monitoring and regulation of pesticide use among farmers was very limited.

The presence of dichlorvos residues in fruits and vegetables, a pesticide banned and/restricted in other countries, is a serious concern and threatens the health of consumers. Violation of codex MRLs in some samples is an indication of deviation from good agriculture practices, which is corroborated by the KIIs. These findings are contrary to results of a study done in Lusaka, Zambia which showed that the average levels of dichlorvos were significantly above the maximum accepted limit as set by Zambian Food and Drugs Act on vegetables (Sinyangwe and Sijumbila 2016). High pesticide concentrations in fruits and vegetables may be caused by a number of factors including differences in pesticide application and other habits such as spraying close to maturity stage due to pest pressure. Other countries that have reported residues of banned or restricted pesticides in analyzed vegetable samples include Ghana, Egypt and

Poland (Darko and Akoto 2008; Szpyrka et al. 2015; Farag et al. 2011).

Several studies have shown that even low level exposure to dichlorvos can cause neuropsychiatric conditions (Ross et al. 2010). Moreover, scientists cannot say for sure that there is ever a "safe level" of pesticide residues in food because many chemical messengers in our bodies function at precisely minute quantities of ppm or even ppb (Boobis et al. 2008). The presence of these pesticides therefore calls for vigilance in monitoring pesticide residue levels in fruits and vegetables.

The use dichlorvos in fruits and vegetables as practiced by farmers around Hachaanga and St Mary's poses a health risk to consumers as pesticides can accumulate on vegetable leaf surfaces and may lead to higher pesticide residues. This was similar to findings of studies done in Northern Malawi and Eastern Zambia and Nigeria were respondents reported using pesticides that are hazardous (class1) according to WHO classification. By this practice the farmers are also continually being exposed to pesticides and this can lead to an array of health effects depending on the pesticide's toxicity and the dose take in by the body (Coble et al. 2005). The rate of application by most of the farmers was on weekly basis whether or not pests are present. Despite such an increased rate of application, the majority of the farmers observed reduced waiting periods; one (1) day after pesticide (dichlorvos) application. These variations on the waiting period observed by the farmers indicate non-compliancy to label instructions and lack of training in pesticide use. This was despite the fact that the majority of these farmers knew the health effects of pesticides and confirmed experiencing common effects like skin and eye irritation, abdominal pains and diarrhea.

The results also clearly indicated lack of knowledge among farmers in pesticide use and handling. This was attributed to lack of training and monitoring by agriculture officers and poor attitude of vegetable growers. The findings are in agreement with findings of a study done in Ethiopia that demonstrated lack of knowledge among small scale farmers as one of the contributing factors of pesticide residues (Mekonnen and Agonafir 2002). The studies suggested poor attitude and behavior of the vegetable producers, lack of awareness in the pesticide management coupled with weakness in the agriculture extension services as drivers of poor handling and use of pesticides. Unlike a study in Ghana where half the vegetable farmers had received training on safe handling and application of pesticide (Bempah et al. 2012), majority of farmers interviewed were not trained in pesticide use and handling. The relatively high number of farmers without formal training and knowledge in pesticide use is an important source of concern as pesticides are hazardous chemicals that require to be handled with caution. The commonest way of disposing of empty pesticide containers cited among the respondent farmers was by burying in dip holes and pit-latrine or by burning. However the finding of the study was that the majority of the respondent farmers left pesticide containers in their vegetable fields buried with either a thin layer of glass or earth. This practice is a serious threat to human health and the environment as improper disposal of empty pesticide containers can lead to contamination of soil, ground and surface water, causing serious problems for the pesticide user, consumers and the environments.

The agriculture sector in Zambia has continued to use restricted/banned pesticide such as dichlorvos without rigorous monitoring and regulation system. In advanced countries, strict pesticide regulation and enforcement mechanism are put in place to ensure their safe use and handling and high standards are set to ensure safety of consumers (Bempah et al. 2012). However, this is not the case in developing countries including Zambia. In Zambia, of serious concern is the fact that a list of banned or strictly restricted pesticides is non-existence (COMACO 2014).

Regulatory agencies also face enormous challenges including poor funding, lack of transport and poor policies that prioritize curative over preventive programs. As a result routine pesticide residue monitoring in fruits and vegetables is not done. The Zambia Food and Drug Act standards for dichlorvos residues in fruits and vegetables are also low (0.5 ppm = 0.5 mg/kg) compared to international standards (0.1 mg/kg by Codex and 0.2 mg/kg by European Union). Setting low standards for banned and/ restricted pesticides in developed countries create problems when evaluating potential human risk. Ultimately high amounts of these hazardous chemicals are allowed in the food chain at the expense of everyone's health. A literature review on pattern and epidemiology of poisoning in East African region reported organphosphorus poisoning as one of the key causes of non-traumatic coma in Zambia (Tagwireyi et al. 2016).

Study limitations

In spite of the fact that a wide range of pesticides are currently used to control pests in fruits and vegetables, this study only focused on analysis for the presence of dichlorvos. The qualitative study however covered practices on the use of both dichlorvos and monocrotophos. Also, the study did not attempt to collect all fruit and vegetable samples from all suppliers in town. As such, results of this study will only apply to these two types of pesticides and will not be generalized to all fruit and vegetable suppliers in Monze. Therefore, these results are an underestimation of exposure to pesticides by consumers. There is need to conduct a comprehensive study to cover multiple pesticides and all fruits and vegetable types in order to wider overview of contamination levels.

Conclusion

It can be concluded that, pesticide residues in fruits and vegetables from Hachaanga and St Mary's were within locally acceptable standards. However, the combination of factors including increased application rate, reduced waiting period, limited knowledge and training of the majority of farmers and absence of monitoring services regarding residual levels of pesticides in fruits and vegetables seriously threaten consumers. Therefore, there is need to strengthen pesticide regulation and monitoring by extension of the regulation to include pesticide end users. Training of farmers in pesticide use and handling should also be greatly improved to increase awareness and encourage right practice for the safe use of pesticides. Development of a list of banned pesticides by ZEMA which should include dichlorvos to stop its importation in the country is justified according to the findings of this study. Furthermore, revision of the Zambia Food and Drugs Act in regards to standards of pesticide residues in fruits and vegetables is necessary to meet international standards.

Abbreviations

CSO: Central Statistics Office; EPA: Environmental Protection Act; FAO: Food and Agriculture Organization; FDA: Food and Drugs Act; GAP: Good agriculture practices; KII: Key Informant Interview; MAL: Ministry of Agriculture and Livestock; MOH: Ministry of Health; MRL: Maximum residue levels; UNZA: University of Zambia; WHO: World Health Organization; ZEMA: Zambia Environmental Management Agency

Acknowledgments

We would like to thank and appreciate our Colleagues at Ministry of Agriculture offices in Monze for granting us permission to carry out the research among their farmers. We also acknowledge the indispensable input of professor Michelo (Head of Department of Public Health, University of Zambia), Dr. Anjali Sharma, Dr. Roma Chilengi.M.D. and Dr. Wilbroad Mutale (CIDRZ) in the perfection of this work. We are greatly indebted to you all. We also would like to thank Mr. Mwanza Bisalom for his input in qualitative data analysis.

Funding

No specific funding was received for this research.

Authors' contributions

The research was originated by MM for the award of Master of Public Health with support from NSM, AM and CJ. With the technical support and guidance of the other co-authors, MM conducted sample and data collection. NSM, AM and CJ were all my supervisors and together we worked out the research idea. They also provided technical support in data analysis and interpretation, and draft of this manuscript. All authors have read and agreed to the final version of this manuscript for publication.

Competing interests

The authors declare that they have no competing interests.

Author details

[1]Environmental Health Department, Monze District Medical Office, P.O. Box 66144, Monze, Zambia. [2]School of Medicine Department of Public Health, Environmental Health Unit, University of Zambia, P.O Box 50110, Ridgeway, Lusaka, Zambia. [3]School of Medicine Department of Epidemiology and Biostatistics Unit, University of Zambia, P.O Box 50110, Ridgeway, Lusaka, Zambia.

References

Alba AF. Chromatographic-mass spectrometric food analysis for trace determination of pesticide residues, Spain. vol. 43. Almeria: Elsevier; 2004.

Armah FA. Assessment of pesticide residues in vegetables at the farm gate: Cabbage (*Brassica oleracea*) cultivation in Cape Coast, Ghana. Res J Environ Toxicol. 2011;5(3):180.

Bempah CK, Asomaning J, Boateng J. Market basket survey for some pesticides residues in fruits and vegetables from Ghana. J Microbiol Biotechnol Food Sci. 2012;2(3):850.

Boland J, Koomen I, JvL d J, Oudejans J. AD29E Pesticides: compounds, use and hazards. Netherland. Agromisa Foundation; 2004.

Boobis AR, Ossendorp BC, Banasiak U, Hamey PY, Sebestyen I, Moretto A. Cumulative risk assessment of pesticide residues in food. Toxicol Lett. 2008; 180(2):137–50.

Bwalya SM. Sound Management of the Chemicals in Zambia. A cost benefit analysis of the Agriculture Chemical use in the Kafue basin. 2010

Coble J, Arbuckle T, Lee W, Alavanja M, Dosemeci M. The validation of a pesticide exposure algorithm using biological monitoring results. J Occup Environ Hyg. 2005;2(3):194–201.

COMACO. Community Markets for the COMACO landscape management project P144254 : Pest management plan (PMP), November 2014. Proc Natl Acad Sci. 2014;29:8–33.

CSO. 2010 Zambia Census of Population and Housing, Preliminary report. Lusaka: Central Statistics Office, Republic of Zambia; 2010.

Damalas CA, Eleftherohorinos IG. Pesticide exposure, safety issues, and risk assessment indicators. Int J Environ Res Public Health. 2011;8(5):1402–19.

Darko G, Akoto O. Dietary intake of organophosphorus pesticide residues through vegetables from Kumasi, Ghana. Food Chem Toxicol. 2008;46(12): 3703–6.

FAO/WHO. Joint FAO/WHO Expert Committee on Food Additives. 2004

FAO/WHO. Food Standards Programme. Geneva: Codex Alimentarius Commission; 2014. 14–18 July 2014

Farag R, Latif A, Abd El-Gawad A, Dogheim S. Monitoring of pesticide residues in some Egyptian herbs, fruits and vegetables. Int Food Res J. 2011;18(2):659–67.

Greish S, Ismail SM, Mosleh Y, Loutfy N, Dessouki AA, Ahmed MT. Human Risk Assessment of Profenofos: A Case Study in Ismailia, Egypt. Polycycl Aromat Compd. 2011;31(1):28–47.

Halimatou A, Kohler L, Taren D, Mofu M, Chileshe J, Kalungwana N. Zambia food consumption and micronutrient status survey report. Lusaka: National Food and Nutrition Commission DRAFT; 2014.

Lawson A (2014) UWE Research Repository annual report Sept 2013-Aug 2014.

Mekonnen Y, Agonafir T. Pesticide sprayers' knowledge, attitude and practice of pesticide use on agricultural farms of Ethiopia. Occup Med. 2002;52(6):311–5.

Mowry JB, Spyker DA, Brooks DE, McMillan N, Schauben JL. 2014 annual report of the american association of poison control centers' national poison data system (npds): 32nd annual report. Clin Toxicol. 2015;53(10):962–1147.

Ntow WJ, Gijzen HJ, Kelderman P, Drechsel P. Farmer perceptions and pesticide use practices in vegetable production in Ghana. Pest Manag Sci. 2006;62(4): 356–65.

PAN G. Pesticides and health hazards facts and figures. Bochum: Pestizide und Gesundheitsgefahren: Daten und Fakten; 2012.

Ross SJM, Brewin CR, Curran HV, Furlong CE, Abraham-Smith KM, Harrison V. Neuropsychological and psychiatric functioning in sheep farmers exposed to low levels of organophosphate pesticides. Neurotoxicol Teratol. 2010;32(4):452–9.

Sinyangwe DM, Sijumbila G. Determination of dichlorvos residues levels in vegetables sold in Lusaka Zambia. Pan Afr Med J. 2016;23(1):1–18.

Szpyrka E, Kurdziel A, Matyaszek A, Podbielska M, Rupar J, Słowik-Borowiec M. Evaluation of pesticide residues in fruits and vegetables from the region of south-eastern Poland. Food Control. 2015;48:137–42.

Tagwireyi D, Chingombe P, Khoza S, Maredza M. Pattern and epidemiology of poisoning in the East African region: a literature review. J Toxicol. 2016;2016:1–26.

Wang J, Chow W, Leung D. Applications of LC/ESI-MS/MS and UHPLC QqTOF MS for the determination of 148 pesticides in fruits and vegetables. Anal Bioanal Chem. 2010;396(4):1513–38.

22

Assessment of sulphonamides and tetracyclines antibiotic residue contaminants in rural and peri urban dairy value chains in Kenya

Joy Deborah Orwa[1*], Joseph Wafula Matofari[1], Patrick Simiyu Muliro[1] and Peter Lamuka[2]

Abstract

Background: Antibiotic residues are drug substances found in food from plants or animals initially exposed to antibiotics. In animal husbandry antibiotics have widely been used for the treatment of animal diseases. These residues have the ability to expose the public to serious health hazards. In Kenya drug residues have not only been related to lack of withdrawal periods but also to intentional addition to extend milk's shelf life.

Results: The aim of this study was to investigate the occurrence of 13 veterinary drugs of tetracyclines and sulphonamides along the dairy sub value chain. The study was carried out in Nakuru County which is the leading milk producer in the country. A total of 229 samples were analysed from rural and 80 samples from peri-urban. These were collected from different nodes of the value chain; the farm, milk transporters and at the bulking centers between January 2014 and November 2015. Screening of samples was done by Charm II Blue -Yellow-test while confirmation was done by HPLC-UV for sulfachloropyradizine (SCL), sulfadiazine (SDZ), sulfadimidine (SMTZ), sulfaquinoxaline (SQ), sulfamerazine (SMR), sulfathiazole (STZ), sulfamethoxazole (SMX), sulfadoxin (SDOX), sulfadimethoxin (SDM), oxytetracycline (OTC), doxycycline hyclate (DC), chlortetracycline hydrochloride (CTC) and tetracycline hydrochloride (TC). In the rural 72 out of 229 (31.4%) samples were positive after screening while none of the samples confirmed the presence of tetracyclines after analysis with HPLC-UV. Sulphonamides confirmed after analysis with HPLC-UV were all above the EU MRL limits. In the peri urban 28.8% (23/80) of the samples were positive for antibiotic residues. Tetracyclines were not detected in confirmation while 60% of the positive samples were positive for sulphonamides out of which 71% were above the regulatory limits. Highest percentage of antibiotics was detected in rural farms (46.7%) and at peri urban bulking centers (50%).

Conclusion: The study concluded that antibiotic residues along the dairy value chain are majorly from the farm due to lack of withdrawal periods followed by intentional addition along the value chain. Value chain actors should also be trained on ways of avoiding antibiotic residues from entering the dairy value chain to protect the public from health effects related to antibiotic residues.

Keywords: Antibiotic residues, Contaminants, Tetracyclines, Sulphonamides, Dairy value chain, Rural, Peri-urban

* Correspondence: deborahorwa@gmail.com
[1]Department of Dairy Food Science Technology, Egerton University, Faculty of Agriculture, P.O. Box, 536–20115, Egerton, Kenya
Full list of author information is available at the end of the article

Background

Antibiotics are defined as the antimicrobial substances that are produced either naturally by living organisms or synthetically by laboratory procedures with the ability to inhibit the growth of microorganisms or kill the microorganisms (Wageh et al., 2013). Antibiotics are manufactured for the purpose of the prevention and treatment of animal diseases such as mastitis, arthritis, brucellosis, gastrointestinal diseases, respiratory diseases and many other bacterial infectious diseases (Tollefson and Miller 2000). In intensified farming antibiotics are also used to improve animal production like increase of growth rate and fattening (Nisha, 2008). When these antibiotics are administered to an animal, they dissolve and distribute rapidly in animal tissues and fluids. Over 90% of these antibiotics bind to plasmic proteins and reach a high concentration between the 3rd and 6th hour of administration (Sulejmani et al., 2012). They are then metabolized in the liver and are excreted through glomerular filtration. If the right procedure is not used in administration and use of these drugs they are left in large amounts i.e., residues, in animal products like milk and meat (Richelle, 2007). Once they are in the milk, there is a carry-over effect along the milk value chains.

The most common antibiotics used in animal husbandry include sulfonamides, tetracyclines, beta lactams, aminoglycosides, lincosamides, macrolides and pleuromutilins (Lee et al. 2001). The most common sulphonamides are sulfadiazine, sulphadimidine sulfamethoxazole, sulfamerazine, sulfadimethoxine, sulphasalazine, sulfisoxazole and silver sulfadiazine, which have the sulfonamide as the base structure (Chung et al., 2009). In countries like Kenya, the prevalence of tetracycline's was recorded to be highest at 55%, followed by sulfonamides at 21% and beta lactams at 6% (Mitema et al. 2001; Shitandi and Sternesjö 2004). According to Aboge et al. (2000), the most common antibiotics used in the treatment of livestock were beta lactams, sulfonamides, tetracyclines and aminoglycosides. Antibiotic residues in milk along the value chain have been reported to be above the maximum residual levels (MRL) in Kenya (Aboge et al., 2000; Shitandi and Sternesjö 2004). The main causes of antibiotics residues in milk have been attribute to by lack of observing withdrawal period, extra label usage of drugs, contamination of animal feed with feces of treated animals, or the use of unlicensed antibiotics (Nisha, 2008). Other studies have also attributed the occurrence antibiotics in animal products to lack of educational training in antibiotic use and their effects (Shitandi and Sternesjö 2004; Okeke et al. 1995).

According to the European union and Codex Alimentarious regulation for maximum residual limits, sulfonamides should not exceed 100 µg/kg, while tetracyclines should not exceed 100 µg/kg (EUR-Lex 2010; Codex

2012). Antibiotics have been identified in other parts of the world falling above the standard residual limits, including Germany (Kress et al., 2007), Netherlands (Abjean et al. 2000), Mexico (Tolentino et al., 2005) Turkey (Alkan, 2007) among others. Other African countries have also been reported to have milk contaminated with Antibiotic residues. Some of these countries include Egypt, Ghana, Ethiopia, south Africa, Nigeria, Tanzania and Sudan (Myllyniemi et al. 2000; Kurwijila et al. 2006; Goudah et al. 2007; Addo et al., 2011; El-tayeb et al. 2012).

The most recent research on antibiotic residues has recorded higher levels in milk. In Kenya in autum of 2010, 2.5% samples tested positive for sulphonamides while 0.6% were reported to contain tetracyclines (Ahlberg et al., 2016). High levels of tetracyclines were reported in Algerian milk and milk products in study by Layada et al. (2016). Chowdhury et al. (2015) reported levels of antibiotic residues above recommended levels in Bangladesh milk. Over 60% of milk samples tested positive for antibiotic in Nigerian milk and other milk products (Olatoye et al. 2016).

When milk and other animal products with high levels of antibiotic residues are ingested by humans, there is occurrence of numerous adverse health effects like permanent gene mutation and liver poisoning (Nisha, 2008). Sulfamethazine has been highly associated with including immunopathological effects, transfer of bacterial resistance to humans, hypersensitivity and carcinogenicity. Mutagenicity and nephropathy have been reported to be caused by gentamycin. Hepatoxicity, reproductive disorders and bone marrow toxicity have been related to the occurrence of some chlorophenical (Wageh et al., 2013). Penicillin however has been reported to be associated with allergy development. Tetracycline's are also capable of staining teeth in little children. Technologically, antibiotic residues in milk inhibit the growth of starter cultures used in the manufacturing of cultured milk products like yoghurt and cheese.

If no control measures are taken against the occurrence of antibiotic residues in animal products, its projected that there will be over 65% increase in antibiotic residue contamination worldwide in animal products between 2010 and 2030 (Van Boeckel et al. 2015). In Kenya reports on drug residues have increased gradually since 1978 where penicillin residues in milk was reported to be 1% (Kang'ethe et al., 2005), in 2000 there was 16% of drug residues reported (Omore et al., 2004; Kang'ethe et al., 2005) while Shitandi and Sternesjö (2004) reported 14.9% of milk to contain penicillin. Further, in 2010 over 24% of milk at the farm level tested positive for antibiotic residues (Ahlberg et al., 2016).

However, in the middle and low income countries, there is limited data on antibiotic usage and levels (Van Boeckel et al. 2015). Also, many studies on milk antibiotic residues in Kenya have majorly focused on tetracyclines and sulphonamide levels at the farm and market levels but no study has focused on the entire informal milk value chain. Therefore, the aim of this study was to assess the presence and levels of sulfonamides and tetracyclines contamination in raw milk from small scale farmers, transporters, and bulking centers who are key actors along the dairy value chain using rapid screening (Charm II Blue-Yellow) and confirmation by HPLC.

Methodology
Study site
The study was carried in Nakuru County which is rated as a high milk production center in the country. It is estimated to produce over 40 million liters of milk per annum (Ministry of Livestock Development Department and Fisheries 2012). The divisions within the county where the study was carried out were; Olenguruone, Bahati and Wanyororo. Olenguruone division represented a rural dairy system which lies at 35° 40′60″E and 0° 34′60″S DMS (degree minute seconds). Wanyororo and Bahati divisions represented the peri- urban locations at 36° 16′ 12″ E and 0° 12′ 0″ S. Samples were collected between January 2014 and November 2015.

Sampling
A nested design (in RCBD) was applied in sample collection where the nodes were nested within the locations. Sampling was done in three visits to the dairy system. The first visit 40 samples were collected from the farm, 35 samples from transporters and the three bulking centers were sampled from. This provided a total of 79 samples per visit in the rural dairy system. In the peri urban dairy system, 17 farms were visited with one cow per farm being sampled from, 7 milk transporters were sampled and the two cooling centers in that diary system were also sampled from. This provided a total of 26 samples per visit. Dairy farming is not a priority source of income for the population in peri urban and hence less than fifty percent of the population carried out dairy farming, hence a relatively smaller sample size was collected from the peri urban dairy system. Sample volumes were 100 ml per sampling point, the samples were then stored at temperatures not higher than 4 °C and were analyzed within 2 months of sampling.

Antibiotic analysis
Screening
Screening was done using the Charm II Blue-Yellow test. The charm test is done using a kit which is provided by the manufacturer (ALDRICH). The kit has 96 wells

containing a media, pre-measured bacterial spores (*Bacillus stearothermophilus var. calidolactis*) and a pH indicator. The Charm II Blue Yellow (Serial number: BlueYellowII.01843CharmInc.) kit has 96wells. The Kit detects 29 antimicrobials including beta lactams, sulphonamides, tetacyclines and aminoglycosides at and above EU maximum residual limits (MRL). The test is based on the principle of growth inhibition of the bacterial spores by antibiotics. The starting color of the bacterial wells is blue. If the sample is positive, the bacteria will be inhibited and there will be no color change hence the blue color will be retained after the test is completed. If the sample is negative, then the bacterial spores will germinate producing acid which will change the pH of the media to yellow. Samples which are not clearly positive will have a yellow to bluish color.

Thawed milk samples at room temperature, were centrifuged and the supernatant (not fat) was used in screening. Fifty microliter of the skimmed milk was measures and transferred into a well. The procedure was repeated until all samples were transferred to individual wells in duplicate. A positive control containing Penicillin G (4 ppb) and a negative control were included before proceeding. Wells with added samples and the controls were then incubated at 64 °C in a humidified incubator for three hours. At the end of the three hours, the wells were removed and observed for color changes which were read using interpretation chart provided by the manufacturer (Charm Blue Yellow II test Manual).

Suspect positive samples were further heated to 80 °C for 10 s to eliminate the presence of other natural antibiotic inhibitors, lysozyme and lactoferrin (Kellnerová et al. 2015; Layada et al., 2016). The boiled samples would then be taken through the charm screening test. Samples that tested negative were eliminated but those which remained positive and caution samples were preceded to the HPLC for confirmation and quantification.

Standards and reagents
The standards used in examination were; Sulfachloropyradizine (Sigma-Adrich 46778), Sulfadiazine (Sigma-Adrich 35033), Sulfadimidine (Sigma-Adrich 46802), Sulfaquinoxaline (Sigma-Adrich 45662), Sulfamerazine (Sigma-Adrich 46826), Sulfathiazole (Sigma-Adrich 46902), Sulfamethoxazole (Sigma-Adrich 46850), Sulfadoxin (Sigma-Adrich 46810), Sulfadimethoxin (Sigma-Adrich 46794), Oxytetracycline (Sigma-Adrich 46598), Doxycycline hyclate (Sigma-Adrich 33429), Chlortetracycline hydrochloride (Sigma-Adrich 26430), Tetracycline hydrochloride (Sigma-Adrich 87130).

Reagents were Acetonitrile (J.T. Baker 9017), Methanol (J.T. Baker 8402), Trichloroacetic acid (J.T. Baker 0344), Disodium hydrogen phosphate (J.T. Baker 0326), Citric acid (Acros 124912500), Sodium-EDTA (J.T.

Baker 1073), Calcium Chloride (Merck 1.2378.0500), Sodium Cetate (J.T. Baker 0258), Ammonium acetate (J.T. Baker 0011). Mobile Phase A was prepared by mixing Na acetate (0.075 M) and Calcium Chloride (0.035 M) to Sodium EDTA (0.025 M) and the pH adjusted to 7.0. Mobile Phase B was prepared by mixing 75% methanol to 25% Acetonitrite.

Equipment

The equipment used in the study was Shimadzu HPLC-Japan. Equipped with a UV-vis detector, SID 20A, Column Oven-CTO-10ASVP, X-Terra^R MS C$_{18}$ (3.5 μm, 2.1 × 150 mm column Waters made in Ireland), an X-Terra Guard column C18 (3.5 μm, 2.1 × 10 mm) solvent delivery module, LC 20AT, degassing unit DGU-20A$_3$, an auto sampler SIL-20AHT and system controller CBM-20A connected to a HP intergrator with LC Solution Version 3.5 Shimadzu Corp-Japan.

Confirmation

Positive suspects were confirmed using HPLC where exact antibiotic quantities were determined. The following antibiotics were sought out based on the high prevalence of Mastitis in both locations. These antibiotics included; Sulfachloropyradizine (SCL), Sulfadiazine (SDZ), Sulfadimidine (SMTZ), Sulfaquinoxaline (SQ), Sulfamerazine (SMR), Sulfathiazole (STZ), Sulfamethoxazole (SMX), Sulfadoxin (SDOX), Sulfadimethoxin (SDM), Oxytetracycline (OTC), Doxycycline hyclate (DC), Chlortetracycline hydrochloride (CTC) and Tetracycline hydrochloride (TC). The antibiotics selected are also wide in spectrum in terms of their activity and are usually used in treating other bacterial diseases among cattle.

The HPLC procedure used had four main steps as described by Mamani and Reyes (2009) and Koesukwiwat et al. (2007). These were; (a) protein precipitation and purification by centrifugation and trichloroacetic acid and McIIvaine-EDTA buffer, (b) Sample extraction with Oasis HLB (200 mg) cartridge, (c) sample evaporation and (d) quantification by HPLC with gradient mode on C$_{18}$ column and UV-detection.

Sample preparation
Protein precipitation and purification

5 ml of presumptive positive sample was measured into a 25 ml centrifuge tube; 2.5 ml of 25% TCA in water was added and mixed for 10 s by vortexing. 10 ml of McIlavine- EDTA buffer was added to the mixture, vortexed (Vortex-Assistant- Reamix 2789) for 10 s and then mixed in a sonicator (Power Sonic405 LUC) for 10 min. This mixture was then centrifuged at 400 rpm at 10 °C. The clear supernatant was poured out to a new 25 ml centrifuge tube with the fat remaining in the tube walls.

To the old centrifuge tube containing the subnatant, 10 ml McIlavine-EDTA was added and mixed by vortexing for 10 s. This was the sonicated for another 10 min. This was then centrifuged (Centrifuge- Heraeus-Labofuge) at 4000 rpm. The resulting supernatant was mixed with old supernatant initially collected from the same sample.

Solid phase extraction

The C$_{18}$ cartridges (Oasis HLB cartridges C18, 6 cc 200 mg, waters Corporation USA) were marked and fixed on the solid extraction vacuum. Additional funnels (20 ml) were fixed on the cartridges. The C18 column cartridge was activated by 5 ml methanol, followed by 10 ml acetonitrile then 5 ml McIlavine-EDTA without letting the cartridge run dry. The clear supernatant was then poured to the cartridge funnels so that it trickles through in approximately 20 min. This was then washed with 5 ml methanol in McIlavine. The cartridge was then dried by using the vacuum drier. After the vacuum was relieved the washes under the cartridge were discarded marked glass tube for sample collection was placed under the cartridges. 5 ml of methanol was added to the dry cartridge and allowed to absorb for 5 min after which the samples were eluded out of the cartridge at a flow rate of 1 ml/min.

Evaporation

Glass tubes containing methanol eluent were placed in a sand bath at 50 °C to evaporate the methanol and leaving a thick fluid at the bottom of the glass tube.

HPLC analysis

Sulphonamides were detected at 265 nm while Tetracyclines were detected at 385 nm. The column temperature was set at 40 °C while the flow rate was at 0.2 ml/min. Used gradient was A:B 90:10 at 0–35 min, 65:35 at 35–36 min, 90:10 at 36–45 min and 90:10 at 45–55 min. sample run time was 45 min while the injection volume was 10 μl. The retention times of Sulphonamides (SDZ, SMX, SMR, SCL, SDOX, SMTZ, SDM, SQ) was 4 min, 7 min, 7 min, 8 min, 8 min, 14 min, 16 min and 17 min respectively. The retention times for tetracyclines (OTC, DC, TC and CTC) were 11 min, 24 min, 33 min and 36 min respectively.

The remaining fluid in the glass tube (from evaporation stage) was added to 200 μl of mobile phase. This was then mixed by vortexing for 15 s. 0.3 μl of mobile phase A was added to the mixture and vortexed vigorously for 15 s. The sample was then filtered through 0.2 μm syringe filter to HPLC vials and put inside the HPLC and results were generated after the run time was completed.

Calibration

Calibration graphs were first determined by preparation of different concentration of the standard solutions. From a stock solution of 1 mg/ml of each standard the following concentrations were prepared; 2,000 ng/ml, 1,000 ng/ml, 500 ng/ml and 50 ng/ml using mobile phase A. The calibration graphs produced by the standards were used to determine the concentration of drugs in the samples. The calibration curves were used to provide information on recovery, retention factor and the standard deviation. The Limits of detection were also provided, but these were equipment and procedure specific and were provided by the manufacturer of the HPLC-UV. Results of calibration are as shown in Table 1. The standard calibration curves used in the generation of results were generated based on a formula by (Sulejmani et al. 2012). The absorbance read became a percentage (%) of optical density relative to zero standards B_0 and it is based on the calibration line assigned to each series of standard solutions and has the following formula:

$$y = a + b * lnX$$

Y-read signal expressed in% of optical density,

X-concentration of the substance and a and b-coefficients.

In every batch of samples analyzed for values of Rr^2, tetracycline's must be at least 0.8278, while the sulphonamides' $Rr^2 > 0.98$. The calibration curve results are presented in Table 1.

The calibration curve provides information on recovery, retention factor and the standard deviation. The Limits of detections are also provided in the table but these are equipment and procedure specific and were provided by the manufacturer of the HPLC-UV. One such calibration curve has been provided in Fig. 1. The figure shows calibrationn curves for SDM and SMTZ.

Method validation

The method was validated by the use of blank samples ($n = 7$) spiked with a concentration of 200 ng/ml of all the sulphonamides (SCL, SDZ, SMTZ, SQ, SMR, STZ, SMX, SDOX, SDM) and tetracycline's (OTC, DC, CTC, TC). The spiked and blank samples passed through sample preparation as other milk samples described above. In the validation procedure Sulfachloropyradizine (SCL) was not eluded in any of the7 spiked samples. However Sulfadiazine was only eluded from one of the spiked sample hence it was not possible to calculate a standard deviation as reported in Table 2.

Results

Out of 229 samples in the rural dairy system, 72 (31.4%) samples tested positive after screening. Caution samples were treated as positive samples since all of them were preceded to confirmation stage by HPLC-UV. Out of the positive rural samples from charm test, 59 (56 confirmed positive and 3 caution samples) were from farm level, 12 samples at the transporters and 1 sample was from the bulking center. In the peri urban dairy system, out of the 80 samples collected 23 samples were positive including caution. Out of which 14 of them were recorded at the farm level, 4 at the transporters node and 2 samples at the bulking center. Samples which did not produce a distinct colour were treated as caution and were

Table 1 Calibration curves results of standards

STANDARS	Regression equation from calibration curves	Rr^2	LOD HPLC-UV	RF	SD (%)
SMR	f(x) = 662.48x + 74645.3	1.0	50	737.1	0.08
SDOX	F(x) = 376.583x + 1846.29	1.0	100	378.4	0.004
SCL-rt30	F(x) = 17.72x − 7393.03	1.0	100	10.3	0.62
SDZ	F(x) = 340.51x + 2232.01	1.0	50	342.7	0.005
SMTZ	F(x) = 321.58x + 4987.23	1.0	50	326.6	0.013
SQ	F (x) = 259.997x + 1480.45	1.0	50	258.5	0.005
SDM	F(x) = 292.90x − 8808.51	1.0	80	284.1	0.027
SMX	F(x) = 386.15x − 297.48	1.0	50	385.9	0.001
STZ-32	F(x) = 65.84x + 18977	1.0	50	84.8	0.19
OTC	F(x) = 43.30x + 48278.9	1.0	40	91.6	0.46
DC	F(x) = 55.08x + 29908.4	1.0	40	85	0.3
CTC	F(x) = 127.8x + 18909.9	1.0	40	146.7	0.11
TC	F(x) = 63.85x − 2396.01	1.0	40	61.5	0.034

Rr^2 recovery ratio, RF: retention factor, LOD:limit of detection, SD:standard deviation, Sulfachloropyradizine (SCL), Sulfadiazine (SDZ), Sulfadimidine (SMTZ), Sulfaquinoxaline (SQ), Sulfamerazine (SMR), Sulfathiazole (STZ), Sulfamethoxazole (SMX), Sulfadoxin (SDOX), Sulfadimethoxin (SDM), Oxytetracycline (OTC), Doxycycline hyclate (DC), Chlortetracycline hydrochloride (CTC) and Tetracycline hydrochloride (TC)

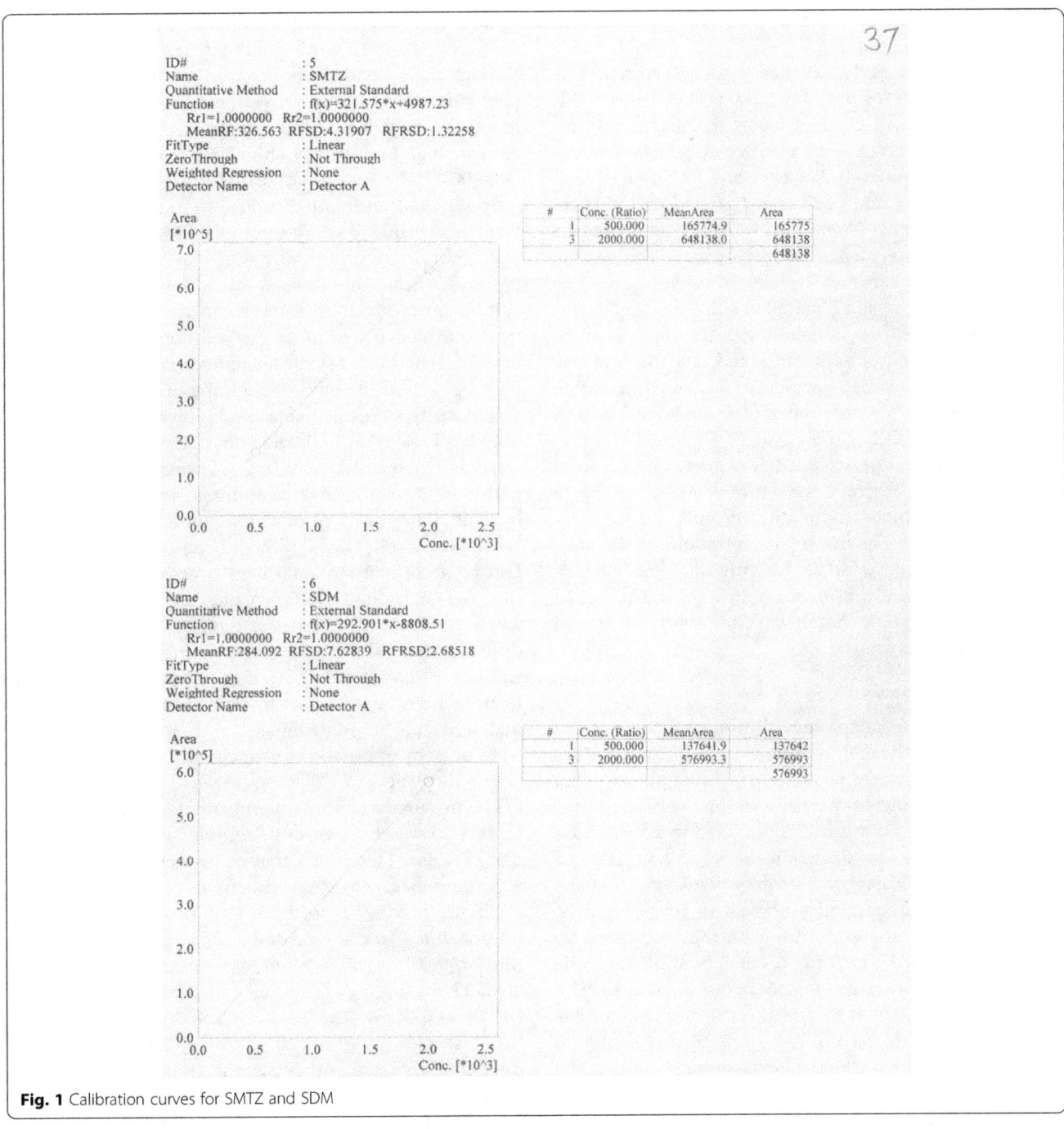

Fig. 1 Calibration curves for SMTZ and SDM

Table 2 Results from method validation using spiked milk samples

Samples	SDZ	SMX	SCL	SMR	SDOX	SMTZ	SDM	SQ	OTC	DC	TC	CTC
Recovery (mean %)	144	92	-	71	71	112	56	42	99	92	70	64
SD	-	6.2	-	5.2	3.9	7.0	3.2	2.7	3.7	5.7	3.2	4.5

recorded as shown in Table 3 and were proceeded to HPLC for identification.

Results from the HPLC showed that none of the samples contained any of the four tetracyclines tested (OT, CTC, DOC, TC). All of the samples that tested positive for sulphonamides at the rural dairy system had values above the EU MRL levels. Positive samples for sulphonamides at the farm node and transporters node also recorded values higher than the EU MRL levels. At the bulking (coolong) center in the peri urban dairy system, SDZ and SDM however recorded values less than 100ug/kg in the positive sample. Only 2 samples recorded presence of sulphonamides in the rural farms, 1 sample at the transporters node and 1 sample at the bulking node recorded presence of sulphonamides. Samples positive for sulphonamides contained SDOX (148.78 µg/kg,), SDZ (90.03 µg/kg), SDM (66.14 µg/kg) and SMX (8,979.59 µg/kg; 8,979.51 µg/kg). A sample of a positive Chromatogram has been provided in Fig. 2 (for SDOX). Results for quantity of antibiotics are recorded in Table 4. The mean concentrations of the antibiotic contaminants were significantly different between locations. The highest mean concentration was recorded at peri urban bulking center which was highest for all the nodes (Fig. 3).

Discussion

After screening, no sample tested positive for any of the tetracyclines in the confirmation stage. Charm Blue-Yellow kit tests for the presence of a wide spectrum of antibiotics including betalactams, sulphonamides, tetracyclines and most of the antibiotics used in animal husbandry. The lack of detection of sulphonamides or tetracyclines in some of the samples would indicate the presence of other antibiotics as well in these samples. These results are almost similar with those obtained by Ahlberg et al. (2016) in study to analyze antibiotic residues in milk from smallholder farms in Kenya. The study used two methods to screen for presence of antibiotics, but the HPLC was not able to detect any antibiotic concentrations at and above the concentration used. The

Charm II Blue Yellow kit eliminates the possibility of analyzing samples with inherent (natural antibiotics) through the second stage of screening. In this stage samples were exposed to heat treatment (80 °C/10 min) treatment to breakdown most natural inhibitors and also emulate high temperature short time pasteurization (Mullan, 2003; Kellnerová et al., 2014; Layada et al., 2016).

Presence of antibiotic residues in milk sampled at the farm from individual farms indicates that farmers are not observing withdrawal periods in lactating animals. A study done in Kosovo (Sulejmani et al. 2012), sulphonamide residue levels were compared based on time and delivery level. It showed that during the first days of delivery (1–4 days), sulphonamide levels remain high up to mid time after which the drug levels reduced significantly to incalculable levels towards the last days (day 5). This is an indication that since sulphonamides were detected in farm milk, the farmer milked the cow within 5 days of drug administration. Some farmers have attributed lack of observing withdrawal periods to harsh economic times (Shitandi and Sternesjö 2004). During treatment, the farmers however have to milk the cow to facilitate letdown but is expected to throw away the milk. Most farmers tend to find this practice difficult since the physical appearance of the milk is similar to that from a cow that is not undergoing any form of treatment. A part from lack of withdrawal, animal feed can be contaminated with antibiotics through feces or poor disposal of treatment kits containing antibiotics (Aboge et al., 2000; Kang'ethe et al., 2005).

The number of positive farm samples in the rural was slightly higher than positive samples in the peri urban dairy system. This shows that consumers in rural setting are more likely to consume milk contaminated with antibiotic residues than those in peri urban. This would be possible if these consumers buy milk directly from the farmers, which is a common practice in the rural area. These findings are similar to those of Aboge et al. (2000) and Kang'ethe et al. (2005) they reported that rural farmers are three times more likely to consume milk contaminated with antibiotic residues compared to their counterparts in the peri urban farms.

Antibiotic residues in transporters milk, shows that the antibiotics may have been intentionally added to milk to extend their shelf life. Transporters collect milk from farms and deliver them to the next value chain node. These include cooling centers which are collection points for dairy processing factories. Transporters face a challenge of milk spoilage since they transport the milk without any cooling facilities. Milk at this node is at a high risk of spoilage due to time taken moving from one farm to the other before reporting to the cooling center. Most transporters however, have been reported to add antibiotics to milk to prevent milk spoilage

Table 3 Screening results from Charm II Blue-Yellow Test

Dairy system	Nodes	N	Positive (n)	Caution (n)	Total
Rural	Cows	120	56 (46.7%)	3 (2.5%)	
	Transporters	105	11 (9.5%)	1 (0.95%)	
	Bulking	4	1 (25%)	0	
Total		229	68	4	72 (31.4%)
Peri urban	Farm	57	11 (19.5%)	3 (5.3%)	
	Transporters	21	6 (28.6%)	1 (4.8%)	
	Bulking	2	1 (50%)	1 (50%)	
Total		80	18	5	23 (28.8%)

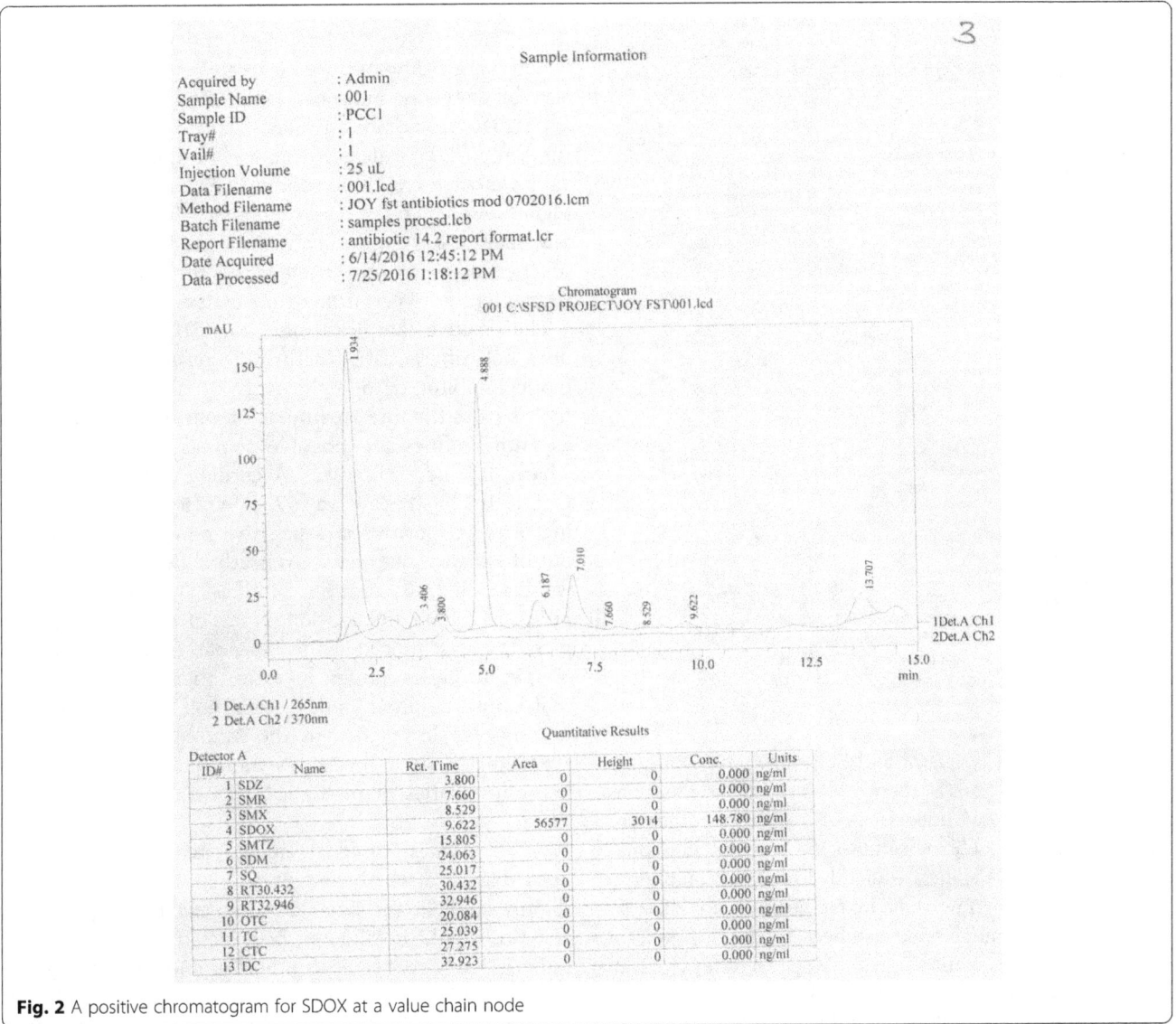

Fig. 2 A positive chromatogram for SDOX at a value chain node

(Aboge et al. 2000). Occurrence of antibiotic residues at the transporters node in peri urban is four times higher than in rural location. Consumers of dairy products and milk in the peri urban are more likely to consume milk with added antibiotics to extend shelf life. The prevalence of antibiotic contaminants was highest at rural farms (Table 3) while the concentration at this node was significantly lower than bulking centers in the peri urban (Fig. 3). This result indicates that the amount of antibiotic contaminants due to lack of observing withdrawal period, is lower than quantities added along the value chain.

The percentage positive samples from screening results in this study was 30%, these results are slightly higher than those identified in the recent studies in Kenya. In 2004 Shitandi and Sternesjo recorded 14%,

Ekuttan et al. (2007); 4% Kang'ethe et al. (2005) 14%. The proportion of sulphonamides groups in milk samples differs sparingly in this study compared to other studies done in Kenya. We found that sulphonamide occurred at 4.1% while tetracycline was at 0%. A study by Ahlberg et al. (2016) recorded sulphonamides at 0.4% and tetracyclines at 2.5%. Mitema et al. (2001) recorded sulphonamids at 24% and tetracyclines at 61%.

Studies from different parts of Africa have reported different results from this study. In Tanzania over 36% of milk supply chain was reported to contain antibiotic residues (Kurwijila et al. 2006). Other studies in Africa have also recorded the presence of antibiotic residues in milk like in Egypt (Goudah et al., 2007), Ghana (Addo et al. 2011), South Africa (Bester and Lombard 1979), Ethiopia (Myllyniemi et al. 2000), Sudan (El-tayeb et al.

Table 4 Quantity of Sulphonamides and Tetracyclines

Residue	RURAL (conc ppb)			PERI (conc ppb)		
	Farm	Transport	Bulking	Farm	Transport	Bulking
SMR	0	0	0	0	0	0
SDOX	0	0	0	0	0	148.78
SCL	0	0	0	0	0	0
STZ	0	0	0	1923	0	0
SMTZ	389.176	0	0	0	0	0
SQ	0	0	0	0	0	0
SDZ	0	0	0	0	0	90.03
SDM	0	0	2305	0	674.83	66.14
SMX	179.026	2389.844	0	0	0	8979.59 8979.51
Positive n	2	1	1	1	1	2
OTC	0	0	0	0	0	0
DC	0	0	0	0	0	0
CTC	0	0	0	0	0	0
TC	0	0	0	0	0	0

Sulfachloropyradizine (SCL), Sulfadiazine (SDZ), Sulfadimidine (SMTZ), Sulfaquinoxaline (SQ), Sulfamerazine (SMR), Sulfathiazole (STZ), Sulfamethoxazole (SMX), Sulfadoxin (SDOX), Sulfadimethoxin (SDM), Oxytetracycline (OTC), Doxycycline hyclate (DC), Chlortetracycline hydrochloride (CTC) and Tetracycline hydrochloride (TC)

2012) and Nigeria (Olufemi and Ehinmowo 2009) among others. In other nations of the world antibiotic residues has been reported in raw milk. In 2007 Kress and other co-authors reported 1.6% of samples to contain sulphonamides in Germany. These are slightly lower than sulphonamides identified in raw milk in Netherlands in 2000 by Abjean. Sulphonamides have also been reported in milk in Mexico 51.3% (Tolentino et al. 2005), Turkey (Alkan, 2007) and Korea (Chung et al. 2009). In the most recent studies, tetracyclines, sulphonamides and other antibiotics have been recorded at levels above recommended limits in milk at different nodes of the value chain (Olatoye et al., 2016; Chowdhury et al., 2016; Layada et al., 2016).

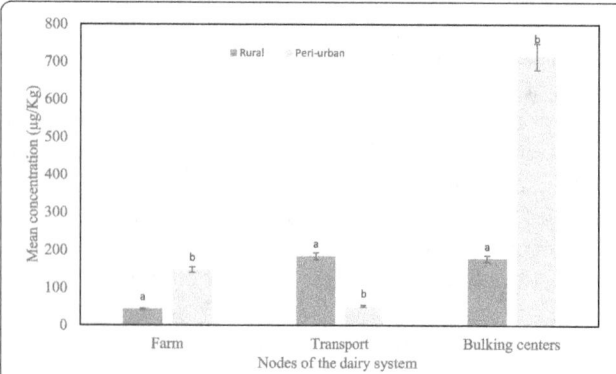

Fig. 3 Comparison of mean concentration of antibiotic residues in milk along the dairy value chains of rural and peri-urban systems

All sulphonamides detected read values above the EU MRL value except Sulfadiazine and Sulfadimethoxine which were below 100 µg/kg in this study. These levels also fall above the European Union Comission Regulation (EU) 2010 of antibiotic residues in raw milk. The rest of the samples were detected above the Limit of detection values of each antibiotic residue. This indicates that even higher levels than the read value might have been present. Sulfadiazine is a common antibiotic used in the veterinary practice in several countries and was recorded at the bulking center in the peri urban dairy system only.

The positive samples from Charm II Blue-Yellow test does not differentiate whether the result is due to antibiotic inhibitor or other growth inhibitors. Growth inhibitors used in the treatment of worm infections such as anthelmintics are possible sources of error for the Charm II Blue- Yellow Kit. According to this study, The Charm II Blue-Yellow was 97.1% efficient in distinguishing between positive and negative samples since only 9 out of 309 samples were not clearly differentiated. These were labeled as caution samples and proceeded to HPLC UV for confirmation. None of the caution samples however, recorded the presence of sulphonamides or Tetracyclines under investigation. It is also likely that some sulphonamides were not detected due to presence of impurities in the sample. Sulphonamides are detected at a lower UV range of 268 nm where many impurities of biological origin can interfere with the analysis.

When milk is stored at ambient temperature, antibiotics degrade (Marth and Steele, 2001). When milk is slightly spoiled, the beta lactamse enzyme is produced and this would breakdown beta lactam antibiotics. The same is likely to occur to other antibiotics (Guay et al. 1987). In this study, samples were stored for a maximum of 2 months at 4 °C (frozen), however cold chain would not be maintained between storage and the point of analysis due to unavoidable circumstances. These challenges could explain the low levels of antibiotics detected by the HPLC-UV.

Conclusion

It can be concluded that presence of antibiotics in the farm is more common in the rural farms that the peri-urban farms. The level of antibiotics in the peri-urban increased through the transporters to the collection center. Lack of observation of withdrawal period might be a common practice given the high level of antibiotics in the farm milk. Addition of antibiotics for shelf life extension may be practiced more in the peri-urban by milk transporters. The study recommends the implementation of rapid tests along the value chain to be able to detect presence of antibiotics since this is not practiced in both locations.

Abbreviations

CTC: Chlortetracycline hydrochloride; DC: Doxycycline hyclate; EU MRL: European Union Maximum Residual Limits; HPLC UV: High Performance Liquid Chromatography with Ultra Violet; LOD: Limit of detection; OTC: Oxytetracycline; RF: Retention; Rr^2: Recovery; SCL: Sulfachloropyradizine; SD: Standard deviation; SDM: Sulfadimethoxin; SDOX: Sulfadoxin; SDZ: Sulfadiazine; SMR: Sulfamerazine; SMTZ: Sulfadimidine; SMX: Sulfamethoxazole; SQ: Sulfaquinoxaline; STZ: Sulfathiazole; TC: Tetracycline hydrochloride

Acknowledgement

We are grateful to the following individuals for their support in sample collection at the field, transportation to the laboratory Ms. Caroline Makau, Mrs. Faith Ndungi, Mr. Muyoma Nato, Mr. Olivier Kashongwe and Ms. Linnet Mwangi. Mr. Nderitu, and Mr. Pancras From Nairobi University are acknowledged for helping in sample analysis. Mr. Nobert Wafula is also acknowledged for data analysis and editing of this manuscript.

Funding

This research was funded by the Federal Government of Germany (BMBF) through ReLOAD (Reducing post harvest losses adding value in East African Food Value Chains) Project.

Authors' contributions

The authors JDO, JWM, PSM and PL participated in the design of the research and location of the study. JDO participated in data collection in the field, sample analysis, result generation and interpretation under close supervision of JWM, PSM and PL. PL was responsible in the analysis in availing the reagents and identification of the method and equipment for sample analysis by JDO. All authors read and approved the final manuscript.

Authors' information

Joy Deborah Orwa is a Masters Student with a Bachelors in Food Science and Technology from Egerton University. She has acquired skills in research including data analysis and interpretation especially in the Food Science. She also has experience working in the food industry. Joseph W. Matofari ia a Professor in Food Microbiology currently a senior lecturer at Egerton University in the Department of Dairy, Food Science and Technology. Patrick Muliro is a PhD holder in Dairy Science and a senior lecturer at Egerton University department of food science and technology. Peter Lamuka is a lecturer at Nairobi University in the department of Food Science, Nutrition and technology with vast knowledge in Food Safety. The three lecturers have supervised a number of Masters and PhD students over the years in their respective Universities.

Competing interests

The authors of this document have not declared any competing interests financial or non-financial in nature.

Author details

[1]Department of Dairy Food Science Technology, Egerton University, Faculty of Agriculture, P.O. Box, 536–20115, Egerton, Kenya. [2]Department of Food Science, Nairobi University, Nutrition and Technology, P.O. Box 29053–00625, Nairobi, Kenya.

References

Abjean JP, Delepine B, Hurtaud-Pessel D. Qualitative or quantitative Methods for residue analysis? A Strategy for drug residue monitoring. Veldhoven: Proceedings of the Conference EuroResidue IV; 2000.

Aboge GO, Kang'ethe EK, Arimi SM, Omore AO, McDermot JJ, Kanja LW, Macharia JK, Nduhiu JK, Githua A. Antimicrobial Agents Detected in Marketed Milk in Kenya. Oral presentation at the 3rd All Africa Conference On Animal Agriculture. Smallholder Dairy Project. 2000.

Addo KK, Mensah GI, Aning KG, Nartey N, Nipah GK, Bonsu C, Akyeh ML, Smits HL. Microbiological quality and antibiotic residues in informally marketed raw cow milk within the coastal savannah zone of Ghana. Trop Med Int Health. 2011;16(2):227–32.

Ahlberg S, Korhonen H, Lindfors E, Kang'ethe E. Analysis of Antibiotic Residues in milk from Smallholder farms in Kenya. African J Dairy Farmin Milk Prodc. 2016;3(4):152–8.

Alkan A. The confirmation of used commercial kits in the detection of antibiotics in milk with HPLC (High Pressure Liquid Chromatography). Izmir: Master thesis. Graduate School of Engineering and Sciences of Izmir Institute of Technology; 2007.

Bester BH, Lombard SH. The effect of the dye-marking of mastitis remedies on the incidence of antibiotic residues in Pretoria's market milk supplies. J S Afr Vet Assoc. 1979;50(3):151–3.

Chowdhury S, Hassan M, Alam M, Sattar S, Bari MS, Saifuddin AK, Ahasanul MH. Antibiotic residues in milk and eggs of commercial and local farms at Chittagong, Bangladesh. Vet World. 2015;8(4):467–71.

Chung HH, Lee JB, Chung YH, Lee KG. Analysis of sulfonamide and quinolone antibiotic residues in Korean milk using microbial assays and high performance liquid chromatography. Food Chem. 2009;113(1):297–301.

Ekuttan CE, Kang'ethe EK, Kimani VN. Randolph TF Investigation on the prevalence of antimicrobial residues in milk obtained from urban smallholder dairy and non-dairy farming households in Dagoretti division, Nairobi, Kenya. East Afr Med J. 2007;84(11):87–91.

El-tayeb A, Barakat S, Marrone G, Shaddad S, Stålsby LC. Antibiotic use and resistance in animal farming: a quantitative and qualitative study on knowledge and practices among farmers in Khartoum, Sudan. Zoonoses Public Health. 2012;59(5):330–8.

European Union Commission Regulation (EU). On Pharmacologically active substances and their classification regarding Maximum Residual Limits in Food stuffs of animal Origin. 2010. EU 37/2010.

Goudah A, Sher S, Shin HC, Shim JH, Abd El-Aty AM. Pharmacokinetics and mammary residual depletion of erythromycin in healthy lactating ewes. J Vet Med A Physiol Pathol Clin Med. 2007;54(10):607–11.

Guay R, Cardinal P, Bourassa C, Brassard N. Decrease of penicillin G residue incidence in milk: A fact or an artefact? Int J Food Microbiol. 1987;4(3):187–96.

Kang'ethe EK, Aboge GO, Arimi SM, Kanja LW, Omore AO. McDermott JJ Investigation of the Risk of Consuming Marketed Milk with Antimicrobial Residues in Kenya. Food Control. 2005;16(4):349–55.

Kellnerová E, Navrátilová P, Borkovcová I. Effect of pasteurization on the residues of tetracyclines in milk. Acta Vet Brno. 2015;83(10):21–6.

Kress C, Seidler C, Kerp B, Schneider E, Usleber E. Experiences with an identification and quantification program for inhibitor-positive milk samples. Anal Chim Acta. 2007;586(1):275–9.

Koesukwiwat U, Jayanta S, Leepipatpiboon N. Solid-phase extraction for multi-residue determination of sulfonamides, tetracyclines, and pyrimethamine in Bovine's milk. J Chromatogr. 2007;1149(1):102–11.

Kurwijila LR, Omore A, Staal S, Mdoe NSY. Investigation of the Risk of Exposure to Antimicrobial Residues Present in Marketed milk in Tanzania. J Food Prot. 2006;69(10):2487–92.

Layada S, Benouareth D, Coucke E, Andjelkovic M. Assessment of antibiotic residues in commercial and farm milk collected in the region of Guelma (Algeria). Int J Food Contam. 2016;3:19.

Lee HJ, Lee MH, Ruy PD. Public health risks: Chemical and antibiotic residues. Asian Australas J Anim Sci. 2001;14(3):402–13.

Mamani MCV, Reyes FGR. Rath S Multiresidue determination of tetracyclines, sulfonamides and chloramphenicol in bovine milk using HPLC-DAD. Food Chem. 2009;117(3):545–52.

Marth EH, Steele JL. Applied Dairy Microbiology, vol. 2. New York: Marcel Dekker, Inc; 2001. p. 320–7.

Ministry of Livestock Development Department and Fisheries. MoLF District Livestock Production Annual Report. Nairobi: Nakuru North; 2012.

Mitema ES, Kikuvi GM, Wegener HC, Stohr K. An assessment of antimicrobial consumption in food producing animals in Kenya. J Vet Pharmacol Ther. 2001;24(6):385–90.

Mullan WMA. Inhibitors in milk. 2003. Available from: https://www.dairyscience.info/index.php/inhibitors-in-milk/51-inhibitors-in-milk.html. Accessed 26 Jan 2017.

Myllyniemi AL, Rannikko R, Lindfors E, Niemi A, Bäckman C. Microbiological and chemical detection of incurred penicillin G, oxytetracycline, enrofloxacin and ciprofloxacin residues in bovine and porcine tissues. Food Addit Contam. 2000;17(12):991–1000.

Nisha AR. Antibiotics residues-A global health hazard. Vet World. 2008;1(12):375–7.

Okeke N, Lamikana A, Edelman R. Socioeconomic and behaviourial factors leading to acquired bacterial resistance in developing countries. Emerg Infect Dis. 1995;5(1):18–27.

Olatoye IO, Daniel OF, Ishola SA. Screening of antibiotics and chemical analysis of penicillin residues in fresh milk and traditional dairy products in Oyo state, Nigeria. Vet World. 2016;9(9):948–54.

Olufemi OI, Ehinmowo EA. Oxytetracycline Residues in Edible Tissues of Cattle Slaughtered in Akure, Nigeria. Internet J Food Safety. 2009;11(1):62–6.

Omore AO, Arimi SM, Kang'ethe EK, McDermott JJ. Analysis of Public Health Risks from Consumption of Informally Marketed Milk in Kenya. Kenya Vet. 2004;27(1):15–7.

Richelle RG. Investigation of Safe-Level Testing for Beta-lactam, Sulfonamide, and Tetracycline Residues in Commingled Bovine Milk. USA: Doctoral thesis, Salve Regina University; 2007.

Sulejmani Z, Shehi A, Hajrulai Z, Mata E. Abuse of Pharmaceutical Drugs-antibiotics in Dairy Cattle in Kosovo and Detection of their Residues in Milk. J Ecosyst Ecogr. 2012;2(19):114–20.

Shitandi A, Sternesjö Å. Factors Contributing to the Occurrence of Antimicrobial Drug Residues in Kenyan Milk. J Food Prot. 2004;67(2):399–402.

Tolentino RG, Perez MA, Gonzales GD, León SV, López MG. Determination of the presence of 10 antimicrobial residues in Mexican pasteurized milk. Interciencia. 2005;30(5):291–4.

Tollefson L, Miller M. Antibiotic use in food animals: controlling the human health impact. J AOAC Int. 2000;83(2):245–54.

Van Boeckel TP, Charles B, Marius G, Bryan TG, Simon AL, Timothy PR, Aude T, Ramanan L. Global trends in antimicrobial use in food animals. Proc Natl Acad Sci. 2015;112(18):5649–54.

Wageh SD, Elsaid AE, Mohamed TE, Yoshinori I, Shouta N, Mayumi I. Antibiotic residues in Food: the African Scenario. Japanese J Vet Resear. 2013; 61(Supplement):13–22.

PERMISSIONS

LIST OF CONTRIBUTORS

Melese Abate Reta
Faculty of Health Science, Department of Nursing, Woldia University, P.O.Box 400, Woldia, Ethiopia

Tesfaye Wolde Bereda
College of Natural Sciences, Department of Biology, Wolkite University, P.O.Box 07, Wolkite, Ethiopia

Ayalew Nigusie Alemu
College of Veterinary Medicine, Department of Veterinary Microbiology and Public Health, Jigjiga University, P.O.Box 1020, Jigjiga, Ethiopia

Ruth A. Oni
Department of Nutrition and Food Science, University of Maryland, College Park 20742, MD, USA

Robert L. Buchanan
Department of Nutrition and Food Science, University of Maryland, College Park 20742, MD, USA
Center for Food Safety and Security Systems, University of Maryland, College Park 20742, MD, USA

Elisabetta Lambertini
Department of Nutrition and Food Science, University of Maryland, College Park 20742, MD, USA
Center for Food Safety and Security Systems, University of Maryland, College Park 20742, MD, USA
Current address: Environmental and Health Sciences, RTI International, Rockville, MD 20852, USA

Md Iftakharul Muhib, Nusrat Jakarin Easha and M. Khabir Uddin
Department of Environmental Sciences, Jahangirnagar University, Dhaka 1342, Bangladesh

Muhammed Alamgir Zaman Chowdhury
Department of Environmental Sciences, Jahangirnagar University, Dhaka 1342, Bangladesh
Agrochemicals and Environmental Research Division, Institute of Food & Radiation Biology, Atomic Energy Research Establishment, G.P.O. Box 3787, Savar 1349, Bangladesh

Md Mostafizur Rahman
Department of Environmental Sciences, Jahangirnagar University, Dhaka 1342, Bangladesh.
Faculty of Environmental Earth Science, Hokkaido University, Sapporo 060-0810, Japan

Mashura Shammi
Department of Environmental Sciences, Jahangirnagar University, Dhaka 1342, Bangladesh
Department of Environmental Pollution and Process Control, Xinjiang Institute of Ecology and Geography, ChineseAcademy of Sciences, Urumqi 830011, Xinjiang, People's Republic of China

Zeenath Fardous and Md Khorshed Alam
Agrochemicals and Environmental Research Division, Institute of Food & Radiation Biology, Atomic Energy Research Establishment, G.P.O. Box 3787, Savar 1349, Bangladesh

Mohammad Latiful Bari
Food Analysis Research Laboratory, Center for Advanced Research in Sciences, University of Dhaka, Dhaka 1000, Bangladesh

Masaaki Kurasaki
Faculty of Environmental Earth Science, Hokkaido University, Sapporo 060-0810, Japan

Simbarashe Samapundo and Frank Devlieghere
Department of Food Safety and Food Quality, Laboratory of Food Microbiology and Food Preservation, Food2Know, Ghent University, Coupure Links 653, Ghent 9000, Belgium

Anh Ngoc Tong Thi
Department of Food Safety and Food Quality, Laboratory of Food Microbiology and Food Preservation, Food2Know, Ghent University, Coupure Links 653, Ghent 9000, Belgium
Department of Food Technology, Faculty of Agriculture and Applied Biology, Can Tho University, 3-2 Street, Can Tho City, Viet Nam

Marc Heyndrickx
Institute for Agricultural and Fisheries Research (ILVO), Technology and Food Science Unit, Brusselsesteenweg 370, Melle 9090, Belgium

Department of Pathology, Bacteriology and Avian Diseases, Faculty of Veterinary Medicine, Ghent University, Salisburylaan 133, Merelbeke 9820, Belgium

José J. Pérez, Rutilio Ortiz and María L. Ramírez
Departamento de Producción Agrícola y Animal, Laboratorio de Análisis Instrumental, Universidad Autónoma Metropolitana, Unidad Xochimilco, Calzada del Hueso No 1100, Colonia Villa Quietud, Delegación Coyoacán C.P. 04960, D.F., México

Javier Olivares, Daniel Ruíz and David Montiel
Departamento de Producción Agrícola y Animal, Laboratorio de Fitopatología, Universidad Autónoma Metropolitana, Unidad Xochimilco, Calzada del Hueso No 1100, Colonia Villa Quietud, Delegación Coyoacán C.P. 04960, D.F., México

Linnet Wanjiru Mwangi, Joseph Wafula Matofari and Patrick Simiyu Muliro
Department of Dairy and Food Science and Technology, Egerton University, P.O. Box 536-20115, Egerton, Kenya

Bockline Omedo Bebe
Department of Animal Sciences, Egerton University, P.O. Box 536-20115, Egerton, Kenya

Betelihem Tegegne
Wollo University, School of Veterinary Medicine, Dessie, Ethiopia

Shimels Tesfaye
Faculty of Veterinary Medicine, Department of Para-Clinical Studies (Veterinary Microbiology), University of Gondar, Gondar, Ethiopia

Osei Akoto
Department of Chemistry, Kwame Nkrumah University of Science and Technology, Kumasi, Ghana

Fredrick Addai-Mensah
Department of Theoretical and Applied Biology, Kwame Nkrumah University of Science and Technology, Kumasi, Ghana

Eric K. K. Abavare
Department of Physics, Kwame Nkrumah university of Science and Technology, Kumasi, Ghana

Barbara Duquenne, Katleen Coudijzer and Jan De Block
Institute for Agricultural and Fisheries Research (ILVO), Technology and Food Science Unit, Brusselsesteenweg 370, 9090 Melle, Belgium

Sophie Marchand
Institute for Agricultural and Fisheries Research (ILVO), Technology and Food Science Unit, Brusselsesteenweg 370, 9090 Melle, Belgium University Hospital Ghent, Metabolic and Cardiovascular Diseases, Ghent University, De Pintelaan 185, 9000 Gent, Belgium

Marc Heyndrickx
Institute for Agricultural and Fisheries Research (ILVO), Technology and Food Science Unit, Brusselsesteenweg 370, 9090 Melle, Belgium Department of Pathology, Bacteriology and Poultry Diseases, Ghent University, Salisburylaan 133, 9820 Merelbeke, Belgium

Wanjala Nobert Wafula, Wafula Joseph Matofari and Masani John Nduko
Department of Dairy and Food Science and Technology, Egerton University, P.O. Box 536-20115, Egerton, Kenya

Peter Lamuka
Department of Food Science, Nutrition and Technology, University of Nairobi, P.O. Box 29053, Nairobi, Kenya

Caroline Mwende Makau, Joseph Wafula Matofari and Patrick Simiyu Muliro
Egerton University, Faculty of Agriculture, Department of Dairy & Food Science and Technology, P.O. Box 536-20115, Egerton, Kenya

Bockline Omedo Bebe
Egerton University, Faculty of Agriculture, Department of Animal Science, P.O. Box 536-20115, Egerton, Kenya

Gloria Nwakego Elemo
Department of Biotechnology, Federal Institute of Industrial Research Oshodi, Lagos, Nigeria

Princewill Chimezie Okorie
Department of Biotechnology, Federal Institute of Industrial Research Oshodi, Lagos, Nigeria Department of Microbiology, Faculty of Science, Lagos State University, Lagos, Nigeria

Nurudeen Ayoade Olasupo
Department of Microbiology, Faculty of Science, Lagos State University, Lagos, Nigeria

Felicia Ngozi Anike and Omoanghe Samuel Isikhuemhen
Department of Natural Resources and Environmental Design, North Carolina Agricultural and Technical State University, Greensboro, North Carolina, USA

Desta Woldetsadik and Heluf Gebrekidan
School of Natural Resources Management and Environmental Sciences, Haramaya University, PObox: 138, Dire Dawa, Ethiopia

Pay Drechsel
International Water Management Institute, Colombo, Sri Lanka

Bernard Keraita
Department of Global Health, University of Copenhagen, Copenhagen, Denmark

Fisseha Itanna
Department of Crop Science, University of Namibia, Windhoek, Namibia

Noel Bakobie, Amponsah Samuel Addae and Samuel Jerry Cobbina
Department of Ecotourism and Environmental Management, Faculty of Natural Resources and Environment, University for Development Studies, Tamale, Ghana

Abudu Ballu Duwiejuah
Department of Ecotourism and Environmental Management, Faculty of Natural Resources and Environment, University for Development Studies, Tamale, Ghana
Department of Biotechnology, Faculty of Agriculture, University for Development Studies, Tamale, Ghana

Solomon Miniyila
Ghana Integrated Water Sanitation and Hygiene World Vision, Tamale, Ghana

Mildred Mwanja
Environmental Health Department, Monze District Medical Office, P.O. Box 66144, Monze, Zambia. School of Medicine Department of Public Health, Environmental Health Unit, University of Zambia, P.O Box 50110, Ridgeway, Lusaka, Zambia

Allan Rabson Mbewe and Nosiku Sipilanyambe Munyinda
School of Medicine Department of Public Health, Environmental Health Unit, University of Zambia, P.O Box 50110, Ridgeway, Lusaka, Zambia

Choolwe Jacobs
School of Medicine Department of Epidemiology and Biostatistics Unit, University of Zambia, P.O Box 50110, Ridgeway, Lusaka, Zambia

Joy Deborah Orwa, Joseph Wafula Matofari and Patrick Simiyu Muliro
Department of Dairy Food Science Technology, Egerton University, Faculty of Agriculture, P.O. Box, 536–20115, Egerton, Kenya

Peter Lamuka
Department of Food Science, Nairobi University, Nutrition and Technology, P.O. Box 29053–00625, Nairobi, Kenya

Janet Irungu, Suresh Raina and Baldwyn Torto

Md. Shahadat Hossain, Md. Samiul Islam, Subrata Bhadra and Abu Shara Shamsur Rouf

Elvis D. Okoffo, Benedicta Y. Fosu-Mensah and Christopher Gordon

Hayford Ofori, Charles Tortoe, Paa Toah Akonor and Jonathan Ampah

Samiha Layada, Djemel-Eddine Benouareth, Wim Coucke and Mirjana Andjelkovic

Grace L. Anderson, Lindsey Garnick, Mai S. Fung and Shannon H. Gaffney

Index